HUMAN SEXUALITY 94/95

Nineteenth Edition

Editor

Ollie Pocs
Illinois State University

Ollie Pocs is a professor in the Department of Sociology, Anthropology, and Social Work at Illinois State University. He received his B.A. and M.A. in sociology from the University of Illinois, and a Ph.D. in family studies from Purdue University. His primary areas of interest are marriage and family, human sexuality and sexuality education, sex roles, and counseling/therapy. He has published several books and articles in these areas.

Cover illustration by Mike Eagle

The Dushkin Publishing Group, Inc.
Sluice Dock, Guilford, Connecticut 06437

The Annual Editions Series

Annual Editions is a series of over 60 volumes designed to provide the reader with convenient, low-cost access to a wide range of current, carefully selected articles from some of the most important magazines, newspapers, and journals published today. Annual Editions are updated on an annual basis through a continuous monitoring of over 300 periodical sources. All Annual Editions have a number of features designed to make them particularly useful, including topic guides, annotated tables of contents, unit overviews, and indexes. For the teacher using Annual Editions in the classroom, an Instructor's Resource Guide with test questions is available for each volume.

VOLUMES AVAILABLE

Africa
Aging
American Foreign Policy
American Government
American History, Pre-Civil War
American History, Post-Civil War
Anthropology
Biology
Business Ethics
Canadian Politics
Child Growth and Development
China
Comparative Politics
Computers in Education
Computers in Business
Computers in Society
Criminal Justice
Drugs, Society, and Behavior
Dying, Death, and Bereavement
Early Childhood Education
Economics
Educating Exceptional Children
Education
Educational Psychology
Environment
Geography
Global Issues
Health
Human Development
Human Resources
Human Sexuality
India and South Asia
International Business
Japan and the Pacific Rim

Latin America
Life Management
Macroeconomics
Management
Marketing
Marriage and Family
Mass Media
Microeconomics
Middle East and the Islamic World
Money and Banking
Multicultural Education
Nutrition
Personal Growth and Behavior
Physical Anthropology
Psychology
Public Administration
Race and Ethnic Relations
Russia, Eurasia, and Central/Eastern Europe
Social Problems
Sociology
State and Local Government
Third World
Urban Society
Violence and Terrorism
Western Civilization, Pre-Reformation
Western Civilization, Post-Reformation
Western Europe
World History, Pre-Modern
World History, Modern
World Politics

Library of Congress Cataloging in Publication Data
Main entry under title: Annual Editions: Human sexuality. 1994/95.
    1. Sexual behavior—Periodicals. 2. Sexual hygiene—Periodicals. 3. Sex education—Periodicals. 4. Human relations—Periodicals. I. Pocs, Ollie, comp.
II. Title: Human sexuality.
ISBN 1-56134-279-3   155.3′05   75-20756

© 1994 by The Dushkin Publishing Group, Inc., Guilford, CT 06437

Copyright law prohibits the reproduction, storage, or transmission in any form by any means of any portion of this publication without the express written permission of The Dushkin Publishing Group, Inc., and of the copyright holder (if different) of the part of the publication to be reproduced. The Guidelines for Classroom Copying endorsed by Congress explicitly state that unauthorized copying may not be used to create, to replace, or to substitute for anthologies, compilations, or collective works.

Annual Editions® is a Registered Trademark of The Dushkin Publishing Group, Inc.

Nineteenth Edition

Printed in the United States of America

Editors/Advisory Board

EDITOR
Ollie Pocs
Illinois State University

ADVISORY BOARD

Janice Baldwin
University of California
Santa Barbara

John Baldwin
University of California
Santa Barbara

Kelli McCormack Brown
Western Illinois University

T. Jean Byrne
Kent State University

Donald Devers
Northern Virginia Community
College

Harry F. Felton
The Pennsylvania State University

Robert T. Francoeur
Fairleigh Dickinson University
Madison

Marylou Hacker
Modesto Junior College

Karen Hicks
Albright College

Kathleen Kaiser
California State University
Chico

Narendra Kalia
SUNY College, Buffalo

Gary F. Kelly
Clarkson University

John Lambert
Mohawk College

Bruce LeBlanc
Black Hawk College

Ted J. Maroun
McGill University

Owen Morgan
Arizona State University

Phillip Patros
Southern Connecticut State
University

Fred L. Peterson
The University of Texas
Austin

Robert Pollack
University of Georgia

Dale Rajacich
University of Windsor

Mina Robbins
California State University
Sacramento

Laurna Rubinson
University of Illinois
Urbana-Champaign

Martin Turnaver
Radford University

Deitra Wengert
Towson State University

Members of the Advisory Board are instrumental in the final selection of articles for each edition of Annual Editions. Their review of articles for content, level, currentness, and appropriateness provides critical direction to the editor and staff. We think you'll find their careful consideration well reflected in this volume.

STAFF

Ian A. Nielsen, Publisher
Brenda S. Filley, Production Manager
Roberta Monaco, Editor
Addie Raucci, Administrative Editor
Cheryl Greenleaf, Permissions Editor
Diane Barker, Editorial Assistant
Lisa Holmes-Doebrick, Administrative Coordinator
Charles Vitelli, Designer
Shawn Callahan, Graphics
Steve Shumaker, Graphics
Lara M. Johnson, Graphics
Libra A. Cusack, Typesetting Supervisor
Juliana Arbo, Typesetter

To the Reader

In publishing ANNUAL EDITIONS we recognize the enormous role played by the magazines, newspapers, and journals of the *public press* in providing current, first-rate educational information in a broad spectrum of interest areas. Within the articles, the best scientists, practitioners, researchers, and commentators draw issues into new perspective as accepted theories and viewpoints are called into account by new events, recent discoveries change old facts, and fresh debate breaks out over important controversies.

Many of the articles resulting from this enormous editorial effort are appropriate for students, researchers, and professionals seeking accurate, current material to help bridge the gap between principles and theories and the real world. These articles, however, become more useful for study when those of lasting value are carefully *collected, organized, indexed,* and *reproduced* in a *low-cost format*, which provides easy and permanent access when the material is needed. That is the role played by *Annual Editions*. Under the direction of each volume's *Editor*, who is an expert in the subject area, and with the guidance of an *Advisory Board*, we seek each year to provide in each ANNUAL EDITION a current, well-balanced, carefully selected collection of the best of the public press for your study and enjoyment. We think you'll find this volume useful, and we hope you'll take a moment to let us know what you think.

Sex lies at the root of life, and we can never learn to reverence life until we know how to understand sex.

—Havelock Ellis

The above observation by one of the first sexologists highlights the objective of this book. Learning about sex is a lifelong process that can occur informally and formally. With knowledge comes the understanding that we are all born sexual, and that sex, per se, is neither good nor bad, beautiful nor ugly, moral nor immoral.

While we are all born with basic sexual interests, drives, and desires, human sexuality is a dynamic and complex force that involves psychological and sociocultural dimensions in addition to the physiological ones. Sexuality includes an individual's whole body and personality. We are not born with a fully developed body or mind, but instead grow and learn; so is it with respect to our sexuality. Sexuality is learned. We learn what "appropriate" sexual behavior is, how to express it, when to do so, and under what circumstances. We also learn sexual feelings: positive feelings such as acceptance of sexuality, or negative and repressive feelings such as guilt and shame.

Sexuality, which affects human life so basically and powerfully, has, until recently, received little attention in scientific research, and even less attention within higher education communities. Yet our contemporary social environment is expanding its sexual and social horizons toward greater freedom for the individual, especially for women and for people who deviate from societal norms: those who are somehow handicapped and those who make less common sexual or relationship choices. Without proper understanding, this expansion in sexual freedom can lead to new forms of sexual bondage as easily as to increased joy and pleasure. The celebration of sexuality today is most likely found somewhere between the traditional, rigid, repressive morality that is our sociosexual heritage, and a new performance-oriented, irresponsible, self-seeking mentality. However, there are many signs of growing public disagreement and controversy as to where that appropriate balance exists.

In trying to understand sexuality, our goal is to seek a joyful acceptance of being sexual, and to express this awareness in the most considerate way for ourselves and our sexual partners, while at the same time taking personal and social consequences into account. This anthology is aimed at helping all of us achieve this goal.

The articles selected for this edition cover a wide range of important topics and were written primarily by professionals for a nonprofessional audience. In them, health educators, psychologists, sociologists, sexologists, and sex therapists writing for professional journals and popular magazines present their views on how and why sexual attitudes and behaviors are developed, maintained, and changed. This edition of *Annual Editions: Human Sexuality* is organized into six sections. *Sexuality and Society* notes historical and cross-cultural views and analyzes our constantly changing society and sexuality. *Sexual Biology, Behavior, and Orientation* explains the functioning and responses of the human body, and a range of common, and not-so-common, sexual behaviors and practices. *Interpersonal Relationships* provides suggestions for establishing and maintaining intimate, responsible, quality relationships. *Reproduction* discusses some recent trends related to pregnancy and childbearing and deals with reproductive topics including conception, contraception, and abortion. *Sexuality Through the Life Cycle* looks at what happens sexually throughout one's lifetime—from childhood to the later years. Finally, *Old/New Sexual Concerns* deals with such topics as sexual hygiene, sexually transmitted diseases (STDs), sexual abuse, harassment, violence and rape, and issues and dynamics of male and female relationships.

The articles in this anthology have been carefully reviewed and selected for their quality, currency, and interest. They present a variety of viewpoints. Some you will agree with, some you will not, but you will learn from all of them.

Appreciation and a thank-you go to Susan Bunting for her work and expertise. We feel that *Annual Editions: Human Sexuality 94/95* is one of the most useful and up-to-date books available. Please let us know what you think. Return the article rating form on the last page of this book with your suggestions and comments. Any book can be improved. This one will continue to be—annually.

Ollie Pocs
Editor

Contents

Sexuality and Society.

Seven selections consider sexuality from historical and cross-cultural perspectives and examine today's changing attitudes toward human sexual interaction.

To the Reader	iv
Article List	1
Topic Guide	2
Overview	4

A. HISTORICAL AND CROSS-CULTURAL PERSPECTIVES

1. **Sexuality and Spirituality: The Relevance of Eastern Traditions,** Robert T. Francoeur, *SIECUS Report,* April/May 1992. — 6

   Although few Americans still associate sexuality with Adam and Eve's original sin, ***Western culture lacks sex-positive views.*** Robert Francoeur examines Eastern spirituality's Tantrism and Taoism, advocating an Eastern-Western merger "for sexual relations as expressions of love, passion, commitment, procreation, playful fun, friendship, as well as nuptial transcendence and spiritual oneness."

2. **Cross-Cultural Perspectives on Sexuality Education,** Jay Friedman, *SIECUS Report,* August/September 1992. — 13

   A European summer vacation provided this author opportunities to observe and compare messages ***European adolescents*** receive with those in the United States. The article considers the similarities and differences in formal ***sex education and sexual messages*** in families.

3. **Family, Work, and Gender Equality: A Policy Comparison of Scandinavia, the United States, and the Former Soviet Union,** Elina Haavio-Mannila, *SIECUS Report,* August/September 1993. — 20

   A look at family and work roles of ***men and women across several cultures*** brings some surprising differences. The author identifies the United States as unique in its reluctance to address family and gender roles and asserts that these policies discourage people from having and rearing children.

4. **Beyond Abortion: Transforming the Pro-Choice Movement,** Marlene Gerber Fried, *Social Policy,* Summer 1993. — 25

   Abortion . . . choice . . . rights . . . have been personal and political topics for over 25 years. But, has the focus narrowed too much? This author says so and asserts that access and other practical considerations are more meaningful for poor, rural, disadvantaged ***women in the United States and around the globe.***

B. CHANGING SOCIETY/CHANGING SEXUALITY

5. **The Brave New World of Men,** Diane Crispell, *American Demographics,* January 1992. — 31

   Although most men shop, do housework, or care for children only because women tell them to, things are changing. Television, films, and advertisers are capitalizing on ***a "new" image of men***—an image that men are more romantic, share shopping and cooking, and help with child care. Read all about who is changing (and who is not).

6. **Sex in the Snoring '90s,** Jerry Adler, *Newsweek,* April 26, 1993. — 36

   What is normal? What do other people do? These are common questions. When it comes to sex, it can be easy to feel intimidated by the media accounts of who does what with whom. The Battelle Human Affairs Research Center interviewed over 3,000 American men about their ***sexual habits*** and found the results less than astounding.

7. **The New Sexual Revolution: Liberation at Last? Or the Same Old Mess?** Jim Walsh, *Utne Reader,* July/August 1993. — 39

   The sexually liberated seventies. The fear of herpes, then AIDS, eighties. What about the nineties? Jim Walsh calls them the ***decade of differences,*** with some return to the unsafe do-it-because-it-feels-good mentality, but also some very positive, responsible, imaginative, technology-aided, and playful sex.

The concepts in bold italics are developed in the article. For further expansion please refer to the Topic Guide, the Index, and the Glossary.

Sexual Biology, Behavior, and Orientation

Seven selections examine the biological aspects of human sexuality and emphasize the importance of understanding sexual hygiene.

Overview   42

A. THE BODY AND ITS RESPONSES

8. **The Five Sexes: Why Male and Female Are Not Enough,** Anne Fausto-Sterling, *The Sciences,* March/April 1993.   44

   True **hermaphrodites** are people who possess both male and female sex organs. Pseudo-hermaphrodites are genetically one gender but have sex organs of the other. Medical treatment of hermaphrodites has sought to make these people fit into a two-gender system. What if these intersexuals were not forced to do so?

9. **What Is It with Women and Breasts?** *Newsweek,* January 20, 1992.   48

   Does our society prize women more highly for their body measurements than for themselves? The ongoing obsession with breasts—by men, women, and society—seems to say so. Recent alarming findings about **breast implantations** illustrate the significant dangers associated with this obsession.

B. SEXUAL ATTITUDES AND PRACTICES

10. **Why Do We Know So Little About Human Sex?** Anne Fausto-Sterling, *Discover,* June 1992.   50

    We agree that we need to stop the spread of AIDS and other sexually transmitted diseases. However, many in the scientific community contend that a **lack of current information about what men and women do sexually** blocks this effort. Unfortunately, fear is keeping human sexuality in the dark.

11. **Sexual Arousal of College Students in Relation to Sex Experiences,** Peter R. Kilmann, Joseph P. Boland, Melissa O. West, C. Jean Jonet, and Ryan E. Ramsey, *Journal of Sex Education & Therapy,* Fall 1993.   53

    Readers are likely to find these findings based on 362 **students in undergraduate human sexuality classes** quite interesting. Males and females, virgins and nonvirgins, those who have sex only with affection versus those who have casual sex varied little on measures of **sexual arousability.**

C. SEXUAL ORIENTATION

12. **Homosexuality, the Bible, and Us—A Jewish Perspective,** Dennis Prager, *The Public Interest,* Summer 1993.   57

    Homosexuality remains one of the most anxiety-raising and heated debates of our time. Few realize that homosexuality was once accepted, even revered, in the world. Dennis Prager's **comprehensive history of homosexuality** and society sheds oft-ignored light on values, rights, preferences, behavior, fears, and discrimination, helping thoughtful people clarify their beliefs.

13. **Homosexuality and Biology,** Chandler Burr, *The Atlantic,* March 1993.   66

    Science has studied homosexuality for decades. First, psychiatry labeled it psychopathology, then reversed itself. Recently, biology, neurobiology, genetics, and endocrinology have studied brain, chemical, hormonal, and prenatal differences between male, female, homosexual, bisexual, and heterosexual animals and humans. Taken together, scientific evidence demonstrates immutability and **biological contributions to sexual orientation.**

14. **The Gay Debate: Is Homosexuality a Matter of Choice or Chance?** Meredith F. Small, *American Health,* March 1993.   76

    Although there is now general acceptance of some **biological and/or genetic contributions to homosexuality,** there is no agreement about the extent of the contributions or how they happen. Questions about early childhood influences, intervening personality traits, social, peer, and family roles continue to muddy the choice-or-chance debate.

The concepts in bold italics are developed in the article. For further expansion please refer to the Topic Guide, the Index, and the Glossary.

Unit 3

Interpersonal Relationships

Five selections examine the dynamics of establishing sexual relationships and the need to make these relationships responsible and effective.

Unit 4

Reproduction

Ten articles discuss the roles of both males and females in pregnancy and childbirth and consider the influences of the latest birth control methods and practices on individuals and society as a whole.

Overview 80

A. ESTABLISHING SEXUAL RELATIONSHIPS

15. **The Mating Game,** U.S. News & World Report, July 19, 1993. 82

    From Romeo and Juliet, to Tristram and Isolde, to Julia Roberts and Lyle Lovett, not much has changed in the mating game, according to scientists. The **chemistry of attraction,** the dances of intrigue, assessment, and negotiation; the patterns of male and female arousal and behavior are age-old.

16. **Ways of Loving: A Matter of Style,** Diane Swanbrow, American Health, May 1991. 86

    Are you a cuddler? Romantic? Easily trusting? Attentive to your partner? Often anxious or distant in relationships? The answers to these and similar questions denote **your loving or attachment style,** according to Diane Swanbrow, and they are partly a set of expectations built on your past and present experiences.

B. RESPONSIBLE QUALITY RELATIONSHIPS

17. **What Is Love?** Paul Gray, Time, February 15, 1993. 88

    Until recently, few scientists studied love. Love was viewed as too trivial, too "mushy," or a cultural delusion. However, recently the volume of scientific research on love has skyrocketed. Identifying a **biological predisposition to love** or confirming the near-universality of love cross-culturally seems unlikely to rob love of its mystery.

18. **The Lessons of Love,** Beth Livermore, Psychology Today, March/April 1993. 90

    As scientists have taken up **the study of love,** they have found many loves. Passionate love has a long evolution as a bonding mechanism. Romantic love is not as universally evident, since it is very culture-bound. Read about other kinds of love and take the Colors of Love self-test.

19. **Forecast for Couples,** Barry Dym and Michael Glenn, Psychology Today, July/August 1993. 97

    The family therapist authors of this article describe the **process of intimate relationships highlighting** endless circles of advance and retreat, conflict and resolution. Differences between couples who survive, even thrive, and those who become frustrated, drift apart, or sever their ties are thought-provoking and helpful.

Overview 102

A. BIRTH CONTROL

20. **Choosing a Contraceptive,** Merle S. Goldberg, FDA Consumer, September 1993. 104

    Even in the 1990s, straightforward, accurate, and **complete information on contraception is often hard to find.** This article meets all of these criteria for the full range of methods available, while addressing effectiveness, risks/safety, and lifestyle factors and disease transmission/prevention.

21. **The Female Condom,** Beth Baker, Ms., March/April 1993. 111

    Two varieties of **a condom for females**—Reality and Femidom—are now available to women and couples. Although its appearance evokes surprise, many predict its heightened disease prevention, female control, and polyurethane's greater strength and heat transmission (compared to the male latex version) may make it a viable choice for some.

The concepts in bold italics are developed in the article. For further expansion please refer to the Topic Guide, the Index, and the Glossary.

22. **The Norplant Debate,** Barbara Kantrowitz and Pat Wingert, *Newsweek,* February 15, 1993.   **113**

Norplant has been welcomed as an effective and relatively *long-term addition to the current array of contraceptive methods.* However, its status as a method recommended or required for young, poor, or nonvolunteer women is hotly debated. Proponents cite hope and a better future, while opponents cry coercion, genocide, and immorality.

23. **New, Improved and Ready for Battle,** Jill Smolowe, *Time,* June 14, 1993.   **116**

A scientific breakthrough in the use of the controversial French "abortion pill" *RU 486* promises to greatly simplify and make quite private the process of ending a pregnancy. Although not yet available in the United States, it should be soon. Battle lines are clearly drawn between proponents and opponents.

24. **Desperation: Before *Roe v. Wade,* After *Roe* Is Reversed,** Linda Rocawich, *The Progressive,* January 1992.   **119**

Women had *abortions* before they were legalized by *Roe v. Wade,* have them while legal under *Roe v. Wade,* and still will, according to many people, including the writer of this article, if (or when) *Roe v. Wade* is repealed. This article does not focus on the law, but on how women feel about the prospect—desperate.

25. **The Global Politics of Abortion,** Jodi Jacobson, *Utne Reader,* March/April 1991.   **123**

This article asserts that *the abortion controversy* is not a conflict between black and white views. It is, instead, "a canvas of social morality painted in every imaginable hue" that involves religious, ethnic, racial, and rationalistic questions as well as male, female, and adult questions.

B.   *PREGNANCY AND CHILDBIRTH*

26. **Making Babies: Miracle or Marketing Hype? Risks, Caveats and Costs,** Elayne Clift, *On the Issues,* Spring 1993.   **128**

As infertility clinics and *technologically assisted fertility* treatments proliferate, questions are mounting. Health, economic, legal, and ethical questions involve a growing number of multiple births often requiring costly neonatal services; the lack of standards or regulations on procedures; and ethical and humanitarian concerns about the rarely publicized high failure rates.

27. **Reproductive Revolution Is Jolting Old Views,** Gina Kolata, *New York Times,* January 11, 1994.   **131**

Some of the latest *technological possibilities for human reproduction* are reviewed in this article; no longer are menopause or infertility impediments to having a baby.

28. **Coping with Miscarriage,** Dena K. Salmon, *Parents,* May 1991.   **133**

A *miscarriage* can be a very devastating experience. Trying to talk yourself, your partner, or a friend out of feeling and grieving is ineffective and undesirable. This article addresses common feelings and experiences, and things that can help women and men deal with this fairly common but very difficult event.

29. **Unnecessary Cesarean Sections: Halting a National Epidemic,** *Health Letter,* Public Citizen Health Research Group, June 1992.   **136**

In this publication of a national health research and advocacy group, figures as high as 48.9 percent are cited for the proportion of *unnecessary cesarean sections.* Because of the risks and costs to both mother and babies of these operations, a call for action is made, especially on the part of childbearing women.

Unit 5

Sexuality Through the Life Cycle

Eight articles consider human sexuality as an important element throughout the life cycle. Topics include responsible adolescent sexuality, sex in and out of marriage, and sex in old age.

Overview   142

A. YOUTH AND THEIR SEXUALITY

30. **Raising Sexually Healthy Kids,** Elizabeth Fishel, *Parents,* September 1992.   144

    All parents want to raise ***sexually healthy children.*** Few parents want to carry on the sexual secrecy of the past, but most parents are anxious about how (and when) to talk to children about sex. Elizabeth Fishel provides an excellent guide with important tips for gauging children's readiness and appropriate behavior.

31. **Truth and Consequences: Teen Sex,** Douglas J. Besharov and Karen N. Gardiner, *The American Enterprise,* January/February 1993.   149

    Studies have demonstrated that every year more teenagers are having more sex. Social costs include rising abortion rates, out-of-control births, welfare, and sexually transmitted diseases. What can (and should) be done is a topic of heated nationwide disagreement. These authors caution that the ***availability of contraception is not equal to its use.***

32. **Single Parents and Damaged Children: The Fruits of the Sexual Revolution,** Lloyd Eby and Charles A. Donovan, *The World & I,* July 1993.   156

    This comprehensive review of ***sociocultural changes in sexual attitudes*** and behaviors over the last 30 years focuses on teenage sexual activity and consequences. Pointing out the failures of efforts to reduce out-of-wedlock births, it concludes with more questions than answers about future costs and possible solutions.

B. SEXUALITY IN THE ADULT YEARS

33. **Let the Games Begin: Sex and the *Not*-Thirtysomethings,** Simon Sebag Montefiore, *Psychology Today,* March/April 1993.   162

    Today's ***30–40-year-olds*** are in the middle in several sexual senses. Safe sex may conflict with "getting it while you can." A strong focus on attractiveness, looking young, greater sexual openness and assertiveness, and ticking matrimonial and biological clocks make this a sexually active decade for most.

34. **25 Ways to Make Your Marriage Sexier,** Lynn Harris, *Ladies' Home Journal,* May 1993.   166

    Keeping the heat in the passion, the playfulness in the sex play, and the excitement in marital sex require both people's attention and involvement. An array of suggestions from experts, as well as husbands and wives who are ***cultivating the sexual side of their marriages,*** offers something for everyone.

35. **Beyond Betrayal: Life After Infidelity,** Frank Pittman III, *Psychology Today,* May/June 1993.   169

    Although most first affairs are accidental and unintended, some people, according to Frank Pittman (who wrote *Private Lies),* intentionally and regularly violate their marital vows. Providing a list of commonly believed ***myths about infidelity*** and a typology of varieties of infidelity, Pittman exposes the realities of this often-damaging script.

36. **What Doctors and Others Need to Know: Six Facts on Human Sexuality and Aging,** Richard J. Cross, *SIECUS Report,* June/July 1993.   176

    A doctor in his late 70s, this author challenges doctors and others to remember that ***older people are sexual beings.*** He informs readers about changes in sexual functioning, but identifies negative social attitudes as most sexually frustrating. He ends optimistically with reasons why older lovers can be better lovers.

37. **Sexuality and Aging: What It Means to Be Sixty or Seventy or Eighty in the '90s,** *Mayo Clinic Health Letter,* February 1993.   179

    Contrasting myths with realities this informative newsletter discusses expected or ***common changes in aging women and men.*** It offers advice for oldsters (and all of us) on staying healthy and sexually active as well as simple, easy-to-understand descriptions of medical and nonmedical treatments for common sexual problems.

The concepts in bold italics are developed in the article. For further expansion please refer to the Topic Guide, the Index, and the Glossary.

Old/New Sexual Concerns

Twelve selections discuss ongoing sexual concerns of sexual hygiene, sexual harassment and violence, and gender roles.

Overview 184

A. *SEXUAL HYGIENE*

38. **The Future of AIDS,** William A. Haseltine, *Priorities,* Winter 1993. 186

    William Haseltine describes **the AIDS epidemic** as illustrative of the strengths and weaknesses of all humankind. Although he paints a bleak picture for finding a vaccine or cure and estimates that 100 million people will be infected by the year 2000, the changes he espouses in individual and collective behavior are very possible.

39. **Campuses Confront AIDS: Tapping the Vitality of Caring and Community,** Richard P. Keeling, *Educational Record,* Winter 1993. 189

    Changes in HIV/AIDS policies involve a shift from theoretical preparedness to prevention efforts and caring for the sick. Among the components of effective campus policies are recognition of HIV as a campus concern, policies to reduce occupational risk, educational programs focused on behavior, and the building of a caring community.

40. **Preventing STDs,** Judith Levine Willis, *FDA Consumer,* June 1993. 195

    This very practical **consumer's guide to condoms** was written as part of a health series for teenagers. However, its easy-to-understand information about selecting and using condoms includes helpful hints for people of all ages.

41. **Syphilis in the '90s,** Jane R. Schwebke, *Medical Aspects of Human Sexuality,* April 1991. 198

    Syphilis, one of the "old" STDs, is on the rise again. Since 1985 infection rates have risen so dramatically that we all need to be reminded of the symptoms of this serious, but treatable, STD.

B. *SEXUAL ABUSE AND VIOLENCE*

42. **A Question of Abuse,** Nancy Wartik, *American Health,* May 1993. 203

    Recovered-memory phenomenon is the name for retrieved memories and incremental flashbacks of childhood trauma or abuse. The psychotherapeutic community is divided about whether these reemerging recollections fit with what is known about memory storage and retrieval. Effects on individuals, families, and legal questions make the exposure of these memories often problematic.

43. **The Sexually Abused Boy: Problems in Manhood,** Diana M. Elliott and John Briere, *Medical Aspects of Human Sexuality,* February 1992. 207

    Child sexual abuse happens to one-fourth of all children. Nearly one-fourth of those abused are boys. Differences in the abuse experienced and the aftereffects of the abuse, especially the boys' reluctance to disclose the abuse, seem to predispose the victims to more sexual dysfunction, somatic and psychological distress, and substance abuse.

44. **Who's to Blame for Sexual Violence?** Diana Scully, *USA Today Magazine (Society for the Advancement of Education),* January 1992. 210

    Rape and sexual violence will continue to be prevalent, according to this author, until we move from considering them to be women's problems to becoming men's problems. Diana Scully provides some insights into how the responsibility and blame have been misassigned, using some convicted rapists' perceptions as examples.

C. *MALES/FEMALES—WAR AND PEACE*

45. **Sex Differences in the Brain,** Doreen Kimura, *Scientific American,* September 1992. 213

    Men and women think differently. They solve problems differently. Both of these differences are significantly affected, according to Doreen Kimura, by hormones. New findings also suggest that individual **men and women may experience fluctuations over their lifetimes,** even seasonal changes, in their brain functioning due to hormones.

The concepts in bold italics are developed in the article. For further expansion please refer to the Topic Guide, the Index, and the Glossary.

46. **Women and Men: Can We Get Along? Should We Even Try?** Lawrence Wright, *Utne Reader,* January/February 1993. 220

According to Lawrence Wright, **animosity between women and men** has hit a boiling point similar to that preceding the birth of the 1970s Women's Liberation Movement. The every-man-a-rapist and androgens-contribute-to-violence-and-criminal-behavior perspectives blame all men and threaten men, women, and children relationships and families.

47. **It's a Jungle Out There, So Get Used to It!** Camille Paglia, *Utne Reader,* January/February 1993. 224

Camille Paglia asserts that **feminists and their position on rape** as an act of violence, not sex, have misled, even endangered, millions of young women, especially college students. She counters that women should remember that, for men, aggression and eroticism are deeply entwined, and women's best response combines self-awareness, self-control, and action.

48. **The Blame Game,** Sam Keen, *Utne Reader,* January/February 1993. 226

Sam Keen asserts that it is time to confront feminist analysis and worldview in order to sort out the healing treasures from the toxic trash. Only by doing so can we move past blame to equality and **new and better ways** of relating to each other.

49. **Ending the Battle Between the Sexes,** Aaron R. Kipnis and Elizabeth Herron, *Utne Reader,* January/February 1993. 229

Men and women are embroiled in anger, resentment, and mistrust. Bitter accusations and a dearth of compassion threaten to widen the gender gap. These authors make a plea for a negotiated peace where women's and men's movement goals become our goals and gender stereotypes give way to **gender understanding and acceptance.**

Glossary 235
Index 241
Article Review Form 244
Article Rating Form 245

ARTICLE LIST

The following list of article titles, article numbers (in parenthesis), and page numbers are listed alphabetically for convenience and easy reference:

Beyond Abortion (4) 25	New, Improved and Ready for Battle (23) 116
Beyond Betrayal: Life After Infidelity (35) 169	New Sexual Revolution (7) 39
Blame Game (48) 226	Norplant Debate (22) 113
Brave New World of Men (5) 31	Preventing STDs (40) 195
Campuses Confront AIDS (39) 189	Question of Abuse (42) 203
Choosing a Contraceptive (20) 104	Raising Sexually Healthy Kids (30) 144
Coping with Miscarriage (28) 133	Reproductive Revolution Is Jolting Old Views (27) ... 131
Cross-Cultural Perspectives on Sexuality Education (2) 13	Sex Differences in the Brain (45) 213
Desperation: Before *Roe v. Wade,* After *Roe* Is Reversed (24) 119	Sex in the Snoring '90s (6) 36
Ending the Battle Between the Sexes (49) 229	Sexual Arousal of College Students in Relation to Sex Experiences (11) 53
Family, Work, and Gender Equality (3) 20	Sexuality and Aging (37) 179
Female Condom (21) 111	Sexuality and Spirituality (1) 6
Five Sexes: Why Male and Female Are Not Enough (8) ... 44	Sexually Abused Boy: Problems in Manhood (43) 207
Forecast for Couples (19) 97	Single Parents and Damaged Children (32) 156
Future of AIDS (38) 186	Syphilis in the '90s (41) 198
Gay Debate: Is Homosexuality a Matter of Choice or Chance? (14) 76	Truth and Consequences: Teen Sex (31) 149
Global Politics of Abortion (25) 123	25 Ways to Make Your Marriage Sexier (34) 166
Homosexuality and Biology (13) 66	Unnecessary Cesarean Sections (29) 136
Homosexuality, the Bible, and Us—A Jewish Perspective (12) 57	Ways of Loving: A Matter of Style (16) 86
It's a Jungle Out There, So Get Used to It! (47) 224	What Doctors and Others Need to Know: Six Facts on Human Sexuality and Aging (36) 176
Lessons of Love (18) 90	What Is It with Women and Breasts? (9) 48
Let the Games Begin: Sex and the *Not*-Thirtysomethings (33) 162	What Is Love? (17) 88
	Who's to Blame for Sexual Violence? (44) 210
Making Babies: Miracle or Marketing Hype? (26) 128	Why Do We Know So Little About Human Sex? (10) 50
Mating Game (15) 82	Women and Men: Can We Get Along? (46) 220

The concepts in bold italics are developed in the article. For further expansion please refer to the Topic Guide, the Index, and the Glossary.

Topic Guide

This topic guide suggests how the selections in this book relate to topics of traditional concern to students and professionals involved with the study of human sexuality. It is useful for locating articles that relate to each other for reading and research. The guide is arranged alphabetically according to topic. Articles may, of course, treat topics that do not appear in the topic guide. In turn, entries in the topic guide do not necessarily constitute a comprehensive listing of all the contents of each selection.

TOPIC AREA	TREATED IN:	TOPIC AREA	TREATED IN:
Abortion	2. Cross-Cultural Perspectives 4. Beyond Abortion 23. New, Improved and Ready for Battle 24. Desperation 25. Global Politics of Abortion	Attitudes/Values (cont'd) Birth Control/ Contraception	48. Blame Game 49. Ending the Battle Between the Sexes 2. Cross-Cultural Perspectives 4. Beyond Abortion 20. Choosing a Contraceptive
Abuse, Sexual/Rape	7. New Sexual Revolution 30. Raising Sexually Healthy Kids 42. Question of Abuse 43. Sexually Abused Boy 44. Who's to Blame for Sexual Violence? 46. Women and Men 47. It's a Jungle Out There 48. Blame Game 49. Ending the Battle Between the Sexes		21. Female Condom 22. Norplant Debate 23. New, Improved and Ready for Battle 24. Desperation 25. Global Politics of Abortion 31. Truth and Consequences
		Gender/Sex Roles	2. Cross-Cultural Perspectives 3. Family, Work, and Gender Equality 5. Brave New World of Men 6. Sex in the Snoring '90s 7. New Sexual Revolution 8. Five Sexes 9. What Is It with Women and Breasts? 12. Homosexuality, the Bible, and Us 13. Homosexuality and Biology 14. Gay Debate 15. Mating Game 19. Forecast for Couples 30. Raising Sexually Healthy Kids 33. Let the Games Begin 44. Who's to Blame for Sexual Violence? 45. Sex Differences in the Brain 46. Women and Men 47. It's a Jungle Out There 48. Blame Game 49. Ending the Battle Between the Sexes
Aging	33. Let the Games Begin 36. What Doctors and Others Need to Know 37. Sexuality and Aging		
AIDS	2. Cross-Cultural Perspectives 6. Sex in the Snoring '90s 7. New Sexual Revolution 21. Female Condom 33. Let the Games Begin 38. Future of AIDs 39. Campuses Confront AIDS 40. Preventing STDs		
Attitudes/Values	1. Sexuality and Spirituality 2. Cross-Cultural Perspectives 3. Family, Work, and Gender Equality 4. Beyond Abortion 5. Brave New World of Men 7. New Sexual Revolution 8. Five Sexes 9. What Is It with Women and Breasts? 10. Why Do We Know So Little About Human Sex? 11. Sexual Arousal of College Students 12. Homosexuality, the Bible, and Us 14. Gay Debate 16. Ways of Loving 17. What Is Love? 18. Lessons of Love 19. Forecast for Couples 20. Choosing a Contraceptive 22. Norplant Debate 24. Desperation 25. Global Politics of Abortion 30. Raising Sexually Healthy Kids 31. Truth and Consequences 32. Single Parents and Damaged Children 33. Let the Games Begin 36. What Doctors and Others Need to Know 38. Future of AIDs 39. Campuses Confront AIDS 43. Sexually Abused Boy 44. Who's to Blame for Sexual Violence? 46. Women and Men 47. It's a Jungle Out There	Health	1. Sexuality and Spirituality 2. Cross-Cultural Perspectives 4. Beyond Abortion 7. New Sexual Revolution 9. What Is It with Women and Breasts? 10. Why Do We Know So Little About Human Sex? 13. Homosexuality and Biology 20. Choosing a Contraceptive 21. Female Condom 22. Norplant Debate 23. New, Improved and Ready for Battle 26. Making Babies 27. Reproductive Revolution 28. Coping with Miscarriage 29. Unnecessary Cesarean Sections 36. What Doctors and Others Need to Know 37. Sexuality and Aging 38. Future of AIDs 39. Campuses Confront AIDS 40. Preventing STDs 41. Syphilis in the '90s 43. Sexually Abused Boy 45. Sex Differences in the Brain
		Homosexuality/ Bisexuality	2. Cross-Cultural Perspectives 7. New Sexual Revolution 8. Five Sexes

TOPIC AREA	TREATED IN:	TOPIC AREA	TREATED IN:
Homosexuality/ Bisexuality (cont'd)	10. Why Do We Know So Little About Human Sex? 12. Homosexuality, the Bible, and Us 13. Homosexuality and Biology 14. Gay Debate 30. Raising Sexually Healthy Kids 38. Future of AIDS 43. Sexually Abused Boy	Parents (cont'd)	31. Truth and Consequences 32. Single Parents and Damaged Children 34. 25 Ways to Make Your Marriage Sexier 42. Question of Abuse 43. Sexually Abused Boy 46. Women and Men
Intimacy, Sexual	1. Sexuality and Spirituality 2. Cross-Cultural Perspectives 5. Brave New World of Men 6. Sex in the Snoring '90s 7. New Sexual Revolution 10. Why Do We Know So Little About Human Sex? 11. Sexual Arousal of College Students 14. Gay Debate 15. Mating Game 19. Forecast for Couples 20. Choosing a Contraceptive 28. Coping with Miscarriage 33. Let the Games Begin 34. 25 Ways to Make Your Marriage Sexier 35. Beyond Betrayal 36. What Doctors and Others Need to Know 37. Sexuality and Aging 38. Future of AIDS 46. Women and Men 47. It's a Jungle Out There 49. Ending the Battle Between the Sexes	Pregnancy	2. Cross-Cultural Perspectives 3. Family, Work, and Gender Equality 4. Beyond Abortion 24. Desperation 26. Making Babies 27. Reproductive Revolution 28. Coping with Miscarriage 29. Unnecessary Cesarean Sections 31. Truth and Consequences
		Rape	See Abuse, Sexual/Rape
		Sex Education	1. Sexuality and Spirituality 2. Cross-Cultural Perspectives 10. Why Do We Know So Little About Human Sex? 20. Choosing a Contraceptive 30. Raising Sexually Healthy Kids 31. Truth and Consequences 36. What Doctors and Others Need to Know 37. Sexuality and Aging 38. Future of AIDS 39. Campuses Confront AIDS 40. Preventing STDs 43. Sexually Abused Boy 49. Ending the Battle Between the Sexes
Legal/Ethical	2. Cross-Cultural Perspectives 3. Family, Work, and Gender Equality 4. Beyond Abortion 7. New Sexual Revolution 14. Gay Debate 22. Norplant Debate 23. New, Improved and Ready for Battle 24. Desperation 25. Global Politics of Abortion 26. Making Babies 27. Reproductive Revolution 29. Unnecessary Cesarean Sections 32. Single Parents and Damaged Children 38. Future of AIDS 39. Campuses Confront AIDS 42. Question of Abuse	Sexual Dysfunction	2. Cross-Cultural Perspectives 6. Sex in the Snoring '90s 8. Five Sexes 10. Why Do We Know So Little About Human Sex? 19. Forecast for Couples 26. Making Babies 28. Coping with Miscarriage 36. What Doctors and Others Need to Know 37. Sexuality and Aging 43. Sexually Abused Boy 46. Women and Men 47. It's a Jungle Out There
Media/Messages	1. Sexuality and Spirituality 2. Cross-Cultural Perspectives 3. Family, Work, and Gender Equality 5. Brave New World of Men 7. New Sexual Revolution 9. What Is It with Women and Breasts? 17. What Is Love? 18. Lessons of Love 19. Forecast for Couples 30. Raising Sexually Healthy Kids 46. Women and Men 48. Blame Game 49. Ending the Battle Between the Sexes	STDs (Sexually Transmitted Diseases)	2. Cross-Cultural Perspectives 7. New Sexual Revolution 20. Choosing a Contraceptive 31. Truth and Consequences 40. Preventing STDs 41. Syphilis in the '90s
Parents	3. Family, Work, and Gender Equality 5. Brave New World of Men 12. Homosexuality, the Bible, and Us 19. Forecast for Couples 26. Making Babies 30. Raising Sexually Healthy Kids	Therapy/Counseling	2. Cross-Cultural Perspectives 7. New Sexual Revolution 8. Five Sexes 16. Ways of Loving 19. Forecast for Couples 34. 25 Ways to Make Your Marriage Sexier 35. Beyond Betrayal 37. Sexuality and Aging 39. Campuses Confront AIDS 42. Question of Abuse 43. Sexually Abused Boy

Sexuality and Society

- Historical and Cross-Cultural Perspectives (Articles 1–4)
- Changing Society/Changing Sexuality (Articles 5–7)

People of different civilizations in different historical periods have engaged in a variety of modes of sexual expression and behavior. Despite this cultural and historical diversity, one important principle should be kept in mind: Sexual awareness, attitudes, and behaviors are learned within sociocultural contexts that define appropriate sexuality for society's members. Our sexual attitudes and behaviors are in large measure social and cultural phenomena.

For several centuries, Western civilization has been characterized by an "antisex ethic" that encompasses a norm of denial and beliefs that sex is bad unless it is controlled or proscribed in certain ways. These certain ways have usually meant that sexual behavior should be confined to monogamous heterosexual pair bonds (marriages) for the sole purpose of procreation. Fear, myth, and lack of factual information or dialogue have maintained antisex, and many say harmful, beliefs, customs, and behaviors. Today, changes in our social environment—the widespread availability of effective contraception, the liberation of women from the kitchen, the reconsideration of democratic values of individual freedom and the pursuit of happiness, and increasing open dialogue about sex and sexuality—are strengthening our concept of ourselves as sexual beings and posing a challenge to the antisex ethic that has traditionally served to orient sexuality.

As a rule, social change is not easily accomplished. Sociologists generally acknowledge that changes in the social environment are accompanied by the presence of

Unit 1

interest groups that confront and question existing beliefs, norms, and behavior. The contemporary sociocultural changes with respect to sexuality are highly illustrative of such social dynamics. Many of the articles in this section document changes in the social environment while highlighting the questions, confrontations, and beliefs of different groups about what are or should be our social policies regarding sexuality. The articles also illustrate the diversity of beliefs regarding what was beneficial or detrimental about the past, and what needs to be preserved or changed for a better future.

The fact that human sexuality is primarily a learned behavior can be both a blessing and a curse. The learning process enables humans to achieve a range of sexual expression and meaning that far exceeds their biological programming. Unfortunately, however, our society's lingering antisex ethic tends to foreclose constructive learning experiences and contexts, often driving learning underground. What is needed for the future is high-quality pervasive sex education to counteract the locker room, commercial sex, and the trial-and-error contexts in which most individuals in our society acquire misinformation, anxiety, and fear—as opposed to knowledge, reassurance, and comfort—about themselves as sexual people.

A view of the past illustrates the connectedness of our values and perceptions of sexuality with other sociopolitical events and beliefs. Cross-cultural perspectives provide common human patterns and needs with respect to sexuality and other interpersonal issues. Several of the articles in this section describe and examine patterns of change in political, economic, medical, and educational spheres as they relate to sexuality, including puberty, orgasm, pregnancy, and childbirth, while exploring and challenging present sociocultural, educational, medical, and legal practices in controversial areas, including education in school and public media, sex roles and expectations, AIDS, abortion, homosexuality, adolescent pornography, and censorship. Although the authors may not agree on the desirability of the changes they describe nor advocate the same future directions, they do emphasize the necessity for people to have information and awareness about a wide range of sexual topics. They also share another belief: We as individuals and as a world society have a vital interest in the translation of social consciousness and sexuality into a meaningful and rewarding awareness and expression for all of society's members.

The first subsection, *Historical and Cross-Cultural Perspectives*, contains four articles on sexuality-related issues in settings as varied as the Netherlands, Belgium, the former Czechoslovakia, Denmark, Sweden, Finland, the former Soviet Union, and the United States. Each article challenges the reader with questions of why cultures dictate and/or proscribe certain beliefs, values, or practices. They also link misinformation and lack of access to accurate or complete information about sexuality to problematic personal, interpersonal, and societal consequences. Each article connects each country's sociocultural beliefs to its sex education, sex role, and prescribed and proscribed sexual behaviors. Each article calls for increased awareness, understanding, and acceptance as avenues for more positive sexual experiences for individuals and improved sexual health.

The three articles that make up the second subsection, *Changing Society/Changing Sexuality*, illustrate the varied, even conflicting, nature of the changes occurring, advocated, and predicted in the sexual arena. The first two articles look at men, male roles, and sexuality. Be ready for some thought-provoking information and a few surprises. The third article describes what some people are labeling the Sexual Revolution of the nineties, where fears have decreased, and imagination and variation have increased. Several of these articles include predictions for the future direction of sociocultural change. Readers will have many opportunities to ponder their own beliefs, experiences, and feelings with respect to sexuality yesterday, today, and tomorrow.

Looking Ahead: Challenge Questions

What do you see as the benefits (and risks) of a more positive view of sex in religion, spirituality, and society in general?

Have you ever spoken to a young person from another culture/country about sexuality-related ideas, norms, education, or behavior? If so, what surprised you? What did you think about their perspective or ways?

What would you change about the sex education you received and why?

What is a "real man" these days? How strong do you think the pressure to prove one's manhood is for today's men? Is it beneficial or detrimental to them, their families, and/or their lovers?

How would you expect your feelings to be different if you were poor, uneducated, and lived in a country with little access to effective contraception or safe abortion? Why?

Sex in the nineties—do you think people are more sexually active than five years ago? How much do you think fear of STDs/HIV affects sexual changes and behavior? Why?

SEXUALITY AND SPIRITUALITY

The Relevance of Eastern Traditions

Robert T. Francoeur, PhD, ACS

Professor of Human Embryology and Sexuality,
Fairleigh Dickinson University, Madison, NJ

In recent years, the age-old association of sex with Adam and Eve's original sin in the Garden of Eden has lost its meaning as individuals increasingly accept sexual desire and pleasure as a natural good. Social turmoil, technological changes, increasing recognition of personal needs, and a sexual revolution have wrecked havoc with the meaning and relevance of the traditional Judeo-Christian sexual images, icons, and myths of the purpose of sex, monogamy and male primacy over female.

Because cultures draw their life blood from their myths and archetypes, human beings are searching for new myths and archetypes.[1] At the same time, Americans in particular are increasingly fascinated by the more sex-positive images of Eastern sexual philosophies. This article outlines two major Eastern sexual and spiritual traditions, Tantrism and Taoism, within the context of Hinduism and other religions and philosophies. After contrasting these Eastern views with Western values, some practical applications that complement Western sexology are discussed.

Eastern Sources

Even when the hidden roots of Eastern sexual traditions can be detected, they are found to be far more tangled than the origins of sexual values in Judaism, Christianity, and Islam. Archaeologists have found 8,000-year-old clay images of feminine power and fertility in the pre-Indus settlements on the northwest edge of India. Similar early expressions of a great Goddess who guarantees fertility have been found, with her subordinate male consort, in regions of ancient Egypt, the Aegean, the Danube, Asia Minor, and western Asia. Between 1800 and 1500 BC, waves of migrating Indo-Aryan people moved from eastern Europe, over the mountains, and into the Indus valley of western India. Their worship of a great Goddess intermingled with the fertility religions of pre-Aryan inhabitants they conquered in the Indus River valleys.[2,3,4] Historian Karl Jaspers calls this the pre-Axial period of human consciousness.[6] In this context, Jaspers is using the term Axial to mean turning point.

According to Jaspers and others, this striking transformation in human consciousness occurred in China, India, Persia, the Middle East, and Greece with the advent of Confucius, Lao Tzu, Buddha, Zoroaster, the Jewish prophets, and the pioneering philosophers of Greece. This opened the first Axial period. Everywhere male consciousness and power gained ascendancy over the female principle. In Christianity and Islam, phallic power virtually subdued the power of the female, except for the veneration of Mary, the Virgin Mother of God. After a male God gave man dominion over nature in Eden and ancient Greece gave priority to analysis and objectification, nature became Western man's toy to control and exploit. Although feminine images of sexual power persisted in the East, they were subordinated to the phallocentric male. But unlike the West, Eastern cultures maintained a respect for nature, emphasizing that health and spirituality are only achieved when humanity respects its place in the cosmos and places itself in harmony with nature.[5,6]

1. Sexuality and Spirituality

Hinduism

In India, the amalgam of pre-Aryan fertility religions with the emerging dominance of male consciousness produced Hinduism, a generic term for the traditional religion of India. Hinduism encompasses a wide range of seemingly contradictory beliefs, including reincarnation or transmigration of souls, atheism, and a pantheon of gods and goddesses who symbolize the many attributes of an indescribable supreme principle or being. Hinduism embraces both monistic and dualistic beliefs, and contains many popular local deities and cults. Thus it is not a religion in the same sense Westerners use that term to refer to a system of clear beliefs about a personal God and a spiritual world apart from this material world.[7]

The ideal life of a Hindu male embraces a wide spectrum of roles, from the student of religion to the householder who produces a son to carry on ancestral tradition, and from the hermit who tries to achieve indifference to everything in the world he previously found desirable to the homeless wanderer who renounces all earthly ties. Passing through these four stages is the *Way of Knowledge*, an expression that denotes the spiritual path, which leads to spiritual union with the Infinite. Along the Way of Knowledge, a Hindu male can pursue four goals: *kama* (sexual love), *artha* (power and material gain), *dharma* (spiritual duty), and *moksha* (liberation).[2] The first two goals deal with desire, the last two extol duty and renunciation. Typical of Axial thinking, Hindu sacred texts explain the paths of desire only from a male viewpoint, as if desire, pleasure, and power play no role in the lives of women whose primary activities are childrearing and household duties.

This mix of desire and duty in Hinduism allows a strong tradition of sexual abstinence by celibate monks to coexist with an equally strong religious celebration of sexual pleasure in all its forms as a path to the Divine. While sexual abstinence is favored at certain stages, Hindu sexual asceticism complements the celebration of sexual desire and pleasure, unlike Christian sexual asceticism, which is rooted in the need for redemption from original sin. Most Hindus, even the ascetics and monks, view sex as something natural, to be enjoyed in moderation without repression or overindulgence.

Hindu sacred writings, devotional poetry, and annual festivals celebrate married love, the fidelity of women, and the religious power of sexual union. Hindu myths of gods and goddesses are symbolic of spiritual powers and energies within and the daily challenges of life faced by all human beings. While monotheistic Western cultures tend to objectify and personalize their God, Eastern cultures view their mythologies as psychological and metaphysical metaphors that reveal the miraculous and natural wonders of human life and its desires.

Mythology provides a key to Hindu sexual views. *Brahma*, the Creator, *Vishnu*, the Preserver, and *Shiva*, the cosmic dancer of the cycle of destruction and rebirth form the basic triad of gods in the Hindu pantheon. Hindu sexual values are expressed in images and rituals associated with Shiva and his consort, the goddess *Shakti*. Shakti has several images, appearing as *Parvati*, the gracious embodiment of sensuality and sexual delights, as *Durga*, the unapproachable, and as *Kali*, the black wild one, the helpful, awesome goddess of sex's transcendent powers.[2,8] The *lingam*, a stone or wood phallus, represents Shiva and the concentration of sexual energy by asceticism. Triangular stone sculptures of the *yoni* represent Shakti and the vulva. Mystical geometric patterns called *yantras* combine the circular lingam with triangular yoni. Used in meditation, yantras reflect the belief that sexual practices can be a way of balancing the male and female energies of one's body and experiencing cosmic unity. The worship of lingam and yoni, of Shiva and Shakti, are a regular part of public and household rituals. *Kama*, the Hindu god of love, is also believed to be present during all acts of love. He represents love and pleasure, both sensual and aesthetic. His wife, *Rati*, is the embodiment of sensual love.

Hindu scriptures include hundreds of treatises on the art of eroticism, allegedly written by the gods and sages. Only three of these manuals, the *Kama Sutra*, *Kama Shastra*, and *Ananga Ranga*, have been translated into English. The *Kama Sutra* (second century BC) discusses the spiritual aspects of sexuality, with advice on positions and techniques for increasing the sensual enjoyment of sexual intercourse. The beautifully illustrated *Ananga Ranga* or *The Theater of God* (15th century AD) describes the sexual organs and erogenous areas of men and women, the cycles of erotic passion, and an encyclopedia of lovemaking positions. This spiritual tradition of erotic love appears in temple art depicting *mithuna*, loving couples in sexual embrace. Such sculptures reached their peak in the sensitive, emotionally warm, and intensely spiritual bas-reliefs celebrating all forms of sexual behavior (except adultery and violence) that cover the 1,000-year-old "love-temples" of Khajuraho and Konarak.[9,10]

Taoist Sexual Traditions

In their quest for spiritual and physical health, including longevity and immortality, the Chinese traditionally turned to Taoism, which originated from the teachings of the sixth century BC philosopher Lao-Tzu.[7] Taoism views nature and spirit as interdependent and mutually sustaining. Tao is "the Way," the "eternally nameless" path followed by the wise, the ever-changing rhythmic source of life, and living in harmony with all things. Taoism advocates a life of simplicity, integration, cooperation, and selflessness, and has no formal dogma or church. It does not recommend asceticism or reject natural desires or cravings. It recommends self-cultivation, healthy living, and the fuller enjoyment of both earthly and heavenly joys.[2,11,12] Harmony in one's sexual desires, passions, and joys is a natural and important aspect of health. Sexuality is considered part of nature and is not associated with any kind of sin or moral guilt. In fact, lovers joined in ecstasy can experience a transcendent union with the cosmos.[13]

1. SEXUALITY AND SOCIETY: Historical and Cross-Cultural Perspectives

"...Eastern cultures maintained a respect for nature, emphasizing that health and spirituality are only achieved when humanity respects its place in the cosmos and places itself in harmony with nature."

Some Taoists have sought the secret of longevity in an alchemical formula. Others have sought longevity by bringing the body and soul into a perfect, harmonious balance,[11,12] or by transforming the male or female essence into the "Elixir of Life."[14]

Taoist sexual traditions emphasize the importance of female satisfaction in all sexual relations. It talks of "a thousand loving thrusts," and the importance of non-genital touch for both the woman and the man.[11] In order to increase the enjoyment of sexual intercourse for both women and men, Taoist exercises help a man gain control over his ejaculation, with simple but sophisticated versions of the Sensate Focus, Stop and Go, and Squeeze Exercises popularized 2,000 years later by Masters and Johnson for treatment of premature ejaculation and inhibited female arousal and orgasm.[11] Taoism teaches that men cannot experience true sexual ecstasy unless they develop the ability to control their ejaculation.

This emphasis on male ejaculation is often misinterpreted. It is not the same as coitus reservatus (withdrawal followed by ejaculation) or the "male continence" practiced by the members of the Oneida Community in the 1800s to prevent unwanted pregnancies. It is not the same as the passive lovemaking of *karezza,* an ancient technique for prolonging sexual intercourse without ejaculation, popularized by Marie Stopes in her 1920s best seller *Married Love*.

Taoism also emphasizes the difference between male orgasm and ejaculation, a distinction rediscovered by modern sexologists. According to Taoism, men deplete their energy when they are driven to ejaculate too frequently. Specific Taoist exercises can enable a man to pleasure his partner and enjoy several "non-explosive" orgasms prior to ejaculation.[13]

The early Taoist traditions recognized the greater capacity of women for sexual pleasure and their vital role in introducing men to the treasures of sexual pleasure and ecstasy. But this mutual, harmonious concern for female and male pleasure did not last. In the Han Dynasty (206-219 BC), male interests began to dominate as Taoist exercises were converted into techniques that focused on men's pleasure, including intercourse with virgins and with numerous women in order to become immortal. Women became the footbound pleasure toys of men in the T'ang Dynasty (618-906 AD). During the Manchu Dynasty (1644-1912 AD), the egalitarian Taoist sexual philosophy practically disappeared in male obsessions.[12,13]

For guidance in the customs and proprieties of society and public life, the Chinese looked to the teachings of Confucius (551-479 BC). Early Confucian thought was quite sex-positive. Only in the last thousand years of imperial rule did Confucianism adopt a negative view of sexuality.

Both Taoism and Confucianism appear to have borrowed the basic idea of two vital energies, Yin and Yang, from earlier Chinese who lived centuries before Confucius and Lao-Tzu. Everything stems from the dynamic interaction of *Yin* and *Yang*.[15]

The polarity of Yin/Yang energies is very different from the body-soul opposition that underlies Western thought. Western thought maintains a very clear split between the body and spirit or soul. In Christian thought, salvation and redemption are achieved by subjugation of the body and its passions to reason and to the spiritual soul. In both Taoism and Confucianism, the vital energies of Yin (earth, dark, receptive, female) and Yang (heaven, light, penetrating, male) are complementary rather than opposing aspects of nature. The challenge of life is to achieve a healthy, dynamic balance between these two energies.[8,12,13]

Since both Yin and Yang coexist in every man and woman, in different proportions, everyone can cultivate, balance, and unite their psychosomatic energies. In sexual play Yin and Yang are aroused and can be channeled from the lower levels to the heart and head. According to some modern interpreters, this can be done in self-pleasuring, and in both heterosexual and homosexual relations.[11,13,16]

Some Taoist masters recommend that a male release his semen according to seasonal changes and infrequently, for example, only two or three times out of ten instances of intercourse, in order to direct and transform the vital life energies. Similarly, women are taught to use proper breathing exercises and meditation as ways of circulating and transforming their Yin energy. The mutual exchange of Yin and Yang essences in intercourse and orgasm is believed to produce perfect harmony, increase vigor, and bring long life.

Tantric Sexual Traditions

Some suggest that Tantric sexual traditions were derived from ancient Chinese Taoism, or that Taoist sexuality was derived from Tantra.[13,14,17,18] Others believe that the earliest Tantric traditions predate Hinduism, Buddhism, and Taoism and that they were derived from the pre-Aryan religion of Indus Valley natives and religious symbols brought from paleolithic Europe by the Indo-Aryan invaders about 1800 BC.[19] Whatever their origins, Tantric ideas are found in Hindu, Buddhist, Jain, and Taoist writings in Nepal, Tibet, China, Japan, Thailand, and Indonesia.[2,19]

Over the centuries, the ecstatic, and at times orgiastic, cults inspired by Tantric visions of cosmic sexuality were attacked by ascetic Hindus and Buddhists, denounced by the invading Muslims, opposed by the British colonial government in India, and outlawed by the Chinese communists.

Tantra is a Sanskrit word meaning "thread" or "continuity." Tantra involves active ways of transforming one's perceptions and energies that plunge one back into the roots of personal identity to nakedly experience the truth and reality of oneself and the world. Tantric rituals are kept highly secret, and require severe discipline and every kind of physical, sexual, mental, and moral effort. Instead of recommending abstinence from the pleasures of life as celibate asceticism in other religious traditions do, Tantra cultivates the realization of an ultimate bliss in order to experience awareness of the true nature of reality, beyond all dualistic conceptions. In Philip Rawson's modern wording, Tantra urges its practitioners to "Raise your enjoyment to its highest power and then use it as a spiritual rocket-fuel."[4] The original Tantras use a cryptic "twilight" language difficult to understand. Some modern books on Tantra such as *Sexual Secrets* by Nik Douglas and Penny Slinger are filled with such symbolic terms, while other writers such as Mantak Chia mix traditional with Western terms to more clearly elucidate the meaning of esoteric terminology.[11,17,20]

Hindu Tantric Doctrine (Shaktism)

In Hinduism, Tantric rituals became associated with the worship of Shakti, Shiva's consort. Hindu Tantra reached its most profound external expression in the "love temples" of northeastern India (700-1100 AD).[7,9,10] Right-handed Shaktism is a refined philosophy that focuses on the benign side of Shakti as the energy of nature and mother-goddess. Left-handed Shaktism focuses on Durga and Kali, the violent side of Shakti, and sweeps one into conventionally forbidden expressions of natural impulses to achieve transcendence. Ritual violation of social taboos against adultery and incest, and coitus for otherwise celibate monks, are an important part of these left-handed Tantric rites.[2,19] In Victor Turner's social dialectics of structure-antistructure, Tantric taboo-breaking (anti-structure) rituals may play a vital role in maintaining the flexibility, dynamism, and creativity of a social structure or culture.[21]

Participants in the *Rite of the Five Essences,* a Tantric love ritual, for instance, use the five forbidden Ms: *madya* (wine), *mansa* (meat), *matsya* (fish), *mudra* (parched grain), and *maithuna* (sexual union) in a kind of holy communion.[2] It includes enhancement of the environment with flowers, incense, music, and candlelight, a period of meditation designed to hasten the ascent of the vital energies of the kundalini (see below), the chanting of a mantra, and the couple's visualization of themselves as an embodiment of Shiva and Shakti, the supreme couple.

Buddhist Tantric Doctrine

In Buddhism, Tantra refers to a series of teachings delivered to humans by the Buddha. According to Buddhist Tantra, the most effective means of awakening to the true nature of reality is not by intellectual pursuits, but by experiencing the state of voidness and bliss through one's own body and mind. The Buddhist Tantrik controls his/her body and its psychic powers to attain Buddhahood by coming face to face with the elemental forces of the world and transcending the desires aroused by them.

In Tibetan Buddhism, devotion to male and female deities stresses the interaction of external and internal energies.[4] *Yabyum* is the Tibetan term for the mystical experience of oneness and wholeness men and women can achieve through sexual intercourse.[22] In mystical sexual union, the male and female principles are combined in an experience that resolves all dualities and reflects the union of wisdom and compassion. Because all natural forces and the deity are a union of male and female elements, the highest and most harmonious energies are experienced in such unions as the realization of the inherent luminosity and emptiness of all phenomena.[22]

Tantra and Yoga

The system known as yoga was first mentioned in the Hindu *Upanishads* (eigth?-fifth? century BC). Yoga, literally translated from the Sanskrit as "union," means being aware without thinking. It is the silence of the mind that is broken by trying to tell another person what one experienced in a yoga meditation/exercise.

Yoga is a highly evolved technique of meditation and concentration for disciplining mind and body and purifying the senses from their bondage to limiting concepts. Yoga combines physiological and psychological methods, which involve postures, breathing, and in some cases the rhythmical repetition of proper sound-syllables or *mantras* that suppress the conscious movement of the mind in body.[23] When the whole body is disciplined to aid the gradual suspension of consciousness, one can experience a state of pure ecstasy that is without thought or sensation. In this ecstasy, the yoga practitioner may use ritual, devotion, meditation, the intellect, or physical pleasure to find a complete freeing of the true self from the external world and natural causation.[24]

Both early Tantra and Taoism adopted yogic exercises to gain access to the spiritual through physical pleasure and discipline. The central concept in sexual yoga is a physiology which conceives of the body as interconnected by many channels, or *nadis,* that are conduits for energy. Two main channels run along either side of the spinal column, connecting power centers known as *chakras,* which correspond to the Taoist *tan tien,* located between the loins and throat. The third conduit, the *susumna,* runs from its base in the perineal region to the crown chakra. The *kundalini,* named for the goddess Kali, is the powerful but latent energy source that lies coiled like a serpent at the lowest chakra. The kundalini is also believed to represent Shakti, the feminine aspect of the creative force, the serpent power or mystical fire in the subtle body. The aim of sexual yoga is to arouse the kundalini or serpent power and channel it upward.[25] Once aroused, the kundalini can be channeled upward through the seven chakras of the subtle body until it merges with the eternal Shiva to confer freedom and immortality. By redirecting the body's most basic and vital generative energies of semen and ovum to the

1. SEXUALITY AND SOCIETY: Historical and Cross-Cultural Perspectives

brain, the yoga practitioner hopes to gain spiritual energy, cosmic consciousness, and salvation, the experience of real self completely freed from earthly bonds and joined with all reality.[24]

In developing the idea of kundalini energy, the Tantriks and Taoists may have adopted earlier Persian ideas, using meditation, breathing control, postures, and finger pressure to prolong sexual intercourse without ejaculation. In the process, they added the goal of transforming and circulating the sexual energy upward in the body and in exchanges with a partner, thereby extending the enjoyment of many orgasms without ejaculation.

Orgasm and ejaculation are two distinct processes and can occur apart or together. William Hartman and Marilyn Fithian, for instance, report that men are capable of experiencing multiple orgasms as long as they do not ejaculate.[26] While most Tantric teachers urge males to avoid ejaculation at all times, Taoist teachers place more emphasis on gaining control of ejaculation rather than eliminating it altogether.[11,13]

The !Kung of Africa, Sufi mystics, and ancient and contemporary practitioners of yoga, Tantra, and Taoism, have cultivated the awakening of kundalini energy. Descriptions of these experiences bear intriguing similarities to reports of spontaneous experiences of Christian mystics and secular contemporaries. Strange as these reports sound in terms of Western physiology, their consistency and persistence over thousands of years deserve serious attention from Western scientists. There are hints in the preliminary research of neurophysicist and author Itzhak Bentov and psychiatrist Lee Sannella that a serious clinical and experimental investigation of the kundalini experience may reveal important new insights, much as modern medicine has benefited from clinical investigations of acupuncture and Ayurvedic herbal medicine.[27,28]

Blending East and West

To understand the Tantric and Taoist sexual systems and appreciate their rich messages, one has to go beyond the surface of sexual acts, rituals, and roles to get in touch with the cosmology, philosophy and world view that frame these exercises. One also has to deal with Eastern erotics, the way the Taoists and Tantriks interpret sexual feelings, ideas, fantasies, excitements, and aesthetics — what is beautiful or ugly, luscious or nauseating, dull or titillating.[29] Unfortunately, too many manuals, especially those presenting Tantric sex, are exotic recipe books or tourist brochures for a sexual Shangri-la. Fang-fu Ruan rightly notes that many books on Oriental sexology, while useful, "...are limited by either concentrating on a specialized topic or presenting a popular treatment of their subject. Some, by treating sexuality as a domain of pleasure independent of the changing contexts of medicine, religion, family life, reproductive strategies, or social control, effectively reinforce stereotypes of exotic Oriental cultures."[12]

Complicating any effort to evaluate the extent to which Westerners, raised with very different, even opposing world views and erotics, can understand, practice, and incorporate these sexual systems into their daily lives, is the fact that, while some proponents rhapsodize about the potential for ecstatic and cosmic experiences in Tantra and Taoism, very little can be actually known about the subjective experiences of men and women who practice these systems.[11,20]

These ancient traditions celebrate the naturalness of sexual pleasure and the spiritual potential of sexual relations, a view that may fit well with many people's sensitivities and yearnings. They also accept female sexuality and women's unlimited sexual potential, a view that is congenial with contemporary feminist awareness. Contemporary sexuality can be enriched and broadened by a reawakening of the experience of sexuality as integral to whole-person connectedness. It can also benefit from seeing sexual satisfaction as a fluctuating, non-goal-oriented, continuum of responses that includes pleasuring, orgasm, and ecstasy.[30] Can these ancient and yet very modern views be translated into the Western consciousness without being trapped by faddism? Advocates of yoga and acupuncture have succeeded in similar challenges.

In Western religions, spirituality refers to a loving, personal union of a human being with the Creator who has no gender or sex, although we are said to be created in "His image and likeness." In the Bible, sexual pleasure is commonly associated with an original sin — a fall from grace. Sexuality tends to be viewed as antagonistic to spiritual liberation.[31,32] In the words of Joseph Campbell, in the West, "eternity withdraws, and nature is corrupt, nature has fallen...we live in exile."[1] Neither Hinduism nor Buddhism have a concept of an original sin or primeval fall. Tantric and Taoist sexual union is viewed as a way to spiritual liberation, a consciousness of and identification with the Divine, and a way of becoming enlightened through one's embodiment and interaction with another. Can Western religious thought incorporate these sex-affirming Eastern views without scrapping much of our religious myths and beliefs? Can the spiritual and cosmic sense of sexuality be expressed in a Western world view without sanitizing or weakening sexual passion, or reducing its playful element?

> *"These ancient traditions celebrate the naturalness of sexual pleasure and the spiritual potential of sexual relations..."*

Despite these questions and challenges, we need to remember that nuclear physicist Werner Heisenberg acknowledged that Indian philosophy helped him make sense of some of the seemingly "crazy" principles of quantum physics. And Western science and medicine increasingly acknowledges the value of ancient traditions, such as Ayurveda, the Hindu system of medicine, and techniques of acupuncture originating from China.

The life cycles of past civilizations clearly suggest

that as they degenerate, their cultures tend to exaggerate the great primordial insights that led to their greatness. Western cultures have overvalued individualism at the expense of the environment, separated human nature from nurturing nature, and turned everything, including the human psyche, into objects to be manipulated, controlled, and exploited. The resultant technological superiority has given humankind dominance in our global village. It has given Western culture the leisure and affluence that has allowed women to regain some of the gender equality they experienced in the pre-Axial era. However, the violent, exploitive extremes of Western intellectual and moral assumptions contain the seeds of self-destruction. History suggests that Western culture may avoid self-destruction and achieve a transformation into a new global consciousness if it can integrate values that will bring forth a more balanced culture, respectful of the unity and harmony of all reality. Jaspers and others see in this renaissance the possible advent of a second Axial Period.[5,6,33]

Many critics have deplored the objectification of sex and the Western obsession with sexual performance. Christianity, for the most part, has not been able to integrate sexuality into a holistic philosophy or see sexual relations, pleasure, and passion as avenues for spiritual meaning and growth. There have been a few prophetic efforts in this direction, but many Christian churches are having difficulties dealing with sexual pleasure, apart from reproduction, and along with the spiritual dimension. For individuals or couples, the Eastern views may have rich meaning, but they will not help with the problem Western religions face in accepting and affirming alternates to heterosexual, exclusive monogamy in today's world.

Eastern sexual and spiritual traditions can help Westerners break out of the prevailing reduction of sexuality to genital activity. Taoist and Tantric sexual practices highlight all the senses and involve the whole energies of both partners in slow, sensual dances that are rich variations of what Western sexologists label the "outercourse" of the Sensate Focus Exercises. In addition, Eastern thought may help refocus our understanding and appreciation of male orgasm. The obsession in sexually explicit films and videos with ejaculation as the affirmation of masculinity leaves the male with an inevitable flaccid vulnerability that requires denial in a vicious cycle of repeated "conquests" followed by inevitable detumescence. Taoist practices can help a male achieve some parity with the multiorgasmic woman by controlling his ejaculation, much to the benefit of both sexes.

Conclusion

Over the centuries, Tantriks and Taoists adopted philosophies and practices involving yoga from others and Yin and Yang from earlier Chinese, and borrowed aspects of the cultures of the pre-Aryans and (possibly) the paleolithic Europeans. Some Americans have already borrowed from the riches of Eastern sexual views. In the future this cross-fertilization may increase and become more sophisticated. The outcome could lead to new icons, archetypes, and meanings for sexual relations as expressions of love, passion, commitment, procreation, playful fun, and friendship as well as mystical transcendence and spiritual oneness.

The Western technological imperative needs a strong antidote to regain its health in the 21st century. Western culture may find a corrective to its highly successful, but dangerously exaggerated technological imperative (Yang) in the ancient Eastern tradition of the nurturing potential of a panerotic sensuality (Yin). The health of Western culture can be improved by learning from key elements of the Taoist and Tantric traditions. At the same time, Eastern cultures are also caught up in the current revolution of human consciousness that some see as the advent of a second Axial Period, which is based on gender equality and a global and cosmic consciousness, sensitivity, and shared responsibility. This requires mutual collaboration and cross-fertilization on all sides.

References

1. Campbell, J. The power of myth. New York: Doubleday, 1988.
2. Bullough, VL. Sexual variance in society and history. Chicago: University of Chicago Press, 1976.
3. Gimbutas, M. The language of the Goddess. New York: Harper and Row, 1989.
4. Rawson, P. Tantra: The Indian cult of ecstasy. New York: Avon Books, 1973.
5. Jaspers, K. The Origin and goal of history. New Haven, CT: Yale University Press, 1953.
6. Cousins, EH. Male-female aspects of the Trinity in Christian mysticism. In B Gupta, ed. Sexual archetypes: East & West. New York: Paragon House, 1986.
7. Noss, DS. A History of the world's religions. New York: Macmillan, 8th edition, 1990.
8. Sivaraman, K. The mysticism of male-female relationships: Some philosophical and lyrical motifs of Hinduism. In B Gupta, ed. Sexual archetypes: East & West. New York: Paragon House, 1986.
9. Watts, A. Erotic spirituality: The vision of Konorak. New York: Collier Macmillan, 1971.
10. Deva, SK. Khajuraho. New Delhi: Brijbasi Printers, 1987.
11. Chia, M & Winn, M. Taoist secrets of love: Cultivating male sexual energy. Santa Fe: Aurora Press, 1984.
12. Ruan, FF. Sex in China: Studies in sexology in Chinese culture. New York: Plenum Press, 1991.
13. Chang, J. The Tao of love and sex: The ancient Chinese way to ecstasy. New York: Viking Penguin Arkana, 1977.
14. Van Gulik, RH. Sexual life in ancient China. Leiden, Netherlands: EJ. Brill, 1961.
15. Srinivasan, TM. Polar principles in yoga and Tantra. In B Gupta, ed. Sexual archetypes: East & West. New York: Paragon House, 1986.
16. Anand, M. The Art of sexual ecstasy: The path of sacred sexuality for Western lovers. Los Angeles: Jeremy Tarcher, 1989.
17. Douglas, N, & Slinger, P. Sexual secrets: The alchemy of ecstasy. New York: Destiny Books, 1979.
18. Needham, J. Science and civilization in China. Cambridge, England: University Press, 1956.

1. SEXUALITY AND SOCIETY: Historical and Cross-Cultural Perspectives

19. Rawson, P. The art of Tantra. Greenwich, CT: New York Graphic Society, 1973.
20. Chia, M & Chia, M. Healing love through the Tao: Cultivating female sexual energy. Huntington, NY: Healing Tao Books, 1986.
21. Turner, V. The ritual process: Structure and anti-structure. Ithaca, NY: Cornell University Press, 1969.
22. Blofeld, J. The Tantric mysticism of Tibet. Boston: Shambhala, 1970.
23. Sharma, PS, & Sharma, Yoga and sex. New York: Cornerstone Library, 1975.
24. Campbell, J. Transformations of myth through time. New York: Harper and Row, 1990.
25. Radha, S. Kundalini yoga for the West. Boston: Shambhala, 1985.
26. Hartman, W, & Fithian, M. Any man can. New York: St. Martin's Press, 1984.
27. Bentov, I. Stalking the wild pendulum. New York: E.P. Dutton, 1977.
28. Sannella, L. The Kundalini experience: Psychosis or transcendence. Lower Lake, CA: Integral Publishing, revised edition, 1987.
29. Herdt, G. Representations of homosexuality: An essay on cultural ontology and historical comparison, Part 1. *Journal of the History of Sexuality,* 1991, 1(3), 481-504.
30. Ogden, G. Women and sexual ecstasy: How can therapists help? *Women and Therapy,* 1988, 7(2,3), 43-56.
31. Lawrence, Jr, RJ. The poisoning of Eros: Sexual values in conflict. New York: Augustin Moore Press, 1989.
32. Rainke-Heinemenn, U. Eunuchs for the kingdom of heaven: Women, sexuality, and the Catholic Church. New York: Doubleday, 1990.
33. Paglia, C. Sexual personae: Art and decadence from Nefertiti to Emily Dickinson. New York: Random House, 1990.

CROSS-CULTURAL PERSPECTIVES ON SEXUALITY EDUCATION

Jay Friedman
Director, The Institute on Relationships, Intimacy & Sexuality, So. Burlington, VT

In 1991 I spent my summer vacation observing the messages European youth were given about sexuality. From this cross-cultural experience I hoped to be better able to answer the controversial question: what should be our message to American teenagers about sexual expression?

One year later, as we approach the 1992 elections, I smirk at an imagined European response to current electioneering hype about "family values." For, at one extreme, young Europeans in multi-generational families must often crowd into single-room apartments where sexual intimacy is virtually impossible; these young people are also often burdened with the economic challenge of helping their families feed themselves. Adding to these pressures is that condoms, particularly those with lubrication, are nearly impossible to obtain, even when money is available.

That's one picture of the family life and sexual realities of European youth. On the other hand, in the more sexually progressive countries, young people speak with pride about a monument for gay men and lesbians, or how their parents encourage them to bring a boyfriend or girlfriend home for the night, an increasingly popular practice in the United States that is prompting intense discussion about sexual morals and values. While the world I studied in 1991 has changed dramatically since my travels, the information I'll share remains relevant. I base these findings upon scholarly research, meetings with sexologists from each country, and discussions with young people, parents and others about their attitudes toward sexuality and sexuality education. I've attempted to paint a portrait of the European sexual youth culture, and to look at each country's political and social attitudes through the sexual expression of its teens. In briefly analyzing six European countries, I necessarily forfeit depth for breadth, choosing to highlight some of my findings, and applying them to a concluding view of sexuality education in the United States.

The Netherlands

My travels began with an invitation to be part of the World Congress of Sexology in the Netherlands, a country hallmarked by its reputation for having the lowest teen pregnancy rate in the world, seven times lower than that of the United States. Hedy d'Ancona, the country's Minister of Welfare, Health and Culture, immediately demonstrated Dutch openness and frankness about sexuality in her opening remarks to the Congress:

> In the Netherlands, prevailing attitudes allow people to speak and write very frankly about sexuality in the broadest sense, and to make television programmes about it. This frankness is not confined to discussion of heterosexuality or sex within marriage. There is much discussion of sex and the elderly, prostitution, homosexuality and so on, and much is written and published on these subjects. Sex has come to be seen as a part of life like any other.

In response to worldwide arguments against sexuality education, Minister d'Ancona asserted that for years her country has believed that, "Prohibitions and taboos on sexuality and sex education are directly responsible for unwanted pregnancies, abortions and the spread of sexually transmitted diseases," suggesting a moral imperative to provide information and counseling campaigns. Frankness about sexual matters in the Netherlands originated in the 1960's with the sexual revolution; Minister d'Ancona quotes Dutch sociologist Paul Schnabel's account of these new ideas and attitudes:

> Birth control came to be seen as being in the interests of women, the institution of marriage and the family, and also of society at large. Unwanted pregnancies were no longer regarded as a disgrace but rather as an unfortunate accident and, in any event, not a good reason to get married. Premarital sex, which had previously been assumed to undermine the foundations for a subsequent happy marriage, came to be seen as the very precondition for one. Masturbation, which had once been assumed to be a disease, was now regarded as innocent, or even healthy: if anyone had a problem, it was not the masturbators but those who were unable to masturbate. Homosexuality likewise ceased to be considered a medical condition: it was now a realistic and reasonable basis for a sexual relationship.

The openness and acceptance of these new ideas, according to d'Ancona, stemmed from a number of factors, including sociological research revealing the gulf between prevailing beliefs about sexuality and actual sexual behavior, and a psychological emphasis on the positive value of sexuality to the individual.

Evening after evening I was treated to insights by local residents, who exulted in their Dutch openness about sexuality. An actress described the ongoing debate in Amsterdam about decriminalizing prostitution, which, though illegal, was tolerated. She expanded the "procreation versus recreation" list of sexual options to include

1. SEXUALITY AND SOCIETY: Historical and Cross-Cultural Perspectives

"vocation," explaining that the vocation of prostitution in the Netherlands provides the benefits of a trade union, health care benefits, and job training, and that many of the women view themselves as human service professionals. A student and his partner introduced me to a government-funded organization that provides social services and activities for gay men and lesbians, and accompanied me to "The Homomonument," also government-supported, which commemorates gay life, past and present, with its message: "Toward Friendship, Such a Longing Desire." And a graphic designer I met explained that "People believe heavily in freedom and personal destiny, so religion and government should not dominate our lives." She anticipates that sexual freedom will grow even stronger with further women's liberation efforts, and commented about abortion that "We'll never return to the days of using the knitting needle."

Clearly, young people in the Netherlands grow up in a climate of sexual tolerance, though a formal sexuality education curriculum is lacking. Doortje Braeken, Head of Education for Rutgers Stichting, the country's family planning association, explained that enforcing something by making it obligatory is "just not a Dutch thing to do." Support for sexuality education is strong, although, as expected, it can wane in the more rural areas.

"Formal and informal learning in the Netherlands promotes an openness about sexuality unlike that of many other countries."

In the Netherlands, young people learn to utilize low-cost, accessible sexual health services; three-quarters of teenage women are on the pill, and condom availability has increased dramatically in response to the AIDS crisis. But, despite evidence of low teen pregnancy and abortion rates, Braeken laments that too much of the sexuality education agenda in the Netherlands deals with prevention issues. In addition, while she opposes the idea of a formal curriculum as a "waste of time and energy," she deplores the inadequate training available to and required of teachers in the field. Like other sexologists I met in the Netherlands, Braeken would like to see sexuality education focus more on pleasure. While sex is considered natural and sexual experimentation is accepted, she says her country is not ready to teach young people "how to do it better." Still, formal and informal learning in the Netherlands promotes an openness about sexuality unlike that of many other countries; Rutgers Stichting, for example, is able to publish and distribute, with government support, a magazine entitled *Making Love* that reaches most of the country's teens. It appears that while pregnancy prevention is the primary goal of sexuality education in the Netherlands, the Dutch are very accepting of teenage sexual expression.

Belgium

Belgium is similar to its neighboring country, the Netherlands, in its lack of a standardized, government-endorsed sexuality education curriculum, partly due to its more complex cultural context. The division of Church and government has resulted in Catholic and state school systems; in both systems, provision of sexuality education depends largely on teacher motivation and local politics.

Historically, the Roman Catholic Church has exerted great influence on the lives of all Belgians; today's family planning organizations saw their start in 1955, after Catholics and others boycotted an earlier movement. Despite the availability of modern contraceptives, most of the population relied upon abstinence and withdrawal until the seventies.[1] In March of 1990, abortion was legalized in Belgium, overturning an 1867 law which banned it, allowing a woman to terminate during the first trimester if two doctors judged her to be in a "state of distress." However, King Baudouin, a staunch anti-abortionist, protested the law by stepping down for a day, the first time in Belgian history a monarch exercised this right as a matter of conscience.[2]

In my meetings with members of Centra voor Geboorteregeling & Seksuele Opvoeding (CGSO), the family planning organization of Flanders (Dutch-speaking Belgium), I learned that Belgium is somewhat more progressive in its thinking about sexuality than all of the literature I surveyed indicates. Vicky Claeys, Executive Director of CGSO, explained that Belgium and the Netherlands are actually very similar in their policies and attitudes related to sexuality, though the Netherlands has received more attention for its success. She described the Netherlands as more sexually open, but boasted that, while Dutch women are entrenched in traditional gender roles, ("they're still in their homes"), Belgian women are more "liberated," with greater opportunities in their lives.

As in the Netherlands, awareness of alarming rates of Belgian teen pregnancies provided the impetus for increased sexuality education in the past two decades, but such education is not mandated by the government. While the recent legalization of abortion has increased the number of family life and sexuality programs in Belgian schools, a universal mandate remains unforeseeable. Despite religion-based resistance to sexuality education, the professionals I met claim that the Belgian teen pregnancy rate is now low enough to rival the much-publicized Dutch rate. While both church and state often preach abstinence, most people respect teen sexual autonomy, and Belgian youth strive to protect themselves from pregnancy and disease through pills and condoms that are readily-available from clinics and pharmacies.

When I inquired about the teaching of sex as positive and pleasurable to teenagers, I received mixed replies. The two Catholic educators I met were uncomfortable with this notion, seeing it as beyond the "duties" of their work. On the other hand, Freddy Deven, a prominent sexologist, saw the inclusion of teaching about sexual pleasure as a part of a future goal for sexuality education in Belgium. As in the Netherlands, he asserted, there is too much emphasis on fertility issues, and not enough on interpersonal issues; he believes that teenagers are more likely to hear prevention messages when they are placed in the context of broader social skills training.

In working with CGSO staff and other sexologists, I learned that the issues Belgian teens confront are strikingly similar to ours. For example, young people rarely talk about sexual issues with their parents, who gladly turn to the schools for help. The media greatly influences attitudes about sexuality though programming and commercials, despite limited broadcast hours. The double standard ("stud" and "slut") of sexual behavior applies, to some degree, especially for adolescents.

As an example of Belgian parental attitudes, my host family in Ghent included three children huddled around a computer playing the popular "Leisure Suit Larry" game, where the goal is to go to a nightclub, win money at the casino, and, after enduring many obstacles, meet a woman and take her to bed. The mother of the house admitted that she feared this "adult" game from the United States would give her young children inappropriate messages about sexuality, but not because the children watched as the couple began to disrobe and become sexual. She felt that way because at a crucial point a large black box containing the word "Censored" appeared on the screen. When the box appeared, Anne said, "See, I told you, this gives children an inappropriate message that sex is negative, like a bad secret." Unable to understand why the software company portrays naked bodies and sexual behavior as abnormal, she told her children she did not agree.

Czechoslovakia

It is impossible to discuss sexuality education in Czechoslovakia without some contextual discussion, especially since the country has recently divided. I arrived in Prague the day that the last Soviet soldier was leaving the country, marking the official end of communism in Czechoslovakia. While the people celebrated, I noticed that despair clouded their expressions of hope. The reality: shortages in the stores, especially of food. Ketchup bottles lined the storefront windows, while little more than bananas, green onions, eggs, bread and potatoes were sold inside.

Politics and national resources in many ways influence issues of sexuality in Czechoslovakia. Under communist rule, people held puritanical beliefs about sexuality, and sexuality education in the schools was practically non-existent, often limited to one or two biology classes that discussed anatomy without referring to the genitals. With the end of communism came the increased influence of the Catholic Church, now waging strong campaigns against sexuality education.

The Church also actively opposes family planning, which it sees as part of a "communist conspiracy," according to Jaroslav Zverina of the newly-formed Sexuologicky Ustav. Abortions, still available at no cost, were expected to reach the price of $200 (the equivalent of about two months' salary) if proposed legislation were approved. This would create a serious economic hardship in a country where contraceptive choices appear to be limited. A growing anti-abortion sentiment thrives on financial support from the government and the perception that liberal abortion law is a relic of Communist rule.

Pill use is low in Czechoslovakia, because of misconceptions about side effects, while the IUD and sterilization are available only after a woman has borne a minimum number of children. Condoms are readily available, but most are the non-lubricated variety; lubrication is scarce and requires a prescription to obtain. A recent study showed that 60% of married couples use withdrawal as their method of birth control. But Zverina was optimistic that positive changes were ahead for his country. He cited the beginnings of a women's movement, and, while frustrated with the lack of public involvement, he recognized that change takes time after many years of oppression. He took me to an "underground" sex shop—some shelves of erotic toys, magazines, and the like in the heart of a gutted building— as an example of the opening attitudes toward sexuality. Finally, Zverina predicted that most people will ultimately be supportive of sexuality education in the schools.

Ondrej Trojan, head of the Czechoslovak Independent Coordinating Agency, explained that his country's teachers traditionally emphasized the negative consequences of sexual behavior and instilled shame in their students. Despite a trend toward including positive elements of sexuality and health in their lessons, teachers today are often ill-prepared to teach this topic, and fearful of not receiving administrative and community support. Those who do provide sexuality education are accused of being immoral, of promoting premature sexual behavior and favoring abortion. Meanwhile, many health professionals frown upon adolescent sexual behavior, saying that such behavior is physically wrong, while silently believing that it is morally wrong as well.

Teenagers in Czechoslovakia have very little exposure to sexuality information. According to Zverina, this has caused problems for those who become coitally active, as only 2 to 3% (compared to 70% in the Netherlands) use birth control during their first coital experience. While this first experience, at an average age of seventeen, occurs later than in many other European countries, there are many early marriages in Czechoslovakia because of unintended pregnancies. Trojan and others hope the argument that sexual activity is an important part of a successful married life will lead more people to support sexuality education.

Denmark

Most Danes view sexuality as a normal, natural part of life, as long as individuals assume enough personal responsibility to avoid negative consequences. Sexuality issues, therefore, tend to be discussed openly from the earliest of ages. Copenhagen's openness reminded me of Amsterdam. My rural experiences in both countries were also similar, so my days in Denmark were somewhat reminiscent of my time in the Netherlands. At Foreningen for Familieplanlaegning (FF), the Danish family planning association, Secretariat Manager Kirsten Kjoelby immediately compared her country to the Netherlands. She explained that Denmark has nearly matched the Netherlands' low abortion rate, but attributes Dutch success in lowering its abortion rate to the existence of a greater number of clinics and more stable pill use, about which misconceptions still linger in Denmark.

Kjoelby described the sexual openness in her country with great pride, saying it exceeds even that of Sweden, whose sexuality curriculum is "preoccupied with feelings of love." She believes that well-established equality between the sexes and a "very relaxed attitude to religion" led to her country's

1. SEXUALITY AND SOCIETY: Historical and Cross-Cultural Perspectives

openness, with one manifestation being the increased popularity of childbirth outside of matrimony, either by cohabiting partners or single mothers. But, despite a divorce rate of 50 to 60%, marriage is again becoming more popular in Denmark because "young people are becoming more romantic," according to Kjoelby.

Accessibility and affordability are two of the priorities of adolescent family planning services in Denmark, which, in 1966, became the first country to allow young people access to family planning services without parental consent. Danish minors become independent health insurance members at age 16, when they are free to choose a general practitioner of their choice, and many school classes from age 14 and up visit family planning clinics to learn about contraceptive methods. Abortion without charge is available to minors under the same conditions as it is for adult women.

> *"Denmark is an illustration of the fact that, even with access to family planning services and quality sexuality education, teens experience their 'sexual debut' later than American teens."*

Denmark is an illustration of the fact that, even with access to family planning services and quality sexuality education, teens experience their "sexual debut" (coitus) later than American teens. Also, Danish teens, according to several surveys, have at least an 80% likelihood of using some method of contraception during their first act of coitus. While pregnancies are low compared to the United States, the government is concerned about what it considers to be a high teenage abortion rate, since most teenage pregnancies are terminated. Instead of trying to curtail sexual behavior in order to lower the figure, the government is stressing increased sexuality education efforts to improve contraceptive effectiveness.

As Danish teens spend more time in other countries as exchange students, they are surprised at the negative attitudes about sexuality they discover. In contrast, I found myself surprised at the extent of the openness reflected in the educational materials at the Danish family planning association. Kjoelby showed me a sample from the "Exact Knowledge" pamphlet series, providing information about puberty in a frank, explicit manner, which is used as part of the compulsory sexuality education curriculum in the schools, and is distributed in cooperation with pharmacists who are trained by FF to be important sexuality educators in their communities. Kjoelby described FF as having a "statutory obligation" to educate Danes about contraceptive information; the Ministry of Health delegated this responsibility to FF, and extended its mission to include HIV/AIDS prevention. FF developed a condom package that includes instructions for use, (the National Board of Health would not approve distribution *without* pictures of the penis in the instructions), and a cartoon book about Oda and Ole that shows the young couple undressing, naked, rolling the condom on the penis, and enjoying a "beautiful, wonderful, lovely night" that includes sexual intercourse.

Most remarkable, however, were the popular animated films financed by FF with support from pharmacists and the National Board of Health, that are commonly used in schools. *So That's How* teaches young children (5-7 years) about reproduction, and shows explicit sexual behaviors that are "enjoyable" for married or unmarried people. The video *Safe for Life* contains explicit depictions of sexual activities between same- and other-gender partners in its messages about safer sex. *Sex: A Guide for the Young*, targeted at middle school students, portrays a young couple preparing for their first sexual experience, complete with pictures and instructions about how to give pleasure to oneself and to a partner. Denmark's formal sexuality education curriculum, made compulsory in 1970 and grounded in the principles of openness, security and the sharing of knowledge, has been under recent attack. Last summer, strong representation from a conservative rural group during a curriculum review process influenced the inclusion of a value stating that sex before marriage is wrong. FF staff were unsure how this would affect sexuality education.

Sweden

Like Denmark, Sweden enjoys a Nordic tradition of sexual openness. Swedes have long striven for an egalitarian lifestyle, with great respect for personal autonomy; cohabitation is generally as accepted as marriage, and Swedish policy is similar to Denmark's in affirming gay and lesbian lifestyles. Although some inequities remain, the Swedes have a comparatively long history of equality between men and women, which often extends to family roles and responsibilities.

Remarkably, Sweden maintains a legal age for intercourse, heterosexual and homosexual, of 15 years. Most Swedish teens abstain from intercourse until then, and their average age of first intercourse is about the same as for teens in the United States, about 16 years. This is basically where the similarity ends, however.

Sweden has had an exceptionally long history of sexuality education, with widespread support from the general public as well as professionals. The Board of Education first endorsed sexuality education in 1921, followed by formal government endorsement in 1942. The demand for sexuality education came from a desire to end the sexual ignorance that often resulted in abortions, STDs, and, perhaps most important, marital breakdown, and an acknowledgement of the importance of education to improve one's ability to relate to others. The early goal of Swedish sexuality education in schools was to remedy the ignorance of bodily functions, anatomy and physiology that led to physiological and psychosexual problems. Thus, students in the fifties and sixties were learning that sexuality is a positive force for pleasure and intimacy.

By 1977, a curriculum titled "Instruction Concerning Interpersonal Relations" was published, with schools directed to include messages about fundamental moral values. Teachers were equipped with a complete justification of the program, teaching techniques, tips on dealing with controversial issues, and a compendium of facts. The guidelines provided moral support for the teachers, who already had overwhelming approval from their communities.

Sexuality education in Swedish schools is far ahead of that in other countries in its almost celebratory outlook on sexual expression. The first component of the latest curriculum is "Promotion of the capacity for intimacy," stating that "Sexuality can take place without intimacy, but with intimacy, sexuality satisfies a profound human need." This positive tone balances accompanying messages about the possible negative consequences of sexual irresponsibility.

Many educational campaigns promote sexual responsibility, especially in this HIV/AIDS era. Most interesting was a passport holder given at one time to train travelers. Included inside are condoms, postcards of condoms arranged in a floral pattern, and the words "flower power" (this and other patterns appear on billboards all over the country), an AIDS referral sheet, travel tips, and a booklet about condoms and sexually transmitted diseases that includes a guide to the translation of phrases such as "I love you," "I want to sleep with you, I have condoms," and "A packet of condoms, please" into six different languages.

I met with Eva Olsson, a midwife at Ungdomsmottagning, a sexual health clinic for young people. The number of these clinics grew as the AIDS crisis escalated; there were 20 in Stockholm when I visited, with a total of 130 scattered throughout the entire country. The clinics, completely state-funded (the government believes young people should have complete information, as "it's not good to have fear of STDs") provide free and confidential care for anyone under 25, though most clients are teenagers.The Ungdomsmottagning is actively involved with community education and, during the ninth year of high school, all students make class visits to the clinic. Olsson mentioned that most students seem well-informed intellectually, but have many questions about the "intimate" aspects of sexuality. She believes that, despite its progressive image, the curriculum fails to address the "heart and soul" questions about relationships, and so is a disappointment to the students.

When asked about what accounts for the sexual openness of Swedish society, Olsson immediately responded that a lack of religious restrictiveness promotes healthy sexuality. She said that the people are spiritual and occasionally go to churches because they are "nice buildings with nice music," but that organized religion does not carry any political influence.

> *"Sexuality education in Swedish schools is far ahead of that in other countries in its almost celebratory outlook on sexual expression."*

In addition, parents are beginning to accept the premise that perhaps they should *not* be the primary sexuality educators of their children; the government mandates school sexuality education because adolescence is seen as a time of developing sexual feelings, when it is easier to talk with someone other than parents about these issues.

Parents smiled with acknowledgment when asked about teenagers bringing home romantic partners for overnight stays. This is difficult for most parents, but when their children become involved in relationships, they challenge themselves to consider what is best for their children's future. Olsson described a colleague who found her daughter and boyfriend kissing and fondling each other on the living room couch. The woman "absorbed her shock" and sat the couple down, explaining that what they were doing was natural and healthy. She advised them to go at their own pace, and if they decided to have sexual intercourse, to use protection and to be home rather than be hurried or hiding elsewhere.

A former American I visited in central Sweden reinforced this openness when he described sexual attitudes as "not as drastic as everyone thinks, but subtle, positive, and even natural." He could not imagine returning back to the restrictiveness of the United States, but refused to see Sweden as extreme. Referring to his ten year-old son, he said that "One day he'll be making decisions about who he wants to be with, and whether that's in five years or in ten years, I can only hope to show him what a relationship with love and respect is all about."

Soviet Union

The Soviet Union does not exist as it did last summer. And although I did not actually travel there, I include what is now the Commonwealth of Independent States in my European study because I was able to meet with a number of professionals from that region during my journey.

At the World Congress of Sexology, for instance, Russian sexologist Igor Kon spoke of a "sexless society," where both communism and the Church are opposed to sexual and sensual images, seeing them as unimportant, distracting, and opposed to the goal of "social liberation." He believes that the lack of research, eroticism, and education since the 1930's is having detrimental effects on a country now experiencing "sexual maturation." Kon outlined a number of sexual issues currently confronting the former Soviet society, including a lack of birth control methods, increased sexual violence, escalation of HIV/AIDS among other sexually transmitted diseases, and rampant homophobia.

Abortion is clearly a major issue in the former Soviet Union. *Komsomolskaia Pravda*, the communist youth league newspaper, reported that 6.5 million abortions, with another 3 million unreported, were performed in the Soviet Union in 1988, one-fifth of them involving teenagers. The average Soviet woman will have about seven pregnancies in her lifetime, and three to five abortions, though many women have eight to ten abortions.[3] Use of abortion as birth control signals a dire problem in the Soviet Union: lack of methods of contraception and prophylaxsis. In a 1987 survey of newly married couples, nearly 60% of men and 70% of women were unaware of the most effective birth control methods, and only 10 to 15% of the former Soviet population uses contraception.[4] Oral contraceptive use suffers from the fear of side effects, while other contraceptive methods are largely unavailable. (The lack of condoms is the result of the appropriation of available rubber by the military.) Prostitution is considered another sexual problem, but, as in other countries, its growing popularity is based in economics. A recent movie's attempt to deglorify prostitution backfired,

1. SEXUALITY AND SOCIETY: Historical and Cross-Cultural Perspectives

and many teenage girls see it as a very prestigious occupation that enables them to make money while exerting some control over their lives.

The status of sexuality in the former Soviet Union is directly related to the country's politics. Opponents of *perestroika* condemn the sexual revolution for its self-indulgence and its contemporary values, while liberals protest the state's inability to meet the actual needs of its people, citing that failure as the cause of sexual problems. Though sexual morality grows as a hotly contested issue in the former Soviet Union, perhaps the greatest problem in the region is that sexuality education continues to be essentially non-existent in the schools. According to the most recent surveys, 87% of former Soviet citizens never had discussions about sexuality with their parents, which may explain why the majority of people recognize a need and support the immediate introduction of school sexuality education. Despite this support, as Kon pointed out, while sexuality education in the United States was once seen as a "communist conspiracy," in the Soviet Union it is seen as a "Jewish conspiracy" that must be contained. In light of recent events in the former Soviet Union and Czechoslovakia, sexuality education in the new Commonwealth of Independent States may also be seen as a communist conspiracy before long.

Before the 1980's, most Soviet youth failed to receive any sexuality education in the schools. Experimental programs were piloted in the early 1980's until sexuality education was mandated for all Soviet eighth graders in 1983, following mounting concern about misinformation and early sexual involvement. But while the weekly 45-minute course, "The Ethics and Psychology of Family Life," purported to provide sexuality education, it failed to meet the students' actual needs and questions, keeping them ignorant and ultimately unsatisfied sexually. A Russian father I met in Germany told me that people believe sexuality education corrupts students' minds, and that he and his wife simply find it impossible to talk to their children about sex. Such conservative thinking has shaped a curriculum that promotes virginity until marriage, and life fulfillment solely in the context of marriage, often through the production of children. One newspaper article from an unknown source describes a sexuality education program that claims "A girl who loses her virginity before marriage is deprived of her charm, becomes less interesting and, most important, loses the belief in beautiful, profound feelings and in herself." Sexuality classes are often separated by gender, often promoting the classic double standard: boys will be boys, while girls should try to control them and avoid promiscuity. Teenagers engage in sexual activity that is primitive, rushed for lack of privacy, and dangerous because of a lack of supplies.

As in Czechoslovakia, the gay and lesbian community in the former Soviet Union has been responsible for advances in HIV/AIDS education and counseling. Homosexuality, however, is officially illegal in the Soviet Union, with government officials often claiming there is no such sexual behavior. Many Russians would extend such claims to include heterosexuals as well, as their society still lacks supplies, privacy, and even a full language for sexual expression. As far as teen sexuality is concerned, their educational needs are vastly unaddressed, and, while teens may become sexually involved in hopes of establishing meaningful connections, society seems to disdain sexual expression outside of marriage.

Conclusions and Implications

We can find much that is useful in this comparative look at sexuality education abroad. One element we see consistently is that teen sexuality has been viewed as having clear social implications during the evolution of sexuality education in these countries. For example, all of the countries seek to reduce the incidence of abortion, for varying levels of medical and moral reasons. Thus, an explicit goal of all programs has been to reduce teenage pregnancies. We also see that the most effective teenage pregnancy prevention programs consist of two components: provision of comprehensive sexuality education, and availability of family planning services. If a country's policy on teen sexuality is to prevent teen pregnancy, we can learn a great deal by examining its program. For ex-

> *"The progressive approaches of Denmark and Sweden illustrate that sexuality can be viewed as a positive part of relationships throughout one's life. These are countries that empower young people to be in control of their sexual lives."*

ample, while Czechoslovakia and the Soviet Union may address contraception in its cursory sexuality education programs, such information is relatively meaningless in the face of limited contraceptive supplies. Furthermore, the climate of social despair provides a new twist to the meaning of sexuality, while decreased secularization means that religious groups will exert a greater influence on sexual learning in the future.

Such is the case in Belgium, which appears to be quagmired in great religious debate about sexual morals. As in the United States, sexuality education is seen as quite controversial, though Belgians seem more practical about the need to prevent teen pregnancy and appear to be achieving success. The Dutch have long controlled religious influence, and they display greater openness about sexual issues than the Belgians. This openness appears to create a climate which does not question teen sexuality, but takes a practical approach to preventing negative consequences for teens. The Netherlands has chosen to concentrate on availability of contraceptives rather than sexuality education, and its incredibly low teen pregnancy rate is a testimony to the country's success.

But why measure the success of sexuality education only in terms of pregnancy prevention? Is a low teenage pregnancy rate symbolic of the successful sexual upbringing of youth?

Why not measure numbers of cases of HIV/AIDS and other sexually transmitted diseases, or the incidences of sexual abuse and assault? To shift away from a focus on danger, why not mark success by increased intimacy, sexual competence, satisfaction, number of orgasms, or other measures of sexual pleasure?

2. Cross-Cultural Perspectives

The progressive approaches of Denmark and Sweden illustrate that sexuality can be viewed as a positive part of relationships throughout one's life. These are countries that empower young people to be in control of their sexual lives. Understanding both the pleasures and consequences of sexual expression, their policies indicate respect for self-autonomy and a trust in teenagers to make reasoned and responsible decisions about their own sexual lives.

What helps me is to envision a continuum of approaches to sexuality education for young people. At one end lies the former Soviet Union and Czechoslovakia. Although the Czechs I spoke with were dismayed at being grouped with the Soviets they worked so hard to expel, both of these countries fall at the "sex negative" end of the range, viewing the goal of sexuality education as a means of preventing teenage sexual expression. Both countries are burdened by political strife, economic difficulties, and the powerful influence of the Catholic Church. Czechoslovakia may be somewhat more progressive than the former Soviet Union, which appears to be in a dismal, desperate sexual state of affairs.

In the middle of the continuum are Belgium and the Netherlands. These two countries, friendly rivals to one another, can be characterized as viewing sexuality education for teenagers as a means of pregnancy prevention. Both countries, pragmatic in their approach, have had enormous success in this endeavor, and in a nonobligatory fashion, are making strides toward more positive sexuality education.

Denmark and Sweden appear to fall at the other end of the continuum, with what I would identify as a "sex positive" approach to sexuality education, one that affirms and seemingly assists teenage sexual expression. Young people in these countries talk most openly about sexual issues and have integrated sexuality as an important part of their lives. Both countries also employ public policy tactics that deliberately go beyond the parents, with the schools being the primary providers of sexuality education.

As I absorbed and reflected upon what I was learning during my journey, I continually asked myself where the United States falls on the continuum. In fact, I asked my European colleagues about their perspectives on sexuality education in my country. Almost everyone placed the United States between the "sex negative" and "neutral" positions on the continuum.

It is interesting to note that, while the eastern European countries would like us to share our expertise and resources with them, they fear our moralistic approach. The western Europeans are proud of their "advanced" standing on the continuum, but they, too, fear that "sex negativism," like everything else from the United States, would soon be exported to their countries.

During the summer of 1992, the International AIDS Conference took place in the Netherlands, whose government, incidentally, just announced the lowering of the age of consent to twelve years. Though the conference was originally scheduled for Boston, organizers moved it to Amsterdam as a way of protesting discriminatory American laws concerning visitors to this country who are persons with AIDS. Encouragingly, Surgeon General Antonia Novella now speaks of being less moralistic, and more realistic, in dealing with HIV/AIDS.

But why such moralism in the United States? Many times during my travels, I was struck by the contrast in sexual expression between my country and those I visited. In explaining the difference, the two terms that my European colleagues mentioned repeatedly were "personal autonomy" and "religious freedom." I felt confused and angry that the United States, a leader in the world, a country that, in its early years, championed such values, could now be so disrespectful of them. How ironic that, as we recognize the 500th anniversary of the arrival of Columbus and the legions who followed him in flight from religious and political intolerance, we, ourselves, have become one of the world's most intolerant, moralistic societies.

Perhaps my greatest challenge met me when I returned to the United States and worked on adolescent sexuality with groups of teenagers, teachers, or other professionals. What should I say? Refreshed by a new sense of teen sexual expression I gained while in Europe, my vision now includes the openness of parent-child relationships, open affirmation of the diversity of sexual relationships, and the powerful practicality of educational materials such as *Sex: A Guide for the Young*, which I've integrated into my professional training to challenge others' visions.

While I found no magic answers to my original questions, I gained many new, unanswered questions from my explorations. My international colleagues were useful in reminding me that when I'm feeling frustrated, demoralized, and confused, I should remember the importance of my vision. They were optimistic about change in the United States, constantly remarking that a young country is prone to mistakes in its development. Finally, they instilled in me a reminder to keep the young people foremost in my mind, for, as Hedy d'Ancona noted in her concluding remarks, encouraging a positive sense of sexual expression can help them enjoy a healthy future in a healthy society:

> Policy-makers must continue to devote specific attention to sex... We remain convinced that continuing sexual openness and attention to the subject are the habits best calculated to enable people to exploit their sexual potential and enjoy their sex lives, as well as avoiding frustration and trauma. It seems reasonable to assume that satisfying sexual contacts and relationships will have a beneficial effect on the mental and physical health of the people concerned and will consequently help them to function better both as individuals and as members of society.

REFERENCES
1. Meredith, Philip and Lyn Thomas, eds. Planned Parenthood in Europe: A human rights perspective. London: Croon Helm, 1986.
2. Planned Parenthood in Europe, April, 1990.
3. Shalin, Dimitri. Glasnost and sex, *The New York Times*, January 23, 1990.
4. Traver, Nancy. Kife: The lives and dreams of Soviet youth. New York: St. Martin's Press, 1989.

FAMILY, WORK, AND GENDER EQUALITY
A Policy Comparison of Scandinavia, the United States, and the Former Soviet Union

Elina Haavio-Mannila, Ph.D.
Department of Sociology, University of Helsinki, Finland

The workplace is not separate from other human institutions. Workers bring into the workplace their values and expectations from home, from the community and from the larger society that bear on their relationships and aspirations at work. They also take home from the workplace feelings of frustration or worth that affect their roles as parents, community members, and citizens. To view work in isolation from other role relationships is to remove it from the normative context that sustains it and gives it meaning.[1]

The sociology of family and gender roles has long tended to link work, the family, and even the community and state. As early as 1956 social scientists presented strategies by which women could combine their traditional family obligations with paid work. Today we recognize that women have many more than two roles, and the contemporary interface of family and work produces many a dilemma. Most women now see the challenge they face not in terms of becoming exactly like men in their work lives or returning to the domestic hearth but in terms of restructuring the family and work roles both for themselves and for men as well.[2]

This article will first compare public work policy among Scandinavia, the U.S., and the former Soviet Union. Traditional work policies focus on norms and standards for safety, occupational health, and working conditions. Labor-market policy also includes workers' participation in management and ownership, as well as rules for collective bargaining. Family policies are a part of all developed nations, designed to give economic support to families. Some countries emphasize cash child allowances, others emphasize tax reductions. Family support -- with the exception of maternity insurance -- does not affect the labor market or the bargaining power of workers.

Background: Comparison of Work Policies

In the United States, large firms provide services that mirror those of government services. For instance, businesses have a social service component to deal with issues of alcoholism, drug abuse, and family needs; a justice component to address issues of equal opportunity; and an educational component to provide training.[3] Nevertheless, only a minuscule proportion of even the large firms have made determined efforts to reform their internal structure to better accommodate workers' needs. Rosabeth Moss Kanter suggests that the "new workplace" is less hierarchical, more egalitarian, and more conducive to the freedom that the new work force seeks.[4] This means bureaucratic control: a system of labor control that weakens bonds of solidarity among workers. In the U.S. auto industry, for instance, men and women who run the factories and staff the offices are said to do surprisingly well in the global competitiveness race — but only when managers give them a chance.[5] At General Motors and Ford, the companies for which we conducted research in 1988, the organization is said to be more authoritarian and patriarchal than in some newer branches of U.S. industry, like computer manufacturing, for example.

Scandinavia has been a forerunner in industrial democracy, management-worker cooperation, and work-safety research and policy. Discussions of the democratization of work life entered the political arena in Scandinavian countries around 1960, and since then many changes have been made.[6] Many Scandinavian organizations assumed new experimental forms with researchers, trade unions, and cooperative employers' organizations working together to bring about change. The Scandinavian countries revised work-safety laws in the 1970s, allowing the state to establish norms, supervise and sanction employer compliance through, for example, occupational inspections. Those inspections rely heavily on medical, chemical, and technical expertise. Since the 1980s, laws dealing with worker/management cooperation and workers' influence on decision making have guaranteed a degree of industrial democracy in all Nordic countries. Dialogue among partners in the labor market is encouraged, even to the point that the state intervenes as a third party in wage and fringe-benefit negotiations. It is considered important for work policy to encourage innovative organizational arrangements and to avoid polarization of work conditions and tasks.

Occupational health -- the prevention and treatment of occupational diseases and accidents -- is a major work-policy issue in Nordic countries. Recently, quality of work, the environment at work, and human relations in the work community have also become concerns of workers' health policy.[7] Work is an important part of the quality of life. A healthy, safe, and comfortable work environment which promotes motivating, meaningful work is believed to promote the health of workers, better productivity, and a higher quality product.

In the Nordic countries, there is at present a trend toward less governmental control of work life. This means that both public and private employers have an increasing "freedom of responsibility" in the development of working conditions. An additional goal of recent Finnish work policy is to increase self-initiation in work organizations.[8]

In the former Soviet Union, state control of individuals was strict. During the politically shaky, economically difficult, and inflationary last years of Communist rule, social policy could not guarantee reasonable welfare to pensioners, the ill, the handicapped, and many other groups in need of social services.

Family Policy in the United States
The United States is unique in its reluctance to address the issues of family roles and women's and men's work in the policy arena. European countries provide support to families, parents, and women in the form of children's allowances, paid parental leaves of absence, and maternity benefits. The United States, while paying lip service to the importance of families and children, has no comprehensive family policy and has been notably disinclined to pay the costs of government benefits and services for families, for working parents, and for women.[9]

Child allowances and tax reductions, which exist in the United States, are only a small part of family policy. Otherwise, the official infrastructure of the country does not encourage people to have and rear children.[10] Working hours are long;[11] paid parental leave is short, if paid at all. Only a portion of employers offer job guarantees for parents staying home without pay to care for a baby, the quality of municipal day-care centers varies (when such facilities are available), and high-quality private child care is expensive.

Some employers have privately developed measures to make the workplace more responsible to the needs of employees. Major companies are providing modest paid disability benefits at the time of pregnancy and childbirth, and some allow female employees brief additional unpaid but job-protected leaves. The 1970s saw a rapid expansion in the establishment of flex-time policies, but the trend has since slowed. Part-time work, in contrast, is growing, as is temporary work. The number of counseling services in the workplace has increased, and that expansion seems likely to continue. Childcare remains the family-related service most discussed in the workplace in the United States.

3. Family, Work, and Gender Equality

For years, a dramatic difference in vacation time between many European countries and the United States has existed. In Scandinavia, for instance, most employees have an annual paid vacation of five to six weeks. Further, daily work hours are considerably shorter than in the United States.

It is puzzling that, in the United States, women have high labor-force participation rates despite no benefit legislation, highly uneven coverage through collective bargaining, and modest benefits for sickness or maternity and parental leave. At the end of the 1980s, the proportion of women in the labor force was similar in the United States (45%) and Scandinavia (45%-48%). In the former Soviet Union, the proportion of women in the labor force at that time was over 50%. It is generally agreed that changes in the relationship between men and women and transformations in family life have had profound social repercussions. As the rate of divorces climbed and the ranks of women in the labor force has grown, a whole new set of issues relating to child support and the problems involved in mixing work and family have been raised in all income ranges in all countries.

Family Policy in Scandinavia
Gender equality is one of the main goals of Scandinavia's social democracies. Actual policy, however, has been developed according to conflicting views about the extent to which public policy should support family functions or substitute for the family through public-service arrangements.[12] According to detailed studies by Kamerman and Kahn, Sweden has the most comprehensive family policy of all capitalist countries. Finland and Denmark also have explicit policies, but they are focused more narrowly.

Policy in the Scandinavian countries supports employed parents in their attempts to cope with both work and family demands. Women, who have carried the main responsibility for care in the family circle and in the multi-generational chain, now receive help from public services in fulfilling the tasks of everyday life. Day care for children, elder care, home health services, and so on, have begun to liberate women from both family and intergenerational dependency.[13] The municipal day-care system covers a large proportion of children under school age (which is seven years).

Parental leave with 80% - 90% compensation for lost salary extends for 15 months in Sweden, 12 months in Finland, and about half a year in Denmark. Fathers as well as mothers are eligible for this leave. In the early 1990s, practically every eligible Swedish father used "daddy days" immediately after the birth of a child, and 34% of Finnish fathers took advantage of "daddy days" in 1989.[14] In addition, parents have their jobs guaranteed by law for several years if they choose to stay at home to care for their children. It is still predominantly mothers who take this longer period of parental leave.

The Nordic countries represent a special tradition of women's economic activity: early integration into the labor market, high labor-force participation rates, high rates of union organization, low unemployment, high segregation, frequent but "safe" atypical employment (part-time and temporary work), and relatively small wage differen-

1. SEXUALITY AND SOCIETY: Historical and Cross-Cultural Perspectives

tials. Most of the stability of women's employment in Scandinavia was achieved by means of general economic policy, not so much by policies targeting gender equality in pay or employment opportunities.[15]

The Nordic countries have laws guaranteeing equality between men and women in society generally and at work. There are also state gender-equality councils and ombudsmen for coordinating public equality policy. Shortcomings and backlashes concerning gender equality still exist in the Nordic countries, however. The Nordic model has supported, perhaps even strengthened, segregation and gender division in the workplace. It has made women increasingly dependent on the state instead of on husbands: the private patriarchy has merely become public.[16]

From the point of view of gender equality in the labor market, Scandinavia and the United States show different developmental tendencies. The Swedish employment structure was, in the 1970s and 1980s, evolving toward two economies: one, a heavily male private sector; the other, a female-dominated public sector. Although there was a small increase in "female" job concentration, women were also moving into traditional male jobs. The share of women in privileged "male" occupations in the United States was twice that in Sweden.[17]

The present cuts in social services in Scandinavia due to economic recession will mean a reduction in employment opportunities for women.[18] Since 1970, women have composed almost two-thirds of the public employees in the Nordic Countries.[19] In 1985, the public sector employed almost every second working woman. In Denmark 45%, in Finland 39%, and in Sweden 55% of employed women were public employees.

Family Policy in the Former Soviet Union

A socialist vision of the future emphasized the importance of public life. Women's drudgery in the home would be replaced by public, collective arrangements in which men and women would participate freely and symmetrically. But the introduction of women into the labor force is not enough to produce change in the gender system. The traditional attitudes of male superiority in the former Soviet Union outweighed the pronouncements of official egalitarian ideology. No real change occurred in the roles of men; they merely looked on while women took more jobs. The main goal of recent Soviet social policy, according to Narusk,[20] was to reconcile the woman's duties in society and at home, to facilitate her ability to perform her dual role, and to lessen the differences in material welfare existing among families based on number of children.

As of the latter half of the 1980s, it was obligatory for both men (aged 16 - 59) and women (aged 18 - 54) to work or study outside the home. Free education (with grants to students in secondary and vocational schools and universities), free medical care (with payments for being on the sick list and also for time spent in the hospital), almost-free day-care centers, cash child allowances, and other social services helped families. There were practically no housewives who stayed at home, and women were supposed to take an active role in social life and politics.

A new stage in Estonian family policy began in the early 1980s, when the rights of state enterprises and collective farms were broadened to include the use of company social funds for the help of workers' families. Working mothers became entitled to paid maternity leave until their child was one year old and to additional unpaid leave until the child was 18 months old. A state grant was established for the birth of a first child, and a double sum for the birth of a second or third.[21] Mothers of small children were entitled to shorter working days or were given the opportunity to work at home, if their managers deemed this possible. The use of these measures depended on the good will of the managers and on the real amount of social funds at their disposal. There were great differences in benefits from enterprise to enterprise.

Lapidus identifies three features of the Soviet system that impede women's work freedom and choices.[22] First, sexual stereotyping of occupations was not eliminated. Second, female occupational choices were profoundly influenced by the continuing identification of both creativity and authority with men. Third, culturally and in legislation, household and family responsibilities were explicitly treated as the primary and proper domain of women. At the same time, shortages of consumer goods and everyday services made household responsibilities especially onerous.

The social policy system was supposed to satisfy the basic needs for food and housing, as well as for health, education, cultural and social services for all citizens without separate charge. However, the continuing lack of consumer goods and housing have made daily life difficult, and the quality of state services has been poor. For example, day-care facilities were often understaffed; children disliked them and got sick, adding to parental stress. In addition, the hours were too long for children, often nine hours a day, since most women worked, and full-time and flex-time did not exist.[23]

Nevertheless, the official Soviet ideology of gender equality was successful in some respects. For example, female-male gaps in the number of third-level science students, in the labor force, and in parliament, were smaller in the Soviet Union than in the capitalist countries studied. After the fall of the Soviet Union, however, women criticized the official gender-equality ideology for its lip-service. They questioned the value of gender equality in science, the labor force, and politics. For example, surveys of Estonian high school students in 1979 and 1990 showed that the value girls assigned to education, particularly technical studies, had diminished in the 1980s. The few girls who were interested in technology had low self-esteem.[24] According to a survey of 921 adult women in 1990, women valued family and children more than economic welfare. They also valued love, hobbies, friends, and relationships with their parents more than their jobs or studies.[25]

In 1990, Estonian women were relatively unconscious of inequality between men and women. More than half of the 921 respondents in the survey could not point out gender inequalities, a situation related to the fact that "'feminist' in Estonia is a word of abuse."[26] Opposition to forced equality during the Communist period resulted in a strengthening of traditional attitudes with regard to the position of women. In 1988-90, the Estonian women's movement proposed that women should concentrate on giving birth to many (Estonian) children and care for them at home.

The high number of women in the Supreme Soviets (parliaments) in the former Soviet Union was achieved by applying a 30% quota for women members in elected bodies at the national, regional, and local levels.[27] After the breakdown of the Soviet Union, the proportion of women sank in the representative political bodies of the Soviet republics. The legislature elected in independent Estonia in 1991, for example, included only seven women among 105 members. Women were simply not interested in running as candidates. Their attitude toward politics was negative and based on frustration and alienation. In the prior Communist period, women legislators were like toys in the hands of their male counterparts, even though some of the women were quite competent.[28] The Soviet Women's Committees were the only officially accepted women's organizations functioning on both the national and local levels during the period from 1940 to 1991, and they were not allowed to discuss women's social, economic, and political inequality, the existence of which was denied.

Summary

In the mid-1980s, the five countries we studied were ranked according to gender equality as measured by a combined index.[29] The comparison included 99 countries, representing 92% of the world's female population, and the index was based on 20 indicators, including health, marriage and children, education, employment, and social equality. The status of the women in the five countries in this study declined in the following order:

Sweden	(with a score of 87)
Finland	(85)
U.S.	(82.5)
Denmark	(80)
USSR	(70)

This comparison indicates that, in practice, there was more gender equality in Scandinavia and the U.S. than in the former Soviet Union.

Gender equality seems to be especially high in Scandinavia, according to another international comparison. The Human Development Index combined indices of longevity (life expectancy at birth); knowledge (adult literacy rate and years of schooling); and decent living standards (adjusted real gross domestic product). In 1990, the overall Human Development Index was on the same level for the U.S. (.976) and Scandinavia (Denmark .967, Finland .963, and Sweden .982), but lower in the USSR (.908). When the male-female disparities were taken into account, Scandinavian countries turned out to have higher scores (.878 for Denmark, .902 for Finland, and .886 for Sweden) than the United States (.809). Finland ranked highest among the countries of the world on this gender-sensitive index.[30]

The high level of social development in Scandinavia is reflected in people's satisfaction with life. At the beginning of 1980s, life satisfaction was higher in Scandinavia than in the U.S.[31] (Life satisfaction is part of a syndrome of positive attitudes toward the world in which one lives.)

Nowhere, however, has equality of economic opportunity for women followed automatically from their higher levels of educational attainment and labor force participation. The basic problem is still lower pay for women's work than for men's work. Men derive greater benefits from educational and occupational attainments, even when women's work experience and levels of current labor-force participation are comparable. The gender gap in wages is larger in the United States, where female wages are 69% of male wages, as compared to Scandinavia, where female wages vary between 76% and 84% of male wages.[32] In 1989, the average full-time earnings of women in the Soviet Union ranged from 65% to 75% of those of men.[33] Scandinavian countries have relatively high economic equality between men and women, but a gender gap in earnings remains.

We found by conducting social research and other comparative work among the three regions that type of society and type of workplace (with its specific gender composition) have strong connections with the organization of work, with the social relations at work and in the family, and with the well-being of women.[34]

Elina Haavio-Mannila is a professor of sociology at the University of Helsinki. She has published nearly 200 sociological books and articles in her main field of interest which is gender roles in the family, at work, and in politics. Haavio-Mannila has also conducted studies on medical sociology and sexual behavior. She has been a visiting scholar at the Center for the Education of Women at the University of Michigan and a visiting professor at the University of Minnesota, among others. Presently she is conducting a large study of sexual behavior in Finland with Osmo Kontula.

Author's References

[1] Epstein, C. The Cultural perspective and the study of work. In Kai Erikson & Steven Peter Vallas (eds.), The Nature of Work: Sociological Perspectives (New Haven and London: American Sociological Association Presidential Series and Yale University Press, 1990).

[2] Moen P. Women's Two Roles — A Contemporary Dilemma (New York: Auburn House, 1992).

[3] Vallas SP. The future of work. In Kai Erikson and Steven Peter Vallas (eds.), The Nature of Work: Sociological Perspectives. (New Haven and London: American Sociological Association Presidential Series and Yale University, 1990).

[4] Kanter, RM Men and Women of the Corporation. (New York: Basic Books, 1990.)

[5] Magnet M. The truth about the American worker. *Fortune Magazine*, May 4, 1992, pp. 48-65.

[6] Gustavsen B. Reforms on the place of work and democratic dialogue. In Jan Odhnoff and Casten von Otter (eds.), Rationalities of Work: On Working Life of the Future (Stockholm: Arbetslivscentrum, 1987).

[7] Komiteanmietinto. Report on the Committee on Working Conditions. (Helsinki: Valtion Painatuskeskus, 1991).

[8] Komiteanmietinto. Report on the Committee on Working Conditions. (Helsinki: Valtion Painatuskeskus, 1991).

[9] Moen P. Op. cit.

[10] Kamerman SB and Kahn AJ. Family Policy: Government and Families in Fourteen Countries (New York: Columbia University Press, 1978); Kamerman SB and Kahn AJ. Child Care, Family Benefits and Working Parents: A study in Comparative Policy (New York: Columbia University Press, 1981); and Kamerman SB and Kahn AJ. The Responsive Workplace (New York: Columbia University Press, 1987).

[11] Schor J. The Overworked American: The Unexpected Decline of Leisure (New York: Basic Books, 1991).

[12] Dahlstrom E. Theories and ideologies of family ideologies, gender relations and human reproduction. In Katja Boh et al

1. SEXUALITY AND SOCIETY: Historical and Cross-Cultural Perspectives

(eds.) Changing Patterns of European Family Life: A Comparative Analysis of 14 European Countries. (London and New York: Routledge, 1989).

[13] Simonen L. Change of Women's Caring Work at the Turning Point of the Welfare State — A Nordic Perspective. Paper Presented at the Fourth Annual International Conference of the Society for the Advancement of Socio-Economics, Graduate School of Management, University of California, Irvine, March 27-29, 1992.

[14] Haavio-Mannila E, Kauppinen, K. Women and the welfare state in Nordic countries. In Hilda Kahne and Janet Z. Giele (eds.) Women's Work and and Women's Lives: The Continuing Struggle Worldwide (Boulder: Westview Press, 1992).

[15] Allen T. The Nordic Model of Gender Equality and the Labour Market. Paper presented at the Fourth Annual International Conference of the Society for the Advancement of Socio-Economics, Graduate School of Management, University of California, Irvine, March 27-29, 1992.

[16] Ibid.

[17] Esping-Andersen, G. The Three Worlds of Welfare Capitalism (Cambridge: Polity Press, 1990).

[18] Julkunen R. The Welfare State at a Turning Point (Tampere: Vastapaino, 1992).

[19] Marklund S. Paradise Lost? The Nordic Welfare States and the Recession, 1975-1985. (Lund: Lund Studies in Social Welfare II, 1988 and Alestalo M. The Expansion of the Public Sector (Vammala: Sosiaalipolitiikka, 1991).

[20] Narusk A. Parenthood, partnership and family in Estonia. In Ulla Bjornberg (ed.) European Parents in the 1990s: Contradictions and Comparisons (London: Transaction Publishers, 1991).

[21] Ibid.

[22] Lapidus GW. Gender and restructuring: The impact of perestroika on soviet women. In Valentine M. Mogdaham (ed.) *Gender and Restructuring* (Oxford: Clarendon Press 1993).

[23] Laas A. The future for Estonian women: Real women are mothers. Family Research Institute of Tartu University, Estonia. (Stencil) 1992.

[24] Hansen H. Attitudes of Girls to Future Occupations. Paper presented at Finnish-Estonian Seminar on the position of women at work and in society, Tallinn, May 14-15, 1992.

[25] Narusk A. Satisfaction of women with work and family life. Paper presented at Finnish-Estonian Seminar on the position of women at work and in Society, Tallinn, May 14-15, 1992.

[26] Meri T. Feminist on Eestis soimusona (feminist is in Estonia a word of abuse) *Naised Eestis* 2(32), 28 February, 8, 1992

[27] Laas A. The future for Estonian women: Real women are mothers. Family Research Institute of Tartu University, Estonia. (Stencil) 1992.

[28] Kilvet K. Interview of Krista Kilvet, president of the Estonian Women's Union, Finnish Radio, July 1, 1992).

[29] Population Crisis Committee. Country rankings of the status of women: Poor, powerless and pregnant. Population Briefing Paper No. 20, June, 1988 (Washington: Population Crisis Committee).

[30] Human Development Report. Published for the United Nations Development Programme. (New York and Oxford: Oxford University Press, 1991).

[31] Inglehart R. Culture Shift in Advanced Industrial Society (Princeton: Princeton University Press, 1990).

[32] Human Development Report. Op. cit.

[33] Lapidus G. Op. cit.

[34] Haavio-Mannila E. Work, Family and Well-Being in Five North and East-European Capitals. (Helsinki: The Finish Academy of Science and Letters B:255, 1992); Haavio-Mannila E. Women's Work in Three Types of Societies (Ann Arbor, Michigan: University of Michigan, Center for the Education of Women, 1993).

Beyond Abortion: Transforming the Pro-Choice Movement

Marlene Gerber Fried

Marlene Gerber Fried is director of the Civil Liberties and Public Policy Program at Hampshire College. She is a long-time reproductive rights activist with the Boston Reproductive Rights Network (R2N2), the Abortion Rights Fund of Western Massachusetts, and the newly formed National Network of Abortion Funds. She is editor of From Abortion to Reproductive Freedom: Transforming A Movement *(South End Press, 1990).*

In the 1970s, Congress battled over and finally enacted the Hyde Amendment, rolling back reproductive rights by prohibiting federal Medicaid funding for abortion only a few years after the Supreme Court decision on *Roe v. Wade*. The success of the Hyde Amendment marked the beginning of a decade and a half of challenging, confining, and generally eroding abortion rights.

Today, the "right to choose" that we find in the United States—without funding, with parental consent and notification laws, mandated waiting periods and other onerous state restrictions—bears little relation to the feminist goal of women controlling their own bodies, a notion that shaped the fight for legalized abortion.

A multi-issue women's liberation movement fought for abortion rights as part of a broader agenda for women's freedom, including women's ability to control their own reproductive lives. Legal abortion was a necessary and significant step toward this goal, but only the first step. And, in 1973, *Roe* itself was a compromise—legal abortion controlled by doctors and the state, protected by the right to privacy.[1]

For the abortion rights movement of the 1980s, however, defending *Roe* became the goal itself. In the face of a formidable anti-abortion movement dedicated to making abortion increasingly inaccessible and ultimately criminal, the pro-choice movement took on a reactive role, defending *Roe v. Wade* rather than pressing for a more expansive vision of reproductive choice.

But narrowing the vision narrows the potential base of the movement. Equating reproductive choice with legal abortion is simply not meaningful to the millions of women who have no access to that right because of poverty, age, geography, language or other barriers. Isolating the abortion struggle from the real conditions of women's lives and struggles for other freedoms is divisive and only weakens the movement.

Today, with a pro-choice President and an end, even if only temporarily, to the period of continual erosion, and with popular attention finally focused on access to both abortion and health care, the pro-choice movement has unprecedented opportunities. It can redefine its goals and help create an inclusive movement committed to fighting for reproductive choice and for the health, rights and freedom of all women.

Third World women activists committed to struggling for reproductive rights have taken

1. SEXUALITY AND SOCIETY: Historical and Cross-Cultural Perspectives

the lead. They place the attacks on abortion rights in the context of sweeping attacks on the health and on the social and economic opportunities of their communities. Forming their own organizations, they initiated a process of transforming the pro-choice movement from one focused on abortion to one concerned with *all* aspects of women's reproductive health and survival. These groups insist that the fight for abortion rights incorporate understandings of race and class oppression.[2]

While the success of this process is critical, mainstream pro-choice groups have been slow to recognize its validity and resist the necessary organizational changes and power-sharing required by this process. The voices of Third World women have been marginalized in the pro-choice movement, which remains focused on the concerns of those who feel they have rights to lose rather than on the needs of women for whom the rights have never been a reality. Changing this situation requires changing consciousness and altering power relationships among organizations. New organizational and political forms must be created that reflect the transformation.

Equating reproductive choice with legal abortion is simply not meaningful to the millions of women who have no access to that right.

Today, many of us struggle with these issues as we try to change our movement. Even large, mainstream pro-choice groups are talking about widening their agendas to include health-care reform. At this time, as a part of the conversation about how we can build a movement responsive to these concerns, I propose to look at some ideas that have emerged from a conference I have organized for several years now.

Expanding the Agenda

In 1987, I initiated an annual conference at Hampshire College on The Fight for Abortion Rights and Reproductive Freedom for campus and community activists. The conference features speakers from a wide range of women's struggles—grassroots, national and international.

Over the years, the goals of the conference have been shaped by the speakers and participants. It is an opportunity for activists in different arenas and national settings to come together; it is a forum for projecting a broad vision of reproductive freedom and for challenging abortion-centric politics; it provides an opportunity for activists drawn into the movement via concerns about abortion rights to broaden their political conception; finally, it provides support for the pro-choice activities of younger activists, especially women of color.[3]

The significance of the conference lies in the breadth of issues covered and the linkages among issues and communities, which over the years the speakers have so powerfully and eloquently articulated. As I listen each year to women who are activists in diverse struggles, I am struck with certain recurrent themes and interconnections.

Breaking Silences

An 18-year-old white working-class student from North Adams, Mass., comes to the conference specifically to participate in the abortion speakout. She joins other women she does not know who also have never before talked about having an abortion. Some have been silent for decades. Denise Palewonsky, an activist from the Dominican Republic who has not planned to participate, is moved by the event to do so. She tells us, "In my country, where abortion is illegal, it is unthinkable that women would get together to speak about their experiences this way."

Describing the work of the Community Health Project in the Occupied Territories, a Hala Salem Atiyeh, researcher and health activist at BirZeit University, says, "We started a program to break the isolation and the silence;

4. Beyond Abortion

to go into the community and share ideas about how to achieve our goal of health for all people."

Across race, culture, age and so many other differences in experience, women talk about breaking the silences in their own communities as a first step to resistance and political change. The fact that women have been silenced about abortion and reproductive health reflects the pervasiveness and power of sexism. While the nature and impact of the silencing differ depending upon the societal context, they always lead to isolation, to women being denied support for their choices or for decisions they feel they must undertake. As Loretta Ross put it, "Our abortion experiences have been invisible.... We have spoken silently with our actions. Black women get abortions at twice the rate of white women. But we have had to do so without community support because of the conspiracy of silence surrounding abortion. A silent community cannot support sisters doing what they need and choose to do."[4]

Women are struggling in their different communities to break these silences—to claim

The pro-choice movement remains focused on the concerns of those who feel they have rights to lose rather than on the women for whom the rights have never been a reality.

their own definitions of reproductive freedom. We need to listen, to hear, and to learn from each other; to make room for experiences and understandings of abortion, birth, mothering, and health that are different from our own. Being pro-choice must mean respecting and supporting each woman's right to make her own reproductive decisions.

Choice without Access is Empty

Mary Chung, a representative of Asian Pacific Islanders for Choice, states, "I want to talk about what 'choice' means for the Asian community. For millions of Asian women the flurry of judicial, legislative and executive activity surrounding abortion has no meaning and brings no relief. Many of these women are Asians and Pacific Islanders; many are poor; many do not speak English."

The demand that the abortion rights movement deal with questions of access is key in a world in which access to abortion and to all rights is largely a function of a woman's economic resources, her race, her age, what language she speaks and where she lives. Until recently, access was neglected by the press, the public, and the pro-choice movement. While hundreds of thousands of women have been mobilized to defend abortion clinics and to demonstrate in defense of *Roe* v. *Wade*, there have been no comparable mobilizations to protect the rights of low-income women. Noting this, Brenda Joyner, of the Tallahassee Feminist Women's Health Center, challenges the movement: "Perhaps the question is not really where are women of color in the abortion rights and reproductive rights movement. Rather, where is the primarily white middle-class movement in our struggles for freedom? Where was a white middle-class movement when the Hyde Amendment took away Medicaid funding of abortions for poor women?"

As a movement concerned with *all* women's rights, we must acknowledge that without access there are no rights, and we must fight for all aspects of access. Even now that public attention is finally being paid to the lack of access to abortion, the tendency is to focus on the decreasing services and the lack of providers rather than on a lack of funding for women who need it. While increasing services is crucial, poor women will still be left out unless they can afford whatever services

1. SEXUALITY AND SOCIETY: Historical and Cross-Cultural Perspectives

exist. One practical step in the right direction is the National Black Women's Health Project's current campaign to repeal the Hyde Amendment.

Campaigns for legislation that purport to protect abortion rights, without securing access, are divisive and unacceptable. For example, the current version of the Freedom of Choice Act does not include public funding, and permits parental involvement laws restricting young women's abortion rights. Pro-choice groups, pursuing such legislation at the federal or state level must seriously re-evaluate this strategy and understand its implications for movement building. We jeopardize potential alliances if, in the rush to protect the rights of some women, we compromise away the rights of our sisters for whom access defines rights.

The Breadth of the Struggle

"When I was first asked to come here I thought, how do I talk about reproductive rights without talking about all of those other rights that people in this country seem to think are just privileges for African-American women. Fighting for rights is something you do from the time you get up in the morning until the time you go to bed at night. We are fighting for the right to a proper education for our children, for the rights to decent housing, for the right just to live safely in our own communities and to be respected as people and as human beings."

This message from Norma Baker, a community activist with Northen Educational Services of Springfield, Mass., is echoed by virtually every woman of color who has spoken about the struggle for reproductive freedom. Repeatedly, activists of color tell us that a woman's reproductive "choices" come in the context of her entire life and the life of her community, and that the struggle for reproductive rights in Third World communities is part of a much broader set of struggles. As Adetoun Ilumoka, a Nigerian health activist, iterates, "I think reproductive rights, even the concept, not just the language, doesn't mean an awful lot to the average Nigerian woman. They are concerned with their health, certainly with their ability to make a living and with their choices in terms of all sorts of things..."

A movement for abortion rights that fails to incorporate struggles against racism and other forms of domination will not make sense to women whose lives are structured by interconnecting systems of oppression. For women of color who are participating in the fight for abortion rights, that struggle is about women's survival in all of its breadth. "It's very hard for Indian women to become part of the reproductive rights movement because we cannot separate reproductive rights from the process of colonization and genocide," says Andrea Smith, from Women of All Red Nations. "Who cares about reproductive rights? We don't have any rights at all. We need to be working on reproductive rights within the broader context of colonization and genocide."

Ultimately, reproductive freedom for all women implies broad social change. As Sundari Ravindron, of the Rural Women's Social Education Center in South India puts it, "For us the fight for reproductive freedom means fighting for another kind of society, because we want many things on our own terms. In the process of organizing for reproductive freedom, we find ourselves fighting for another society."

The political agenda of these activists flows from a fundamentally different way of conceptualizing problems and goals. The fact that women of different races and classes have different choices need not ultimately divide us, as long as our movement's immediate and long-range activities reflect an understanding of the breadth of rights that we need in order to make reproductive choice a reality for all women. That understanding should define our fight.

Our Struggles are Global

- In Sri Lanka, women are offered the equivalent of a year's wages if they agree to be sterilized.
- In the United States, 25 percent of Native American women have been involuntarily sterilized.
- Several state legislatures in the US are considering bills to offer cash stipends to women who have Norplant inserted.
- New Jersey passed a law that capped welfare payments so that women who have additional

children while on welfare will not receive increased funding.
- Women in the US who take drugs while pregnant are charged with fetal neglect.
- In the Third World, 500,000 women die each year from illegal abortions.

These examples remind us that the fight for legal and accessible abortion is only a part of the global struggle for women themselves to determine the conditions under which they will have sex and children. In addition to seeing the common ideological, social, and political barriers, we come to understand the variations. For some women, fighting for reproductive choice involves resisting social conditions and having the children they want. For others, it means opposing policies that force them to risk their health lives or have children they do not want.

"As women of color, our community limits are not drawn at national boundaries. We watch with special concern when women of color abroad are bullied and face aggression, or are subjected to dehumanization and policing. We know that if sisterhood is global, so are many other kinds of relationships in which most of us are involved, even if unwillingly. The clothing we wear is made by women in sweat shops in Third World countries; the contraceptives we use were tested on women in Third World countries with massively larger dosages than what is considered acceptable today...

"The movement for reproductive freedom not only needs to expand beyond the theme of abortion, it has to expand beyond its national borders. It has to become internationalist in its approach because our oppressors are internationalist in their approach," states Ana Ortiz, of Women of Color for Reproductive Freedom.

Although militarism and multinational corporations are the primary causes of poverty and environmental degradation worldwide, blame is deflected by focusing on Third World women's fertility. And solutions that involve coercion and ignore women's health, status and general well-being, but attempt to control their fertility, also cross national boundaries.

The US plays a leading role in creating oppressive conditions and in pushing coercive

4. Beyond Abortion

The US plays a leading role in creating oppressive conditions and in pushing coercive population-control policies worldwide.

population-control policies worldwide. It is therefore especially important that the US pro-choice movement educate itself about these issues and incorporate into its agenda demands for the reproductive health and general well-being of women in developing countries. As in the domestic setting, in order to achieve the goals of reproductive freedom and social justice, we must refuse reproductive choices for more privileged women in the US that come at the expense of the rights of women in other countries. Again, quoting Ana Ortiz, "When advocates of abortion rights come to the front line in the struggles of labor, disability rights, the sharing of wealth equitably across national borders; when the reproductive freedom movement takes its rightful place in the creation of a just and comprehensive human rights agenda, then, strengthened by the spiritual and political power of sisters of color in, out and around national boundaries, it will be an unstoppable force."

Struggling for Survival

At these annual conferences, we have heard many activists—from countries where abortion is totally illegal, Latina women working in a battered women's shelter, AIDS educators, housing advocates, South Asian women working against domestic violence in New York City, an anti-Klan activist from Georgia, women working with pregnant and parenting teens, lesbians struggling for their freedom, labor organizers, disability rights activists, and many more.

The extent to which violence and death pervade women's lives is overwhelming—women's bodies truly are battlegrounds for

1. SEXUALITY AND SOCIETY: Historical and Cross-Cultural Perspectives

sexual violence, hate violence, rape, colonialism, military occupation, AIDS or other sexually transmitted diseases, illegal abortion, childbirth, religious fundamentalism, unsafe contraception, lack of food... the list goes on and on.

Each year, as hundreds of us sit in an auditorium at Hampshire College in rural Massachusetts, we experience collective rage at the vastness of the oppression. But ultimately we do not leave overwhelmed. Instead, these incredible activists inspire us, call us to action, and allow us to experience the potential power in the interconnections among our communities of resistance. We are given a glimpse of the kind of political vision and unity that we are capable of creating, and the experience is inspirational.

Notes

[1] Marlene Fried and Loretta Ross, "Reproductive Freedom: Our Right to Decide," in Greg Ruggiero and Stuart Sahulka (eds.), *Open Fire* (New York: The New Press, 1993), p. 94.

[2] Such organizations include, for example, Asian Pacific Islanders for Choice, Native American Women's Health Education Resource Center, The National Black Women's Health Project, The National Latina Health Organization, Women of All Red Nations, Women of Color Partnership Program of RCAR (Religious Coalition for Abortion Rights), and Women of Color for Reproductive Freedom.

[3] Although a majority of the speakers are women of color, they remain a minority of the nearly 300 activists who attend the conference. While there have been efforts to make the conference accessible to as many women as possible (there is no conference fee, child care, housing and food are provided at no cost and travel stipends are available), we need to do more. We plan to expand the networks through which the conference is publicized, and we are exploring the possibility of providing free buses from a few major cities.

[4] Loretta Ross, "Raising Our Voices," in Marlene Gerber Fried (ed.), *From Abortion to Reproductive Freedom: Transforming A Movement* (Boston: South End Press, 1990), p. 139.

The Brave New World of Men

Men are doing more shopping and housework, but only because women are making them change. Women still decide most household purchases, unless they're high-priced items. Women are disillusioned by the "new" man, but there's hope for young and affluent men, who are more likely to share housework and value romance. Knowing how men are changing—and how they aren't—is the key to targeting them in the 1990s.

Diane Crispell

Diane Crispell is editor of The Numbers News *and a contributing editor of* American Demographics.

Here's a portrait of the American man, circa 1992. He is romantic and self-centered, family-oriented and individualistic, hard-working and leisure-loving. In other words, he is a mass of contradictions. And he is just as mysterious to businesses as men have always been to women.

Women have spent the last three decades behaving more like men in certain ways. In the 1990s, men are adapting to these changes. Women have enjoyed significant advances in educational attainment, labor force participation, career involvement, and economic independence. They have also endured significant increases in smoking, divorce, and single-parent families. The overwhelming result has been stress, as women try to play multiple roles, and increasing conflict, as women demand more from men.

"Men were thrown a curve ball," says Michael Clinton, publisher of *GQ* magazine. "The patterns they had established in terms of interacting with women were no longer relevant."

In response, men started helping out more with housework and child care. Nurturing became a manly thing in the 1980s; it was celebrated in movies like *Three Men and a Baby* and television programs like "Full House." Advertisers became infatuated with the image of a handsome towelled man dandling a baby on his knees. But many women see these images of men as desirable though unattainable fantasies.

In reality, men can't move fast enough to meet women's expectations. As men have come to respect women more, women have become less satisfied with men. In 1970, 40 percent of men said they thought women were better respected than they had been in the past, according to the Roper Organization. In 1990, this share had risen to 62 percent. In 1970, women were most likely to describe men as "basically kind, gentle, and thoughtful." In a similar Roper survey 20 years later, women were most likely to say that men only value their own opinions, that men find it necessary to keep women down, that they immediately think of getting a woman in bed, and that they don't pay attention to things at home.

As torrents of negativity wash over them, men are beginning to reach out to one another—just as women learned to do with the women's movement, social networks, and magazines like *Working Mother*. The first issue of *Full-Time Dads* was published in April 1991. This bimonthly publication serves men who are isolated, in both a physical and psychological sense, as full-time parents. Their stories about the joys and frustrations of child-rearing are not new to women, but they are new to many men.

Men who are full-time homemakers are still rare. But for whatever reason, men are taking on more of the everyday responsibilities of running a household. That includes shopping, child care, and cooking.

FORCED TO CHANGE

Men act as consumers in two ways—as individuals and as members of a household. These two levels of purchase behavior can become intertwined. For example, only 46 percent of men buy all of their own personal items, according to a survey conducted for *American Demographics* by

1. SEXUALITY AND SOCIETY: Changing Society/Changing Sexuality

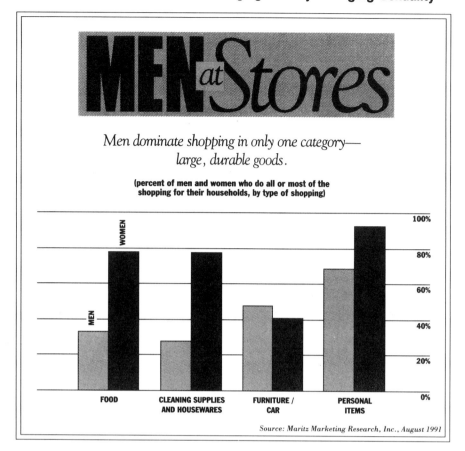

MEN at Stores

Men dominate shopping in only one category—large, durable goods.

(percent of men and women who do all or most of the shopping for their households, by type of shopping)

Source: Maritz Marketing Research, Inc., August 1991

"progress" in men's shopping and housework behavior has been made among men who have no wife. According to the Census Bureau's March 1990 Current Population Survey, 10 percent of men aged 15 or older now live alone, and 8 percent live with others in nonfamily groups. Twenty-two percent of men now live in families but are not householders or spouses of householders, and four-fifths of these men live with their parents. Fifty-seven percent of men head a family with their spouse, and another 3 percent head other kinds of families.

Men's living arrangements largely determine their consumer behavior. For example, men who live alone provide for all of their daily needs. But single men do not act like women who live alone. Single men spend 61 percent of their food budgets away from home, compared with 41 percent for single women, according to the Bureau of Labor Statistics' 1988-89 Consumer Expenditure Surveys.

Some of this gap is due to demographic differences: single men are younger and more affluent than single women. The median age of men who live alone is 43, compared with 66 for women who live alone. The median income of men who live alone was $19,167 in 1989, compared with $12,190 for women.

But women and men are still different, even if they share the same age and income levels. Single men aged 25 to 34

Maritz Marketing Research. Thirty-five percent of men buy half or most of their own things, and 18 percent buy just some or none. In contrast, 82 percent of women control all of their personal purchases, and 10 percent buy most of their own goods.

Men shop almost as frequently as women, but their habits are different. They are slightly more likely than women to shop every day (11 percent do, versus 7 percent of women), and they are also more likely than women to shop less than once a week (28 percent versus 23 percent). The majority of both sexes go shopping at least once a week but not every day.

Retailers like to say that while women shop, men buy. They mean to say that men don't enjoy shopping. It may be true, because men do act differently than women in stores. In some ways, men are better targets for in-store promotions and advertising; in other ways, they're worse.

When shopping for food, men are less likely than women to refer to shelf offers or brochures, according to Simmons Market Research Bureau. But they're more likely than women to read shopping-cart ads and to see overhead aisle markers.

Since 1960, millions of women have been forced into the job market without skills after a divorce. The flip side is that an equal number of unprepared men are being thrust into the consumer marketplace to fend for themselves. Both sexes seem to be adapting, however. The share of ties bought by women is now a mere 60 percent, down from 80 percent in the past, according to *GQ*'s market research.

Most men do at least some of the food shopping for their households. The Maritz poll shows that one-third of men do all or most of the food shopping; this agrees with the share found in a survey conducted for *Men's Health* magazine. But more than one-third of men (39 percent) are not currently married. Men may shop and cook, but only as a last resort.

HIGH-PRICED ITEMS

Married men are picking up some of the shopping and housework their working wives leave behind. But much of the

> **Men may shop and cook, but only as a last resort.**

spend 65 percent of their food budget away from home, compared with 55 percent for young single women. Affluent single men spend 75 percent of their food money away from home, compared with 48 percent for affluent single women. Clearly, men are better restaurant customers.

Married couples spend more of their food budget at home than singles do, 58 percent versus 50 percent in 1989. But even though more food is prepared in a

5. Brave New World of Men

married-couple home, the husband may not contribute. Just 12 percent of married men do all of the food shopping for their household, compared with 41 percent of unmarried men.

Men also shy away from household management chores. Just 28 percent of men in the Maritz poll buy all or most of the cleaning supplies and housewares for their home, compared with 77 percent of women. Twenty-nine percent buy none of them, compared with 7 percent of women. But most men have at least a passing acquaintance with household goods. In a *GQ* study, 51 percent of men personally purchased sheets or towels in the past three years, and 44 percent bought cookware.

Men care more about consumer decision-making when the decision concerns larger household purchases, such as furniture and cars. The Maritz poll shows that 47 percent of men do all or most of this kind of shopping, compared with 41 percent of women. Shopping for big-ticket items is more of a joint decision than other kinds of shopping—29 percent of men and women say they do half of it. Just 11 percent of respondents equally divide the shopping for household supplies, and 14 percent share food shopping equally.

If men are less involved than women in routine shopping chores, it stands to reason that men will spend less than women. The average household spends a total of $172 a week on household shopping, according to the Maritz poll. About one-third of men spend all or most of the weekly total, compared with two-thirds of women. Eighteen percent of men say they spend half of the weekly money, and 39 percent spend some of it. Only 5 percent do not spend any of it.

THE "NEW" MAN

Men have mixed feelings about all of the changes going on in their lives. Some welcome the challenges and appreciate the choices, while others fight them. "Men now see that they have choices to live different kinds of lives," says Michael Clinton of *GQ*. "This is healthy not only for them but also for the women in their lives. Men are no longer following prescribed patterns and getting boxed into preordained behavioral habits."

A 1990 *GQ* study divides men into different attitudinal groups, based on their attitudes toward change. Men who like choices are called "Change Adapters." They are younger, better-educated, and more affluent than "Change Opposers." These two groups shop for different items, and they shop in different ways.

Opposers have less money than Adapters, so they are cautious spenders. They are more likely to shop at discount stores and to look for sales and imitation products. They are not greatly influenced by advertising, but they do lots of background research before making a big purchase. They are not brand loyal.

Change Adapters are more likely to try new things. Within this group are two subgroups: the pragmatic "Tradition-Oriented Adapters," who roll with the changes going on around them, and the adventurous "Self-Oriented Adapters," who are the biggest spenders.

Self-Oriented Adapters are also the youngest generation of the *GQ* group. They are experimenting with new ideas while delaying marriage and the assumption of family responsibilities. Many of them are also having trouble finding a good job. One result is that the majority (58 percent) of 18-to-24-year-old men still live with their parents. Such men are less independent as consumers in some ways, but they also have more discretionary income and time to hone their skills as recreational shoppers. Self-Oriented Adapters are more likely than other groups of men to shop for their own clothing.

Young men may also help heal the wounded state of relations between men and women. The Roper Organization has dubbed today's young men as "new romantics," based on research it has done for *Playboy* magazine. Nearly 70 percent of men aged 18 to 29 consider themselves "romantic," while the definition of that term varies from person to person.

Men and women in their 20s are most likely to agree on the timing of their last

Men at Home

The vast majority of men live with spouses or parents.

(men aged 18 or older in thousands, living arrangements in percent, 1990)

	18 and older	18 to 24	25 to 34	35 to 44	45 to 54	55 to 64	65 and older
MEN	86,872	12,450	21,462	18,331	12,292	10,002	12,333
PERCENT	100.0%	100.0%	100.0%	100.0%	100.0%	100.0%	100.0%
LIVING IN FAMILY GROUPS	81.4	81.0	75.3	83.1	85.8	86.1	82.0
Householder or spouse	63.5	14.8	55.9	74.9	81.6	82.0	76.2
Child of householder	13.8	58.1	15.0	6.0	2.4	1.5	0.3
Other	4.0	8.2	4.4	2.2	1.8	2.6	5.5
LIVING IN NONFAMILY GROUPS	18.6	19.0	24.7	16.9	14.3	13.9	18.0
Living alone	10.4	5.4	11.1	10.0	9.5	10.4	15.8
Other nonfamily	8.2	13.6	13.6	6.9	4.8	3.6	2.3

Source: Bureau of the Census, *Current Population Reports*, Series P-20, No. 450

> The majority of 18-to-24-year-old men still live with their parents.

1. SEXUALITY AND SOCIETY: Changing Society/Changing Sexuality

romantic interlude, according to Roper. This may point the way to increased harmony between the sexes, along with growing markets for clothing, jewelry, and other romantic items.

Looking good is important to many men, and romance is only part of the reason. Nearly half of men strongly agree that they owe it to themselves to look their best, according to *GQ*, and 39 percent say they have a strong sense of personal style. One out of four men feels that dressing for success is a necessity for his job, and nearly half are pleased when others notice and comment on their appearance.

The *GQ* survey found a significant increase in the amount of time men spend grooming themselves. Men spent an average of 44 minutes a day grooming in 1990, up from 30 minutes in 1988. Men under the age of 25 spent the most time (53 minutes a day, on average) arranging their hair and clothes, and otherwise working on their outward appearance.

WHAT CHANGES AND WHAT DOESN'T

Men are changing the ways they shop, work at home, and dress. But one male mindset cuts across all demographic and socioeconomic boundaries, and it has not changed. It is a thirst for knowledge.

"Must-Know" men are do-it-yourselfers whose interest in how things work extends to their understanding of the products they buy, according to a survey sponsored by *Popular Mechanics* and conducted by Yankelovich Clancy Shulman. They are one-fourth of men with household incomes of $20,000 or more.

Must-Know men are a key consumer group because they influence the buying behavior of people around them. These men know and give advice about automobiles, home repair projects, and other traditionally male domains. They also dominate high-tech items like video equipment.

In general, men's tastes in media follow the same desire to search for and acquire factual knowledge. Men spend more time than women reading newspapers and less time reading books and magazines, according to the Americans' Use of Time survey of the University of Maryland. The books men prefer tend toward nonfiction. The novels men like are action-oriented tales, science fiction, and westerns.

Men are more likely than women to watch adventure or science fiction programs on television. They are also more likely to watch all kinds of sports, according to Mediamark's spring 1991 survey. But both sexes are equally interested in news programs, documentaries, entertainment specials, and mystery/police shows. Men are less likely than women to watch award shows and pageants, daytime dramas, feature films, game shows, prime-time dramas, and situation comedies.

The magazines men like show a similar skew. Four of the ten magazines most read by men are also on the top ten for women, according to Mediamark. These are *Modern Maturity*, *People*, *Reader's Digest*, and *TV Guide*. The rest of men's favorite magazines are mostly sports and news-related, while women turn to magazines that focus on home and family.

There are comparatively few general men's or boys' magazines on the market. Only one—*Playboy*—is among men's ten most-read magazines. "One of the problems confronting marketers is that there

---

Why Do Men Love Logos?

Most men are easy-going on matters of style and fashion. But the San Francisco 49ers learned the hard way that men are passionate about their team's logo. When the football team introduced a new symbol for its helmets last February, fan reaction was swift, brutal, and negative. The Niners reversed themselves within a week.

"You're not just talking about it as apparel," says Ralph Barbieri, host of a sports talk show on San Francisco's KNBR radio station. "You're changing the uniform of these people's team."

Many men see a team's logo as an important part of their own identity. "Sport is the glue that bonds many Americans to their place," says John Rooney, a geography professor at Oklahoma State University. "This is more true of men, because they have participated in sports since they were children." Wearing the home team's colors "is another way of saying, 'This place is important to me, and I'll demonstrate it with this jacket.'"

This urge to boast has spawned a big business. Major League Baseball products reaped $1.5 billion in sales from more than 3,000 logo products in 1990. National Football League products grossed $1 billion, and National Basketball Association products topped $750 million in sales this year.

Buying a logo can be a fan's way of voting for the franchise. Barbieri says that many 49ers fans were particularly angry that the team wanted to take "SF" off its helmets. "People were as insecure and paranoid as they could be" at the thought the 49ers might be preparing for a move, he says. But even if the unimaginable happens and the hometown team leaves, the memories of fans still leave behind a profitable market. Major League Baseball's Cooperstown Collection sells reproductions of caps and other equipment from defunct teams like the St. Louis Browns. This product line reaped over $200 million in 1990.

Fashion-conscious people are now wearing the athletic-oriented apparel, even if they don't care about sports. The Chicago White Sox adopted snazzy new black uniforms last year, which quickly rose to the top of major league sales. "Black is definitely in," Barbieri says, referring to the Atlanta Falcons' uniform switch from red to black. "It's in vogue to be bad."

Logos do indeed convey an image. Some street gangs have adopted black and silver as their own colors, in homage to the Los Angeles Raiders' "bad boy" reputation. As an anti-gang measure, a Florida shopping mall recently banned the wearing of Raiders jackets. The ban was quickly rescinded after an outburst of protest from Raiders fans. If a shopping mall outlawed Yves St. Laurent, would anybody care?
—*Dan Fost*

Men spent an average of 44 minutes a day grooming in 1990, up from 30 minutes in 1988.

5. Brave New World of Men

simply is no *Seventeen* or *Teen* for boys," says Peter Zollo, president of Teenage Research Unlimited in Northbrook, Illinois. That may be why Long Communications (creators of *Sassy*) recently launched *Dirt*, the first lifestyle magazine for teen boys.

The best way to reach men with advertising may be to focus on role- or subject-oriented media, rather than on media geared specifically to either sex. *Parenting* and *Parents'* magazines try to encompass both mothers and fathers, for example, because men and women are equally likely to be parents. Men make up only one-fifth of both magazines' readers, according to Mediamark. That share is likely to rise, however. A new generation of active fathers has arrived.

When *GQ* asked baby boomers what significant event had happened to them in the past year, 4 percent said they had become a father for the first time. Men and women both value their family more highly than anything else. This orientation grows stronger with age, and it catches fire when babies arrive.

The true nature of fatherhood can be a revelation to men. When Gordon Rothman of CBS News testified about paternity leave at a Congressional hearing, he echoed a truth that women have known about for a long time: "I don't know why people talk about a wimp factor. This is the hardest work I've ever done in my life."

The huge baby-boom generation has always been responsible for the greatest changes in men's consumer behavior. Today, baby-boom men are increasing the time they spend on household and childcare duties. As a result, they are becoming more savvy about household products.

Meanwhile, more and more women are learning how to buy cars, program VCRs, and use power tools. But some bastions of masculinity remain. Eighty-five percent of beer drinkers are men, for example. And men can pick up women' unique behaviors too. A study conducted for Kiwi Brands finds that 40 percent of professional men sometimes wear sneakers back and forth to work. It's a brave new world.

T A K I N G   I T   F U R T H E R

For information about *GQ's* "American Male Opinion Index," contact Michael Clinton at (212) 880-8800. More on shopping behavior is available by calling Phil Wiseman of Maritz Marketing Research, Inc., at (314) 827-1949. *Full-Time Dads* is published by Chris Stafford at P.O. Box 120773, St. Paul, MN 55122-0773; telephone (612) 633-7424.

Sex in the Snoring '90s

A new survey of American men shocks the nation with what it didn't find

JERRY ADLER

They're the sorts of questions any average American male—Alexander Portnoy, say, or Holden Caulfield—might have asked himself: How many times a night should I be having sex? In what grade should I start? Is 20,000 women too many? But there is nowhere to turn for answers. If you wanted to know something the government considered important, like the size of the sardine catch, you could look it up in a minute. That's because people pay money for sardines, but they usually get sex free—or, in any case, they don't pay taxes on it, so the government has no reason to keep track of it. The Statistical Abstract of the United States, in its relentless emphasis on the material at the expense of the sublime, skips over sex, going directly from "Sewers" to "Sheep." That gap in our national education has been filled by the media, which give us Amy Fisher (Joey Buttafuoco, she says, could copulate six or eight times a night, easy), the Spur Posse (some of whose members presumably had to be driven to their assignations by their moms) and Wilt Chamberlain (who claimed in 1991 to have packed into his career the equivalent of 400 years of lovemaking by ordinary mortals). This is like keeping track of the fishing industry by interviewing someone who says he once caught a 12-pound sardine. What the country needs is data.

Now, at last, science is providing data, even if it's not the kind that would get the scientists onto "Donahue." For several days last week, the most talked-about article in the United States was a technical paper in the March/April issue of Family Planning Perspectives, the journal of the Alan Guttmacher Institute. In what the authors call the first scientifically valid survey of its kind, 3,321 American men in their 20s and 30s were questioned about their sexual practices, and for the first time in the history of sex research, the results are not astounding. The median number of sex partners over the whole lifetimes of these men is 7.3. This is the equivalent of a six-day road trip for Chamberlain. David Koresh has more *wives* than that. More than a quarter of the men have had three partners or fewer (although almost a quarter have had 20 partners or more). The median age at which the white men in the survey lost their virginity is 17.2 years—in other words, they were seniors in high school. (For black men it is 15.) The median frequency of sexual intercourse is slightly less than once a week overall, slightly more than once a week for married men. You'd think that would constitute the ultimate nonstory: *Man Makes Love to Wife*. It is hard to think of another

---

THE REPORT

In the most complete study since the Kinsey report, a new survey charts American sexual activity. The results are—yawn—striking.

1
is the median number of times men said they had sex per week

7.3
is the median number of sexual partners per man

43%
of black men reported performing oral sex while 62% reported receiving it

79%
of white men reported performing oral sex while 81% reported receiving it

21%
of white males have had anal sex

14%
of black males have had anal sex
(Guttmacher, 1993)

---

area in which such mundane findings could be front-page news.

Actually, the authors—four researchers at the Battelle Human Affairs Research Centers in Seattle—did come up with one remarkable statistic. Of the men they surveyed, only 2.3 percent reported any homosexual contacts in the last 10 years, and only half of those—or just over 1 percent of the total—said they were exclusively gay in that period. That number was surprising in light of the widespread belief, dating to the Kinsey report of 1948, that homosexuals accounted for 10 percent of the population. But that was a misreading of what Kinsey actually found (page 37), and more recent surveys had typically put the number much lower (NEWSWEEK, Feb. 15). The 1 percent figure, though, was the lowest yet, and it will be controversial for a long time. It raised the prospect of having to rethink yet again some of the most contentious social issues of the last few years—from the likely future course of AIDS to the much overhashed question of homosexuals in the armed services. Gay-rights groups assumed that their power would diminish if they were seen as speaking for a much smaller constituency, and conservative Christian groups were quick to endorse the Battelle findings. On the other hand, if the 1 percent figure is accurate, then an infantry company, on average, would contain approximately two gay soldiers—hardly enough to cause the dire disruptions predicted by the generals.

Rock groupies: Will the names of the Battelle researchers—John O.G. Billy, Koray Tanfer, William R. Grady and Daniel H. Klepinger—go down in history alongside those of Kinsey, Masters and Johnson, and their illustrious predecessors dating to Ovid? Hard to say. The four set out to investigate male sexual behavior mostly in search of information that might be useful in combatting AIDS. But perhaps their most valuable contribution was to uncover the vast gap between what Americans actually do in bed, as a statistically verifiable fact, and what they assume goes on in most people's lives, based on extrapolating the autobiographies of rock groupies and the letters to the advice columns in girlie magazines. "I don't think they're interviewing the right people," said David Hoffman, 27, a vice president of a Chicago store called Condoms Now. "What about guys on sports teams in colleges?" Many average adults found it hard to believe that the median figure for sexual partners could be as low as 7.3. "I think men lie," said a Boston paralegal who gave only her first

INFIDELITY

How faithful are the women of America? It depends on whom you ask. Two studies produced very different answers.

26%
Married women who have had extramarital affairs
(Janus, 1993)

39%
Married women who have cheated on their husbands
(Cosmopolitan Readers Survey, 1993)

---

approximately as many men with seven or fewer partners as with eight or more; therefore, a man with 100 partners got the same weight as one with 10. In a "mean," or "average," the extreme values would count for more. The mean number of sex partners for the men in the survey is 16.2.

The only people who found nothing surprising in the results were the investigators themselves. "I can't find them surprising or nonsurprising," said Billy, with the magisterial detachment of Rutherford announcing the discovery of the electron. "We have no basis for comparison except popular wisdom—which is often inaccurate." Other researchers on men's sexual behavior generally found nothing astonishing in the report either, except for the data on homosexuality. The data on sex partners "isn't surprising to me," said University of Washington sociologist Pepper Schwartz, who published a 10-year study of American sexual attitudes. "Four to seven [partners] is what the informal studies say." She also wasn't surprised that married men reported hav-

name, Veronica. "They hide things more than women do." Veronica, 30 and married, thought the right number was around 15. Perhaps, like many laypeople, she was confused by the mathematical terms used. A "median" of 7.3 means that there were

ing sex around once a week, although she thinks you have to make allowance for inflation. "People don't have sex every week;

TALL TALES

Some of us are still living the wild life—or at least telling survey takers we are.

Have had sex with more than 100 partners

4% women
10% men
(Janus, 1993)

Have sex more than once a day

6% married
12% living together
13% going steady
(Cosmopolitan Readers Survey, 1993)

The Impact on Gay Political Power

The "1 percent thing" wasn't mentioned at President Clinton's historic Oval Office meeting with gay and lesbian leaders last week. But the question hung in the air outside: if gays really represent such a tiny fraction of the population, will that stall the political momentum the gay-rights movement has built in recent years? Leaders of anti-gay groups were giddy over the prospect. "Tremendous political impact!" exclaimed the Rev. Lou Sheldon, chairman of the Traditional Values Coalition, which represents some 27,000 churches. "Thank you, Alan Guttmacher Institute."

The new report comes at a delicate time for gay political power. Clinton's bid to overturn the ban on gays in the military faces a tough fight in Congress and with military brass. Gay-rights initiatives are under attack in several states. Gay and lesbian leaders, meanwhile, have been striving to liken their cause to the black civil-rights movement of the 1960s. They hope to draw as many as 1 million people to a march on Washington this weekend, demon-

HOMOSEXUALITY

The Kinsey report and the new study provide different counts. Both figures are disputed.

1%
Men who have had same-gender sex exclusively during the past 10 years
(Guttmacher, 1993)

10%
Men more or less exclusively homosexual, for at least three years, between the ages of 16 and 55
(Kinsey, 1948)

---

strating to the rest of the country that gays are part of the mainstream. The new report could cost them a key element of that strategy: their claim of vast, secret numbers in the U.S. population.

Kinsey's 10 percent figure has actually been in dispute for years. Most recent studies put the percentage much lower. Annual surveys by the University of Chicago's National Opinion Research Center have consistently found that about 2 percent of sexually active men said they had had sex only with other men during the past year.

Last week many gay leaders insisted that the 1 percent figure was just as flawed. Some charged that since the researchers had relied on face-to-face interviews, and asked subjects for their names, addresses and the names of relatives, gays might have been afraid to answer honestly. But Koray Tanfer, one of the report's authors, said that those questions did not bias the answers since they were asked at the end of the interviews. And he noted: "For the 10 percent figure to be correct, it would mean that nine out of 10 of the gay people we talked to had lied to us. Is that what the gay activists believe?"

Meanwhile, Stephanie A. Sanders, assistant director of the Kinsey Institute for Research in Sex, Gender and Reproduction, in Bloomington, Ind., said that Alfred Kinsey's notorious findings have been misread. He reported that 10 percent of white males age 16 to 55 said they'd had predominantly gay contacts for at least three years; only 4 percent said they were lifelong homosexuals. The Battelle study found that 1 percent of men age 20 to 39 said they'd been exclusively homosexual for the last 10 years. "The real question is, what are we talking about when we say 'homosexual'?" Sanders said. "Is three years sufficient? Do you have to spend a lifetime?"

To many gay leaders, the numbers are irrelevant. "I don't care if there are only 10 of us in the whole country. Do we have equal rights or not?" demanded Roger MacFarlane, a founder of Gay Men's Health Crisis in New York. But there is strength in numbers, especially in politics. "If a million people go to the march, that'll mean a lot more than a silly survey," says gay author Michelangelo Signorile. But even that won't settle the statistical debate. Will it mean that roughly half of all the gays in America trekked to Washington—or that many nongays also turned out to show support?

MELINDA BECK *with* HOWARD FINEMAN *in Washington and* DANZY SENNA *in New York*

1. SEXUALITY AND SOCIETY: Changing Society/Changing Sexuality

they have good weeks and bad weeks. But they think, 'I'm married and should be having sex more,' and round it off to about once a week." Debra Haffner, executive director of the Sex Information and Education Council of the United States, agrees: "It usually comes out to 57 times a year . . . which I figure means once a week, and three times on vacation."

Appalling slander: But at the same time, there was a body of opinion that the Battelle group had perpetrated an appalling slander on the virility of American men. A one-night stand is by definition forgettable, and you have to guard against the tendency to undercount them. Warren Farrell, author of "Why Men Are the Way They Are," says that in workshops he asks participants to write down—privately—their total number of sexual partners. After a period of discussion he asks them to write it down again, and finds that on average men recall twice as many partners as they did at first (and women three times as many). Therefore he suspects the figure of 7.3 partners should actually be doubled.

One thing the Battelle study did reveal was the huge philosophical gap between two schools of sex research. One group conducts personal interviews with subjects chosen on the basis of census data to represent a scientific sample of the population, which is supposed to provide better statistical validity. The other usually uses anonymous mail-in questionnaires, which seem to elicit juicier responses. The former generally publish papers in scientific journals, while the latter typically write articles in women's magazines and best-selling books. From them comes such illuminating data as the fact that 67 percent of men would not have their penises enlarged, even if it didn't hurt or cost anything (Glamour, 1992). "These kinds of reports are highly suspect," says Tom W. Smith of the University of Chicago's National Opinion Research Center. Shere Hite, the famous bombshell sex researcher of the 1970s, used this technique, as did board-certified sexologist Samuel S. Janus and his wife Cynthia, whose "Janus Report on Sexual Behavior" got a lot of attention just a few weeks ago. Janus, who distributed his surveys through graduate students and in piles left in doctors' offices, asserts that people are more likely to be truthful on an anonymous questionnaire than in face-to-face interviews. The Battelle group's interviews were all conducted in the respondents' homes, by female interviewers who knocked on the door with no previous introduction. Thirty

---

DIVINE GIFTS

A connection between secular and holy love—
or proof that prayers are answered.

Couples over 60 that engage in sex
at least once a week

45%
of those who pray together

33%
of those who don't pray together

("Sex After 60: A Report," Andrew Greeley, 1992)

---

percent of the men contacted in this way refused to participate. Janus suggests that the missing homosexuals are probably in that 30 percent. On the other hand, the people motivated to pick up, fill out and mail in a questionnaire on sex represent a self-selected group—selected, in part, by being interested in sex. Janus's data also show that his subjects were considerably richer, less Protestant and more Jewish than the nation as a whole. And, for what it's worth, they seemed in some ways to be having a lot more sex than the ones surveyed by the Battelle group. In his study more than half the men age 18 to 38 said they had sex either "daily" or "a few times weekly."

So, in the famous phrase, you pays your money ($24.95 for "The Janus Report," $28 a year for Family Planning Perspectives) and you takes your choice—either most men are having sex three to seven times a week (Janus), or the median is slightly less than once a week (Battelle). In favor of the Battelle thesis, you might want to consider what Curtis Pesman, author of "What She Wants: A Man's Guide to Women" had to say: "A lot of men will feel better about their real lives. There are a lot of men who think everybody is having more sex than they are, and they'll feel better being close to the average . . . and I know a lot of wives and girlfriends who will be happy that the average is not up in the teens or twenties."

On the other hand, as Jennifer Knopf, a Chicago sex and marital therapist, points out, it's only reassuring if you're having sex once a week; "if you're having sex twice a month, it's probably frightening."

Which suggests the only possible conclusion on which all sides will undoubtedly agree. We need more data.

With TESSA NAMUTH *in New York,*
PAT WINGERT *in Washington,*
KAREN SPRINGEN *in Chicago and bureau reports*

The New Sexual Revolution

Liberation at last? Or the same old mess?

Jim Walsh

Special to Utne Reader
Jim Walsh is a writer living in Minneapolis, Minnesota.

Five years ago in a cover section entitled *"Remember sex? Are you too tired, too bored, too mad, too scared to enjoy sex these days?"* Utne Reader *concluded that sex for pleasure was about as popular as a spring vacation in Antarctica. Well, as the saying goes, time is a great healer. Mindful of—but undaunted by—AIDS and the increase in other sexually transmitted diseases, today's sexual revolutionaries have pioneered new philosophies, techniques, political movements, and equipment to restore our battered sense of eros. To hear the writers in the following section tell it, the new sexual revolution is primarily about appreciating an imaginative range of options. So relax, and get as lusty as you want. As long as nobody gets hurt, whatever gets you through the night really is all right.*

New Year's Even, 1973–74. People are running naked through the streets, stadiums, and shopping centers of America; tonight the streakers take Times Square. *Roe v. Wade*'s first anniversary is three weeks away. A New York judge has deemed *Deep Throat* "indisputably and irredeemably obscene," though even as he speaks, people are lining up to see *Last Tango in Paris*. In San Francisco, the burgeoning new homosexual capital of the world, a camera shop owner by the name of Harvey Milk has just failed in his first bid to become the city's first openly gay city council member. The number one song in the country is Marvin Gaye's "Let's Get It On."

Me, I'm 13 years old. I'm at my first make-out party, and I'm not making out. I've got braces and zits and already I've been turned down by two girls. The first I want because she's a total fox, the second because she's carrying a copy of Alice Cooper's *Billion Dollar Babies*. I don't know who Alice Cooper is, but I know I love records; therefore, I love this girl. The feeling isn't mutual.

It's winter and freezing outside. Inside, a five-alarm fire of teenage hormones is raging. We're in a guy named Jerry's basement; his parents are gone for the evening, and he's got full access to the liquor cabinet and his dad's collection of dirty magazines. I'm on the floor with a girl by the name of Lynne, though I don't discover this bit of trivia until we've been locking lips for over an hour. There are no first or last names here, no rules, no commitment. Just warm bodies, getting warmer by the minute.

Suddenly, a noise comes from upstairs and the basement lights go on. Parents. A collective. "Oh . . . !" goes up, and kids are jumping up, buttoning shirts, zipping up zippers. Then, just as suddenly, the lights go down again and laughter from a couple of bored partnerless partiers tumbles down the stairs. False alarm. The making out resumes. With the threat of getting caught now in the mix, the hots get hotter. Lynne and I lay back down on the floor and . . .

Today it's sex with machines, sex with cathode rays, sex with latex gloves.

Whoa. Zipping up zippers? This is a *make-out* party. I get up from the floor and tell Lynne I'm going to the bathroom, which is true. But I'm also ducking for cover from the ton of bricks that's just fallen on my head. On my way, I run into Jerry "I . . . mine," he says in a tone that suggests he's already bored with girls, 1974, and adolescence in general. "Are you gonna . . . yours?" I close the door and lock it carefully. He giggles a venomous, singsongy giggle that oozes its way under the bathroom door. "Hey Waaalsh—*are you gonna go all the waaay?*"

Happy new year. I come out of the bathroom and find Lynne on the couch, the same couch where Jerry and little Alice Cooper just had their magic moment. At 1:15, Jerry unpuckers from Alice long enough to announce that his folks are gonna be home soon. Lynne and I are off and smooching again. She's fast, I'm slow, but the ride's been fun: First base. Lovely. Second base. Fabulous. Third base, my first time. Pow! Like going to the moon. And hey, I'm more than happy to stop there. Mission accomplished. Curiosity sated. Lust abated.

Next to me, Jerry's overcome his boredom and he and Alice are doing it for the second time. On the floor, the total fox and a guy I went to grade school with are doing it. I know this because

1. SEXUALITY AND SOCIETY: Changing Society/Changing Sexuality

the guy tells me he's doing it while he's doing it. Everyone, it seems, is doing it. Everyone but Lynne, and she knows it. Time's running out. She unzips my jeans, presses up next to me, and coos, "C'mon . . ."

It's the chance of a lifetime. Opportunity knocking. I don't know what to do, so I keep my lips pressed hard on hers. A minute goes by. I kiss her on the neck and run my hand through her hair. Three minutes later—an eternity—I glide my lips across her cheek and put my mouth to her ear. I don't know what I'm going to say, but when I say it, it . . . squeaks.

"We *can't.*"

What I mean to say was *I* can't. What I wanted to say was, I'm not ready I'm sorry I don't know you I like you well enough but, but . . . But what? But like I said, I'm not ready. I'm saving it, that's what. Not for marriage or because I think I'll go to hell or because I think I'll get a disease. I'm saving it for some other time. For someone else. Someone I don't know yet.

The next Monday in school, thanks to Jerry and a kiss-and-tell chat with Lynne, the word is out: Walsh had a chance to go all the way with a superfox and he blew it. In gym, a guy I barely know comes up to me and asks, "Why?" like I'd just won the lottery and thrown all the money off a bridge. Then there's the guys who simply cluck and flap their arms as they pass me in the hall. It lasts about a week.

It was no big deal, really. This isn't a scar. I don't sit around preoccupied with how I was tormented by those kids that first week of 1974. But just as I felt I was on the outside looking in then, I find myself 20 years later sitting on the very same sidelines of the so-called new sexual revolution. I don't have toys or tapes or even anything all that out of the ordinary when it comes to a fantasy life.

What I've got is a wife, the same woman I've been with for 14 years. Mind you, I didn't trot this fact out with any holier-than-thou vanity; fact is, sometimes I'm almost embarrassed to admit it. See, I write about rock and roll for a living, and I spend a lot of time in bars with people who've been through the sex wringer. And while we're able to see eye to eye on any number of artistic and political levels, when they discover I've been happily married for so long, a certain wall goes up. They can't relate to my monogamous existence, just as I can't connect emotionally with their dating war stories.

But I digress. The memory of that New Year's Eve make-out party came rushing back to me when I read an item about this Spur Posse thing. Have you heard about it? Named after the San Antonio Spurs basketball team, a group of California high schoolboys have been keeping a running tally of their sexual conquests, à la ex-basketball star Wilt Chamberlain (over 20,000 served!) and KISS guitarist Paul Stanley (30,000 and more on the way!).

For sure, the members of the Spur Posse never say "*I can't.*" For these punks, "going to the hole" ("driving to the basket" in playground basketball parlance) has undoubtedly taken on a new meaning. Cute, huh? But I've always had an almost sociological curiosity about guys like the Spur Posse, because ever since that night I was made to feel like the wimpiest 13-year-old on the planet, sex with a partner and without love has pretty much been foreign turf for me.

In my worst moments, the Spur Posse makes me feel like an uptight, hyper-cautious square. In my best, I'm happy to simply not be them. Suffice it to say that those pseudostuds and I part ways when it comes to the definition of sex. In fact, I part ways with the rest of the world when it comes to the definition of sex, and that's the way it should be. Personal. My sex life is just that—my sex life—and yours is yours.

Although sex-as-sport escapades (not to mention rape) are nothing new, maybe the sheer excess of the Spur Posse is a reaction—albeit an extreme one—to sexuality during the AIDS era. After surviving a decade of paranoia and arts censorship that cast a chill over the nation's collective sexual identity, folks everywhere are now making up for lost time. Swinging and partner swapping are back. Unsafe hetero and homo sex is on the rise. Sex clubs are all the rage in New York. And alternative arrangements like polyfidelist households are becoming more common.

Part of this mass reframing—indeed, this coming-out party—is the sex positive movement, which embraces various manifestations of sexuality and sensuality from massage and holistic healing to pornography and technosex. It's all part of the healing process, and one of sex positive's most eloquent healers has been Susie Bright (a.k.a. "Susie Sexpert"), the founder of the lesbian erotica magazine *On Our Backs*.

"We live in a 'just say no' culture," says Bright. " 'Sex positive' was a reaction to Nancy Reagan and the work of Andrea Dworkin in fundamentalist feminism. It was saying, 'Look, we believe sex is integral to our well-being and our progressive politics.' And now, as it becomes more widely used, 'sex positive'—I don't even like to use the term anymore—is in danger of becoming another piece of useless rhetoric. Nevertheless, a younger generation has embraced this as their sexual identity and it explains something fundamental about who they are. So sometimes it's a term, sometimes it's a movement, sometimes it's a generation, and now it's about to become a cliché."

But before it does, it serves the crucial purpose of rehabilitation. Writing about Joe Kramer's New York sexual healing school in the *Village Voice* (April 21, 1992), Don Shewey asks, "What is 'sexual healing,' anyway, besides the name of Marvin Gaye's last great record? Partly it has to do with healing the wounds to the spirit and the flesh caused by sexual abuse, addiction, and AIDS. It also has a lot to do with acknowledging that the fun and the pleasure, the vitality and the divine mystery of sex have nourishing properties in and of themselves—a message that can easily get overwhelmed in a culture where 'sex appeal' is routinely exploited to sell products, but sexuality (read: actual . . .) is usually discussed *only* in the context of abuse, addiction, or AIDS transmission."

As a result, there now exists a burgeoning cottage industry of highly hygienic flesh options: computer sex (software that produces erotic on-screen images); computer bulletin board sex (in which subscribers trade fantasies through the magic of E-mail); and our old friend phone sex and our new friend virtual reality sex, which have recently been wed on the new *Cyberorgasm* CD. Produced by the editor of *Future Sex* magazine, Lisa Palac, the "3-D sound" of *Cyberorgasm* is so in-your-libido vivid, it's like being a fly on the wall of some of your best friends' bedrooms, bathrooms, or backseats. An extremely aroused fly, I might add.

As Bright, Palac, and others contend, at the heart of all this technology is sexuality's single most important tool: the human imagination. But is automated sex getting in the way of the real thing? The phenomenon prompted this comment from comedian Jerry Seinfeld in a recent episode of *Seinfeld*: "This whole talking during sex business—I mean, what are we doing here? The question is, does the talking really improve the sex, or is the sex act now there just to spice up the conversation? Of course, eventually I'm sure people will get too lazy even for phone sex, and they'll just have phone machine sex: 'Yeah, yeah, I want you really bad, just leave it on the tape.' And then the phone company will come out with 'sex waiting.'"

It's a joke, but it's a really smart joke. I mean, what *are* we doing here? The great sex *explosion* of the '60s has turned into the great sex *implosion* of the '90s. That is, free love had its obvious failings, but at least the hippies were doing it with human beings. Today it's sex with machines, sex with cathode rays, sex with latex gloves, sex with vibrators, sex with floppy disks, and, coming soon, thanks to the wonder of "teledildonics," sex with robots, which begs the question: Is it live or is it memo-sex? Better still, is it even sex? Absolutely yes, says Bright.

"The fear that people have about phone sex or virtual reality sex is the same fear they had about vibrators and masturbating," she says. "The fear is that if you have too much sex—once sex is easily available and you don't have to suffer for it and your fantasy becomes easy for you—you will be sucked into this whirlpool of sexual excess and go straight to hell.

"Incorporating technology into people's sex lives may seem strange at first, but once you pick up the phone and dial, or get on E-mail or whatever, once you plug it in, it's a very intuitive, physical process. It doesn't have anything to do with communicating with a machine. I was on *The Joan Rivers Show*, and that was her problem. She's like, "My God, we're going to stay home, hooked up to a helmet in some sort of addictive frenzy and not have any need for personal communication anymore.' That's like saying, 'What if the music or a record player ruins [live] music?'"

It's worth repeating: If there is one, the new sexual revolution is about celebrating difference. For me, the sexiest scenario imaginable is Jerry's apartment on *Seinfeld*—a co-op where neighbors drop by spontaneously, flirt with each other, borrow money and clothes and just . . . hang. It's what everyone is ultimately searching for: human contact and a few laughs. In the fast-paced, fragmented '90s, where neighborhoods and families have become ancient history and/or idealized myth, people have an innate need to connect with each other. And the surest, quickest way to feel good about yourself and your species is by having sex. Real or unreal. Slow or fast, careful or careless.

And if none of the above describes your sexual persona, don't worry about it. Bright says she's sick to death of being called a "lesbian" because of all the baggage that goes with it. Likewise, I'm not down with the "straight male" label, which lumps me in with everybody from Jesse Helms to the Spur Posse. The fact of the matter is, it only says what I've *been,* and then only partially so; it doesn't begin to scratch the surface of my sexuality at any given moment.

The point is, the wimpiest 13-year-old on the planet still has more options than he probably has ever imagined. And more than a few questions. Such as: Does the fact that I have occasional soft crushes on my colleague Andrew (who is gay) and the ultra-dishy male owner of the French deli up the street (who is not) mean I'm gay? To some degree, yeah. Does the fact that I get a hard-on when I'm doing a phone interview with Susie Bright mean I'm a lesbian? One can only hope. Finally, does the fact that I put most of my sexual energy into my marriage mean that I bomb abortion clinics in my spare time, or that because I'm monogamous, I'm a dud when it comes to doing the wild thing?

For more on that last one, you'll have to ask my wife. I mean, there are reasons why it's been 14 years—but who's counting?

Sexual Biology, Behavior, and Orientation

- The Body and Its Responses (Articles 8–9)
- Sexual Attitudes and Practices (Articles 10–11)
- Sexual Orientation (Articles 12–14)

Human bodies are miraculous things. Most of us, however, have less than a complete understanding of how they work. This is especially true of our bodily responses and functioning during sexual activity. Efforts to develop a healthy sexual awareness are severely hindered by misconceptions and lack of quality information about physiology. The first portion of this unit directs attention to the development of a clearer understanding and appreciation of the workings of the human body.

As you read through the articles in this section, you will be able to see more clearly that matters of sexual biology and behavior are not merely physiological in origin. The articles in the first two subsections demonstrate clearly the psychological, social, and cultural origins of sexual behavior as well. Why we humans choose to behave sexually can be quite complex.

A final area in this unit is *Sexual Orientation*. Perhaps no other area of sexual behavior is as misunderstood as this one. Although experts do not agree about what causes our sexual orientation—homosexual, heterosexual, or bisexual—growing evidence suggests a complex interaction of biological or genetic determination, environmental or sociocultural influence, and free choice. In the early years of this century, sexologist Alfred Kinsey's seven-point continuum of sexual orientation was introduced. It placed exclusive heterosexual orientation at one end, exclusive monosexual orientation at the other, and identified the middle range as where most people would fall if society and culture were unprejudiced. Since Kinsey, many others have added research, findings, and theories to what is known about sexual orientation, but the case remains that there is probably more not known than known, and the subject is no less controversial.

That the previous paragraph may have been upsetting, even distasteful, to some readers, emphasizes the connectedness of psychological, social, and cultural issues with those of sexuality. Human sexual function is biology and far more. What behaviors and choices are socially and culturally prescribed and proscribed are also part of human sexuality. This section attempts to address all of these issues.

The subsection *The Body and Its Responses* contains two thought-provoking articles illustrative of the complex perceptual and cultural issues as they relate to human sexuality. "The Five Sexes" provides historical and scientific examples where the two-gender male and female classification system is woefully inadequate. The second article asks an intriguing question: "What Is It with Women and Breasts?"

The *Sexual Attitudes and Practices* subsection opens with a question: "Why Do We Know So Little About Human Sex?" and decries the still-present fear of learning more. The last subsection article reports on research involving 362 male and female college undergraduate students and explores their attitudes, sexual arousability, and history of sexual activity.

The *Sexual Orientation* subsection includes articles that dramatically demonstrate the changes that have occurred during the last decade with respect to homosexuality in the United States. It is clear that there is diversity within the gay community and conflict with what is often called a fundamentally homophobic (or homosexuality-fearing) broader society. The articles also present the dilemma many thoughtful people feel between tolerance for diversity, individual freedoms, and our American norms for a married, heterosexual, procreating lifestyle. What impact the new research findings about the role of genetic factors in determining sexual orientation will have is uncertain, but they suggest a new phase in understanding human sexuality.

Looking Ahead: Challenge Questions

Have you ever felt or been told that you are not acting the way someone of your gender is supposed to act? Have you ever felt constricted or frustrated by gender role expectations? Would you like to have more openness in gender identification and roles? Why or why not?

What makes us so focused on physical attractiveness or attributes, especially as linked to sexiness? How can this individual- and relationship-harming practice be reduced?

What benefits and harm do you see in learning as much as possible about human sexuality?

What are the prerequisites for your sexual arousability? What proportions are physical? Do they seem to have changed in the last few years? If so, how, and to what do you attribute this change?

Unit 2

It is rare for people to wonder why someone is heterosexual in the same ways that we wonder why someone is homosexual. What do you think has contributed to your sexual orientation?

How do you respond when people around you talk about homosexuality and/or gay rights? What do you do when family, friends, or coworkers make jokes about gays or lesbians?

Article 8

THE FIVE SEXES

Why Male and Female Are Not Enough

ANNE FAUSTO-STERLING

ANNE FAUSTO-STERLING is a developmental geneticist and professor of medical science at Brown University in Providence. The second edition of her book MYTHS OF GENDER: BIOLOGICAL THEORIES ABOUT WOMEN AND MEN, *[was] published [in 1992] by Basic Books. She is working on a book titled* THE SEX WHICH PREVAILS: BIOLOGY AND THE SOCIAL/SCIENTIFIC CONSTRUCTION OF SEXUALITY.

IN 1843 LEVI SUYDAM, a twenty-three-year-old resident of Salisbury, Connecticut, asked the town board of selectmen to validate his right to vote as a Whig in a hotly contested local election. The request raised a flurry of objections from the opposition party, for reasons that must be rare in the annals of American democracy: it was said that Suydam was more female than male and thus (some eighty years before suffrage was extended to women) could not be allowed to cast a ballot. To settle the dispute a physician, one William James Barry, was brought in to examine Suydam. And, presumably upon encountering a phallus, the good doctor declared the prospective voter male. With Suydam safely in their column the Whigs won the election by a majority of one.

Barry's diagnosis, however, turned out to be somewhat premature. Within a few days he discovered that, phallus notwithstanding, Suydam menstruated regularly and had a vaginal opening. Both his/her physique and his/her mental predispositions were more complex than was first suspected. S/he had narrow shoulders and broad hips and felt occasional sexual yearnings for women. Suydam's "feminine propensities, such as a fondness for gay colors, for pieces of calico, comparing and placing them together, and an aversion for bodily labor, and an inability to perform the same, were remarked by many," Barry later wrote. It is not clear whether Suydam lost or retained the vote, or whether the election results were reversed.

Western culture is deeply committed to the idea that there are only two sexes. Even language refuses other possibilities; thus to write about Levi Suydam I have had to invent conventions—*s/he* and *his/her*—to denote someone who is clearly neither male nor female or who is perhaps both sexes at once. Legally, too, every adult is either man or woman, and the difference, of course, is not trivial. For Suydam it meant the franchise; today it means being available for, or exempt from, draft registration, as well as being subject, in various ways, to a number of laws governing marriage, the family and human intimacy. In many parts of the United States, for instance, two people legally registered as men cannot have sexual relations without violating anti-sodomy statutes.

But if the state and the legal system have an interest in maintaining a two-party sexual system, they are in defiance of nature. For biologically speaking, there are many gradations running from female to male; and depending on how one calls the shots, one can argue that along that spectrum lie at least five sexes—and perhaps even more.

For some time medical investigators have recognized the concept of the intersexual body. But the standard medical literature uses the term *intersex* as a catch-all for three major subgroups with some mixture of male and female characteristics: the so-called true hermaphrodites, whom I call herms, who possess one testis and one ovary (the sperm- and egg-producing vessels, or gonads); the male pseudohermaphrodites (the "merms"), who have testes and some aspects of the female genitalia but no ovaries; and the female pseudohermaphrodites (the "ferms"), who have ovaries and some aspects of the male genitalia but lack testes. Each of those categories is in itself complex; the percentage of male and female characteristics, for instance, can vary enormously among members of the same subgroup. Moreover, the inner lives of the people in each subgroup—their special needs and their problems, attractions and repulsions—have gone unexplored by science. But on the basis of what is known about them I suggest that the three intersexes, herm, merm and ferm, deserve to be considered additional sexes each in its own right. Indeed, I would argue further that sex is a vast, infinitely malleable continuum that defies the constraints of even five categories.

NOT SURPRISINGLY, it is extremely difficult to estimate the frequency of intersexuality, much less the frequency of each of the three additional sexes: it is not the sort of information one volunteers on a job application. The psychologist John Money of Johns Hopkins University, a specialist in the study of congenital sexual-organ defects, suggests intersexuals may constitute as many as 4 percent of births. As I point out to my students at Brown University, in a student body of about 6,000 that fraction, if correct, implies there may be as many as 240 intersexuals on campus—surely enough to form a minority caucus of some kind.

In reality though, few such students would make it as far as Brown in sexually diverse form. Recent advances in physiology and surgical technology now enable physicians to catch most intersexuals at the moment of birth. Almost at once such infants are entered into a program of hormonal and surgical management so that they can slip quietly into society as "normal" heterosexual males or females. I emphasize that the motive is in no way conspiratorial. The aims of the policy are genuinely humanitarian, reflecting the wish that people be able to "fit in" both physically and psychologically. In the medical community, however, the assumptions behind that wish—that there be only two sexes, that heterosexuality alone is normal, that there is one true model of psychological health—have gone virtually unexamined.

THE WORD *hermaphrodite* comes from the Greek names Hermes, variously known as the messenger of the gods, the patron of music, the controller of dreams or the protector of livestock, and Aphrodite, the goddess of sexual love and beauty. According to Greek mythology, those two gods parented Hermaphroditus, who at age fifteen became half male and half female when his body fused with the body of a nymph he fell in love with. In some true hermaphrodites the testis and the ovary grow separately but bilaterally; in others they grow together within the same organ, forming an ovo-testis. Not infrequently, at least one of the gonads functions quite well, producing either sperm cells or eggs, as well as functional levels of the sex hormones—androgens or estrogens. Although in theory it might be possible for a true hermaphrodite to become both father and mother to a child, in practice the appropriate ducts and tubes are not configured so that egg and sperm can meet.

In contrast with the true hermaphrodites, the pseudo-hermaphrodites possess two gonads of the same kind along with the usual male (XY) or female (XX) chromosomal makeup. But their external genitalia and secondary sex characteristics do not match their chromosomes. Thus merms have testes and XY chromosomes, yet they also have a vagina and a clitoris, and at puberty they often develop breasts. They do not menstruate, however. Ferms have ovaries, two X chromosomes and sometimes a uterus, but they also have at least partly masculine external genitalia. Without medical intervention they can develop beards, deep voices and adult-size penises.

No classification scheme could more than suggest the variety of sexual anatomy encountered in clinical practice. In 1969, for example, two French investigators, Paul Guinet of the Endocrine Clinic in Lyons and Jacques Decourt of the Endocrine Clinic in Paris, described ninety-eight cases of true hermaphroditism—again, signifying people with both ovarian and testicular tissue—solely according to the appearance of the external genitalia and the accompanying ducts. In some cases the people exhibited strongly feminine development. They had separate openings for the vagina and the urethra, a cleft vulva defined by both the large and the small labia, or vaginal lips, and at puberty they developed breasts and usually began to menstruate. It was the oversize and sexually alert clitoris, which threatened sometimes at puberty to grow into a penis, that usually impelled them to seek medical attention. Members of another group also had breasts and a feminine body type, and they menstruated. But their labia were at least partly fused, forming an incomplete scrotum. The phallus (here an embryological term for a structure that during usual development goes on to form either a clitoris or a penis) was between 1.5 and 2.8 inches long; nevertheless, they urinated through a urethra that opened into or near the vagina.

By far the most frequent form of true hermaphrodite encountered by Guinet and Decourt—55 percent—appeared to have a more masculine physique. In such people the urethra runs either through or near the phallus, which looks more like a penis than a clitoris. Any menstrual blood exits periodically during urination. But in spite of the relatively male appearance of the genitalia, breasts appear at puberty. It is possible that a sample larger than ninety-eight so-called true hermaphrodites would yield even more contrasts and subtleties. Suffice it to say that the varieties are so diverse that it is possible to know which parts are present and what is attached to what only after exploratory surgery.

The embryological origins of human hermaphrodites clearly fit what is known about male and female sexual development. The embryonic gonad generally chooses early in development to follow either a male or a female sexual pathway; for the ovo-testis, however, that choice is fudged. Similarly, the embryonic phallus most often ends up as a clitoris or a penis, but the existence of intermediate states comes as no surprise to the embryologist. There are also uro-genital swellings in the embryo that usually either stay open and become the vaginal labia or fuse and become a scrotum. In some hermaphrodites, though, the choice of opening or closing is ambivalent. Finally, all mammalian embryos have structures that can become the female uterus and the fallopian tubes, as well as structures that can become part of the male sperm-transport system. Typically either the male or the female set of those primordial genital organs degenerates, and the remaining structures achieve their sex-appropriate future. In hermaphrodites both sets of organs develop to varying degrees.

INTERSEXUALITY ITSELF is old news. Hermaphrodites, for instance, are often featured in stories about human origins. Early biblical scholars believed Adam began life as a hermaphrodite and later divided into two people—a male and a female—after falling from grace. According to Plato there once were three sexes—male, female and hermaphrodite—but the third sex was lost with time.

Both the Talmud and the Tosefta, the Jewish books of law, list extensive regulations for people of mixed sex. The Tosefta expressly forbids hermaphrodites to inherit their fathers' estates (like daughters), to seclude themselves with women (like sons) or to shave (like men). When hermaphrodites menstruate they must be isolated from men (like women); they are disqualified from serving as witnesses or as priests (like women), but the laws of pederasty apply to them.

In Europe a pattern emerged by the end of the Middle Ages that, in a sense, has lasted to the present day: hermaphrodites were compelled to choose an established

2. SEXUAL BIOLOGY, BEHAVIOR, AND ORIENTATION: Body and Its Responses

gender role and stick with it. The penalty for transgression was often death. Thus in the 1600s a Scottish hermaphrodite living as a woman was buried alive after impregnating his/her master's daughter.

For questions of inheritance, legitimacy, paternity, succession to title and eligibility for certain professions to be determined, modern Anglo-Saxon legal systems require that newborns be registered as either male or female. In the U.S. today sex determination is governed by state laws. Illinois permits adults to change the sex recorded on their birth certificates should a physician attest to having performed the appropriate surgery. The New York Academy of Medicine, on the other hand, has taken an opposite view. In spite of surgical alterations of the external genitalia, the academy argued in 1966, the chromosomal sex remains the same. By that measure, a person's wish to conceal his or her original sex cannot outweigh the public interest in protection against fraud.

During this century the medical community has completed what the legal world began—the complete erasure of any form of embodied sex that does not conform to a male–female, heterosexual pattern. Ironically, a more sophisticated knowledge of the complexity of sexual systems has led to the repression of such intricacy.

In 1937 the urologist Hugh H. Young of Johns Hopkins University published a volume titled *Genital Abnormalities, Hermaphroditism and Related Adrenal Diseases*. The book is remarkable for its erudition, scientific insight and open-mindedness. In it Young drew together a wealth of carefully documented case histories to demonstrate and study the medical treatment of such "accidents of birth." Young did not pass judgment on the people he studied, nor did he attempt to coerce into treatment those intersexuals who rejected that option. And he showed unusual even-handedness in referring to those people who had had sexual experiences as both men and women as "practicing hermaphrodites."

One of Young's more interesting cases was a hermaphrodite named Emma who had grown up as a female. Emma had both a penis-size clitoris and a vagina, which made it possible for him/her to have "normal" heterosexual sex with both men and women. As a teenager Emma had had sex with a number of girls to whom s/he was deeply attracted; but at the age of nineteen s/he had married a man. Unfortunately, he had given Emma little sexual pleasure (though *he* had had no complaints), and so throughout that marriage and subsequent ones Emma had kept girlfriends on the side. With some frequency s/he had pleasurable sex with them. Young describes his subject as appearing "to be quite content and even happy." In conversation Emma occasionally told him of his/her wish to be a man, a circumstance Young said would be relatively easy to bring about. But Emma's reply strikes a heroic blow for self-interest:

Would you have to remove that vagina? I don't know about that because that's my meal ticket. If you did that, I would have to quit my husband and go to work, so I think I'll keep it and stay as I am. My husband supports me well, and even though I don't have any sexual pleasure with him, I do have lots with my girlfriends.

Y ET EVEN AS YOUNG was illuminating intersexuality with the light of scientific reason, he was beginning its suppression. For his book is also an extended treatise on the most modern surgical and hormonal methods of changing intersexuals into either males or females. Young may have differed from his successors in being less judgmental and controlling of the patients and their families, but he nonetheless supplied the foundation on which current intervention practices were built.

By 1969, when the English physicians Christopher J. Dewhurst and Ronald R. Gordon wrote *The Intersexual Disorders*, medical and surgical approaches to intersexuality had neared a state of rigid uniformity. It is hardly surprising that such a hardening of opinion took place in the era of the feminine mystique—of the post–Second World War flight to the suburbs and the strict division of family roles according to sex. That the medical consensus was not quite universal (or perhaps that it seemed poised to break apart again) can be gleaned from the near-hysterical tone of Dewhurst and Gordon's book, which contrasts markedly with the calm reason of Young's founding work. Consider their opening description of an intersexual newborn:

One can only attempt to imagine the anguish of the parents. That a newborn should have a deformity . . . [affecting] so fundamental an issue as the very sex of the child . . . is a tragic event which immediately conjures up visions of a hopeless psychological misfit doomed to live always as a sexual freak in loneliness and frustration.

Dewhurst and Gordon warned that such a miserable fate would, indeed, be a baby's lot should the case be improperly managed; "but fortunately," they wrote, "with correct management the outlook is infinitely better than the poor parents—emotionally stunned by the event—or indeed anyone without special knowledge could ever imagine."

Scientific dogma has held fast to the assumption that without medical care hermaphrodites are doomed to a life of misery. Yet there are few empirical studies to back up that assumption, and some of the same research gathered to build a case for medical treatment contradicts it. Francies Benton, another of Young's practicing hermaphrodites, "had not worried over his condition, did not wish to be changed, and was enjoying life." The same could be said of Emma, the opportunistic hausfrau. Even Dewhurst and Gordon, adamant about the psychological importance of treating intersexuals at the infant stage, acknowledged great success in "changing the sex" of older patients. They reported on twenty cases of children reclassified into a different sex after the supposedly critical age of eighteen months. They asserted that all the reclassifications were "successful," and they wondered then whether reregistration could be "recommended more readily than [had] been suggested so far."

The treatment of intersexuality in this century provides a clear example of what the French historian Michel Foucault has called biopower. The knowledge developed in biochemistry, embryology, endocrinology, psychology and surgery has enabled physicians to control the very sex of the human body. The multiple contradictions in that kind of power call for some scrutiny. On the one hand, the medical "management" of intersexuality certainly developed as part of an attempt to free people from perceived psychological pain (though whether the pain was the patient's, the parents' or the physician's is unclear). And if one accepts the assumption that in a sex-divided culture people can realize their greatest potential for happiness and productivity only if they are sure they belong to one of only two acknowledged sexes, modern medicine has been extremely successful.

On the other hand, the same medical accomplishments can be read not as progress but as a mode of discipline. Hermaphrodites have unruly bodies. They do not fall naturally into a binary classification; only a surgical shoehorn can put them there. But why should we care if a "woman," defined as one who has breasts, a vagina, a uterus and ovaries and who menstruates, also has a clitoris large enough to penetrate the vagina of another woman? Why should we care if there are people whose biological equipment enables them to have sex "naturally" with both men and women? The answers seem to lie in a cultural need to maintain clear distinctions between the sexes. Society mandates the control of intersexual bodies because they blur and bridge the great divide. Inasmuch as hermaphrodites literally embody both sexes, they challenge traditional beliefs about sexual difference: they possess the irritating ability to live sometimes as one sex and sometimes the other, and they raise the specter of homosexuality.

BUT WHAT IF things were altogether different? Imagine a world in which the same knowledge that has enabled medicine to intervene in the management of intersexual patients has been placed at the service of multiple sexualities. Imagine that the sexes have multiplied beyond currently imaginable limits. It would have to be a world of shared powers. Patient and physician, parent and child, male and female, heterosexual and homosexual—all those oppositions and others would have to be dissolved as sources of division. A new ethic of medical treatment would arise, one that would permit ambiguity in a culture that had overcome sexual division. The central mission of medical treatment would be to preserve life. Thus hermaphrodites would be concerned primarily not about whether they can conform to society but about whether they might develop potentially life-threatening conditions—hernias, gonadal tumors, salt imbalance caused by adrenal malfunction—that sometimes accompany hermaphroditic development. In my ideal world medical intervention for intersexuals would take place only rarely before the age of reason; subsequent treatment would be a cooperative venture between physician, patient and other advisers trained in issues of gender multiplicity.

I do not pretend that the transition to my utopia would be smooth. Sex, even the supposedly "normal," heterosexual kind, continues to cause untold anxieties in Western society. And certainly a culture that has yet to come to grips—religiously and, in some states, legally—with the ancient and relatively uncomplicated reality of homosexual love will not readily embrace intersexuality. No doubt the most troublesome arena by far would be the rearing of children. Parents, at least since the Victorian era, have fretted, sometimes to the point of outright denial, over the fact that their children are sexual beings.

All that and more amply explains why intersexual children are generally squeezed into one of the two prevailing sexual categories. But what would be the psychological consequences of taking the alternative road—raising children as unabashed intersexuals? On the surface that tack seems fraught with peril. What, for example, would happen to the intersexual child amid the unrelenting cruelty of the school yard? When the time came to shower in gym class, what horrors and humiliations would await the intersexual as his/her anatomy was displayed in all its non-traditional glory? In whose gym class would s/he register to begin with? What bathroom would s/he use? And how on earth would Mom and Dad help shepherd him/her through the mine field of puberty?

IN THE PAST THIRTY YEARS those questions have been ignored, as the scientific community has, with remarkable unanimity, avoided contemplating the alternative route of unimpeded intersexuality. But modern investigators tend to overlook a substantial body of case histories, most of them compiled between 1930 and 1960, before surgical intervention became rampant. Almost without exception, those reports describe children who grew up knowing they were intersexual (though they did not advertise it) and adjusted to their unusual status. Some of the studies are richly detailed—described at the level of gym-class showering (which most intersexuals avoided without incident); in any event, there is not a psychotic or a suicide in the lot.

Still, the nuances of socialization among intersexuals cry out for more sophisticated analysis. Clearly, before my vision of sexual multiplicity can be realized, the first openly intersexual children and their parents will have to be brave pioneers who will bear the brunt of society's growing pains. But in the long view—though it could take generations to achieve—the prize might be a society in which sexuality is something to be celebrated for its subtleties and not something to be feared or ridiculed.

What Is It With Women and Breasts?

A less-than-magnificent obsession reveals shaky self-esteem

She's a squat little thing, with huge breasts, bulbous hips and belly, and fat thighs. No arms, though, and not much of a face. Apparently the Venus of Willendorf (circa 15,000 B.C.) was revered for the unambiguous symbols of fertility that make up her small person; for sure nobody ever loved her for her mind. It would be nice to believe that in terms of evaluating women we have all progressed beyond the Stone Age, but a quick glance at present-day California is not reassuring. Here is Regina, a Los Angeles homemaker, who had her breasts surgically enlarged for cosmetic reasons, going from an A cup to a C cup. She found that men began talking to her chest instead of her face. "That's all they look at if you're big," she says. "It really did help my self-esteem."

Eighty percent of the 2 million American women with breast implants chose the surgery for nonmedical reasons, most often a reason like Regina's—to enhance the kind of self-esteem that relies on breast size. "So much of women's self-esteem is based on appearance," says Jennifer Knopf, director of the Northwestern University Sex and Marital Therapy Program in Chicago. "But they're very critical of themselves. Women undress in front of a mirror, and they start to do pathology patrol." Bonnie, a Chicago office manager, says she had been thinking about new breasts since she was a teenager; she finally had surgery seven years ago when she was 34. "There are a lot of flat-chested women out there who have wonderful relationships and marriages," she says. "But my opinion is you're more likely to turn someone's head or get a date if you have large breasts. Men are so superficial. That's what prompted me."

Self-confidence: Some husbands and boyfriends give women the implants as gifts (and may even pick out the size they like best). Surveys indicate, however, that most women feel they are making the decision on their own and for themselves, not for the man in their lives. "It's a self-confidence thing," says Bonnie. "That's what it boils down to—feeling like a woman."

Our society, like many before it, prizes women for their packaging

It *is* possible to feel like a woman without buttressing your body with artificial secondary sex characteristics, but our society—like many before it—prizes women most highly for their packaging. The pressure on women to look more feminine than female, more symbolic than real, goes back centuries. Throughout most of the 1800s, writes historian Valerie Steele in her 1985 book "Fashion and Eroticism," stylish women were shaped like two cones whose points met at a tightly constricted waist. Corsets made it hard for women to draw a deep breath and may have caused a broken rib or two, but they displayed a bosom as if it were an ice-cream sundae on a platter. Early in this century the platter effect gave way to the long, straight Empire look, and by the '20s a slim, boyish figure was a fashion necessity. Breasts came back after World War II, this time linked with dumb blondes in the most regrettable partnership since the sweet potato met the marshmallow. During the '60s and '70s the boyish look returned with a vengeance when Twiggy became an icon. Today breasts are back—reportedly even Jane (how to marry a millionaire) Fonda sports them—but there's a difference. "A woman's appearance is now dependent on how closely she approximates Barbie," says Pauline Bart, a sociologist at the University of Illinois at Chicago. "You're supposed to be very thin with large breasts. This is difficult."

It's always difficult. What these changing demands on women's bodies have in common is that the ideal is impossible for most women to achieve. Inadequacy is virtually built into us. Women are equated with their body parts in a way that men simply are not; even now there aren't many women in public life who are nationally known for something besides their looks. Those women who locate their self-esteem in their bra size are accurately reading their culture.

9. What Is It with Women and Breasts?

Braless look: Sad to say, they're also surrendering to their culture. It's unlikely that larger breasts became fashionable in the '80s simply because the braless look went out of style. For the last 20 years women have been joining the work force in unprecedented numbers, choosing to postpone marriage and motherhood longer than any other generation in American history, and grappling for power with all the weaponry and wits they can muster. We've embarked on one of the most far-reaching revolutions conceivable, and it's made a lot of people uncomfortable and frightened—not all of them men. What better symbol for times like these than large breasts? They conjure up just about everything that makes women dependent on men, including sex, maternity and bimbohood.

Many women insist that cosmetic breast surgery is a choice they have a right to make, dangerous or not. "I don't want this to become a debate on whether we're an overly breast-conscious society," says Sharon Green, executive director of Y-ME, a Homewood, Ill.-based support program for breast-cancer patients that receives many queries from healthy women seeking information on breast implants. But a debate on society's grotesque expectations for women is exactly what they need, far more than they need surgery. To "choose" a procedure that may harden the breasts, result in loss of sensation and introduce a range of serious health problems isn't a choice, it's a scripted response. And it's worthy of the Stepford wives.

LAURA SHAPIRO *with* KAREN SPRINGEN *in Chicago and* JEANNE GORDON *in Los Angeles*

Why Do We Know So Little About Human Sex?

Even in the best of times it's hard to study human sexuality. Now it's almost impossible.

Anne Fausto-Sterling

Because I am a biologist, my friends have assigned me the role of resident expert on any news, rumor, or fad having even the slightest connection to the life sciences. Questions about human sexuality—that topic of eternal interest to us all—certainly top the list of queries my buddies shoot hopefully in my direction. What causes homosexuality? Do farm boys really experiment with animals? Are kids reaching puberty at ever younger ages? I'm as eager as anyone to talk about these questions, even at some length, but deep down I realize that we know far too little to answer them accurately.

Last summer, for example, the newspapers and weekly magazines buzzed with reports of a finding by Simon LeVay that the brains of male homosexuals and male heterosexuals differed. LeVay had found a structural variation located in the hypothalamus, a part of the brain involved with the regulation of hormones and some sexual activities. Many people think he has obtained evidence of a biological cause of homosexuality. But when, inevitably, my friends asked me what I thought, I came back again and again to what I see as a central flaw: LeVay had no specific information about the sexual behavior of the men in his study.

In LeVay's study either you're gay or you're straight. He thinks that the men from his heterosexual group, whose brains he obtained at autopsy, were straight. But he doesn't really know for sure. They might have been gay but in the closet, or they might have lived straight, married lives yet had an occasional liaison with a man. He also thinks that his homosexual group represented men who engaged in sex with high frequency. But here too he doesn't know what kinds of sex acts they did nor how often they did them.

The presumption that most people are heterosexual and the idea that heterosexuality and homosexuality represent sharply distinct behaviors seem reasonable to most of us. But human behavior is far more complex than that. One survey of gay men and women in San Francisco, for example, used the plural word *homosexualities* to emphasize the diversity of behaviors subsumed under the term *homosexual*. Self-defined homosexuals turned out to live very varied lives. Some were monogamous and in long-term relationships; others engaged in frequent sex with total strangers. Male and female homosexuality were different in practice. They may eventually be shown to differ in origin as well. LeVay surmised that his sample of homosexual men had frequent sex; could the brain differences he found, if confirmed, correlate with frequency of sexual activity rather than with orientation?

We can't understand the origins of human sexual expression without knowing more about how we actually behave. But sexology, the study of human sexual behavior, began only in the twentieth century. The first, and the most famous, modern scientific survey appeared a mere 44 years ago under the title *Sexual Behavior in the Human Male,* coauthored by Alfred Kinsey, Wardell Pomeroy, and Clyde Martin. It underwent nine reprintings in the first year and a half of publication.

Kinsey and his co-workers discovered a continuum of sexuality. They developed a heterosexual-homosexual rating scale, which they divided into seven categories, from exclusively or predominantly heterosexual to exclusively or predominantly homosexual. Where sex is concerned, it turned out, you find all sorts of shades of gray. They found that 37 percent of the male population surveyed had some overt homosexual experience, that most of these experiences occurred during adolescence, and that at least 25 percent of adult males had more than incidental homosexual experiences for at least three years of their lives.

Not surprisingly, howls of protest met the Kinsey study's conclusions. When it comes to the scientific investigation of sexuality, European-American culture does not have a good track record. Pioneer sexologists included the German Richard von Krafft-Ebing and the Englishman Havelock Ellis, and many found their work dangerous. In 1897 Ellis published a book that treated homosexuality in neutral, scientific tones. His British publisher quickly faced criminal prosecution for issuing a "lewd, wicked, bawdy, scandalous, and obscene" book. In *The Well of Loneliness,* written a couple of decades later, novelist Radclyffe Hall describes the lesbian protagonist as she comes across one of Krafft-Ebing's books, which her father has forgotten to return to its locked cabinet. Trembling, she reads for the first time a description of her "condition." Hall's fictional revelation of these secrets led to a 1928 court declaration banning the novel as obscene. The judge ordered all copies seized.

In 1919 in Germany, Magnus Hirschfeld founded the Institute for the Study of Sexual Behavior, which housed more than 20,000 volumes, numerous photographs, and archival material. Interest in the topic grew, and by the 1930s about 80 sex-reform organizations opened clinics in which professionals and laypeople offered medical and sexual information. The flourishing of such knowledge didn't last long. In 1933, months after their rise to power, the Nazis attacked Hirschfeld's institute and burned its books and papers in the street.

Nor did Kinsey escape unscathed. In 1954 the American Medical Association attacked him for contributing to "a wave of sex hysteria." Conservative congressman Louis Heller called for an investigation, urging that Kinsey's work be barred from the U.S. mails. He accused him of contributing to "the depravity of a whole generation" and "the spread of juvenile delinquency." Under political pressure from the House Committee to Investigate Tax-Exempt Foundations, the Rockefeller Foundation, which had funded Kinsey's work, withdrew its support. Kinsey died of a heart attack two years later, a death some say was hastened by the vilification of his work.

Regrettably, this story turns out to have a contemporary echo. Over the past three years conservative House members and senators have again intervened to halt studies and a new national sex survey that would have provided us with the first truly comprehensive accounting of sexual behavior in this country since the Kinsey report. Current epidemiological estimates of the spreading patterns of sexually transmitted diseases still rely on Kinsey's data even though they are badly out of date. Behavior has certainly changed in the interim, and Kinsey's sample, large as it was, did not represent a cross section of the American population.

Indeed, such is our state of ignorance that in 1989 scientists from the National Research Council warned in a report that we don't know enough to win the war against sexually transmitted diseases, including AIDS. To devise a sensible strategy, they reported, we need to know the prevalence of sexually risky behaviors associated with AIDS transmission in low-risk as well as in high-risk groups. We also have to understand the social contexts promoting risky behavior, and the relationships among sex, drug use, and alcohol consumption.

The report called for longitudinal studies—following the behavior of groups of teens for several years, for example—and for research on how to accurately gather information about behavior that many feel squeamish discussing. There are still many puzzles about AIDS transmission. Among them is that in developing countries the virus is spread mostly by heterosexual sex. But in the United States heterosexual sex seems a minor means of transmission, at least so far, compared with homosexual sex or intravenous drug use. What makes our country so different? At the time of the National Research Council report, studies to answer some of these questions were just getting under way, but since then projects have ground to a sudden and discouraging halt.

Here's what happened. About four years ago the National Institute of Child Health and Development (NICHD) awarded a contract to Edward Laumann, dean of the division of social sciences at the University of Chicago. He was to plan an update of the outmoded Kinsey report and amass the kind of information that would guide public health decisions. Laumann assembled a national team to design a survey that would throw light on practices relating to contraception, fertility, and disease prevention. The researchers wanted the science in the study to be beyond reproach, so they worked hard to solve difficult methodological problems. They asked very basic questions: How would they get the answers they needed from people without violating their privacy? Could they control for interviewer bias—would a male respond differently to a female interviewer than to a male interviewer? How would they check the validity of the answers?

But Laumann and his co-workers wanted to do more than simply gather raw statistics. They wanted to know how social networks influence behavior. Most epidemiological models of the spread of sexually transmitted diseases use estimates of the average number of sexual partners a person has in a given population. For example, statistics might show that on average, women born after 1950 have ten partners in their lifetime. Traditional models would presume that their mating is more or less random. Laumann thinks this is a poor way to model the spread of sexually transmitted diseases. People engage in different patterns of sexual activity at different times of their lives. For instance, after a burst of experimentation in her teens and twenties, a woman might remain monogamous for 20 years, divorce, and have several sexual partners before returning to monogamy. Hence you might expect a different pattern of disease spread in a population where most women are under age 30 than in one where most women are older. Social and economic status also play a role. Sex between partners of the same age, ethnic group, and economic standing may be more openly negotiated, thereby lessening the risk of unprotected sex.

So Laumann proposed a model of sexual behavior as something done by couples living in social networks, rather than by randomly acting individuals, as most epidemiological models presume. His proposal was of such high quality that the NICHD was set to launch the national study when, in 1989, Senator Jesse Helms (Republican of North Carolina) and Congressman William Dannemeyer (Republican of California) caught wind of it. When they were done, the Office of Management and Budget and the House Appropriations Committee had withdrawn funding for the project.

> *Why were the surveys canceled? It seems Helms and Dannemeyer's deep-seated fear and hatred of homosexuality are at issue.*

Meanwhile, researchers at the Carolina Population Center (part of the University of North Carolina) had designed a longitudinal study of teenage sexuality of the sort called for by the National Research Council. They planned to study teens' sexual behavior, ranging from contraceptive use to homosexual activities, taking into account education, religion, and family and peer-group interactions. Their proposal was submitted to the NICHD (the appropriate branch of the National Institutes of Health), was peer-reviewed, and received a high-priority score; in 1991 the researchers received their first year of funding for a five-year survey of 24,000 teenagers and their parents. In July 1991, however, Helms and Dannemeyer reentered the scene and successfully pressured Secretary of Health and Human Services Louis Sullivan into canceling the project.

The story doesn't end there. In October 1990 Laumann applied for a grant from the NICHD to do a more limited adult sex study, using the approaches developed for his ill-fated national survey. His application received rave reviews and a funding priority placing him in the top 2 percent of grants reviewed at the time. Funding seemed all but assured. The following year NICHD director Wendy Baldwin told him it would be "political suicide" to award him the money.

It so happened that on September 12, 1991, Senator Helms had introduced an amendment to the NIH appropriations bill, which determines the money allotted to projects and individual institutes within the NIH. Helms proposed that the money earmarked for sex surveys be removed from the NIH budget. Instead, he wanted the same dollar amount transferred to that portion of the Adolescent Family Life Act devoted to encouraging premarital celibacy (that is, just say no). Voting on the amendment, he argued, would "provide senators with a clear choice between right and wrong." In a one-two punch Congressman Dannemeyer introduced the same amendment in the House. The amendment passed in the Senate, failed in the House, and the House-Senate Conference Committee later dropped it from the final bill. Nevertheless, the debate on both the House and Senate floors had the desired effect. Funding for Laumann's research is on indefinite hold.

Why were the surveys canceled? A broad cross section of the scientific and medical community felt they represented cutting-edge research. The quality of the science had never been in question. Instead, it seems that Senator Helms and Congressman Dannemeyer's deep-seated hatred and fear of homosexuality are at issue. Last August Dannemeyer told a *Los Angeles Times* columnist that he believes the sex surveys are the idea of a conspiratorial cell of homosexuals who operate inside the Department of Health and Human Services. The following month Helms took the Senate floor to say: "The NIH funds these sex surveys... to 'cook the books,' so to speak, in terms of presenting 'scientific facts'—in order to do what? To legitimize homosexual life-styles, of course.

"Mr. President," he went on, "let me just say that I am sick and tired of pandering to the homosexuals in this country."

The surveys, Helms argued, are not really intended "to stop the spread of AIDS. The real purpose is to compile supposedly scientific facts to support the left-wing liberal argument that homosexuality is a normal, acceptable life-style.... As long as I am able to stand on the floor of the U.S. Senate," he added, "I am never going to yield to that sort of thing, because it is not just another life-style; it is sodomy."

Helms concluded by gay-baiting both of Laumann's coinvestigators—distinguished social scientists and acknowledged homosexuals—repeating that "the surveys are part and parcel of the homosexual movement's agenda to legitimize their sexual behavior."

And so the battle rages on. On one side stands the social science and medical community, which wants to know what people do behind closed doors. These researchers don't want to gain prurient pleasure from it, or judge it, or encourage it. They wish merely to devise sound public health policies aimed at stopping the spread of a deadly disease. On the other stand two powerful legislators and their conservative constituencies. Their approach to stopping the spread of sexually transmitted diseases is simple: just say no to any kind of sexual activity other than heterosexual relations within the confines of marriage. During the past three rounds of the fight, the top managers at Health and Human Services have put their money on the conservatives. Who will be left standing at the end of the fight remains to be seen.

Sexual Arousal of College Students in Relation to Sex Experiences

Peter R. Kilmann, Ph.D., M.P.H., Joseph P. Boland, Ph.D., Melissa O. West, Ph.D., C. Jean Jonet, Ph.D., and Ryan E. Ramsey, M.A.

University of South Carolina

This study explored whether casual sexual intercourse experiences would be associated with sexual arousability. Three hundred and sixty-two never-married undergraduates were divided into four subgroups based on their history of sexual activity: virgins, sexual intercourse with affection only, relatively less casual sex experiences (1–5), and relatively more casual sex experiences (6 or more). While males rated themselves more easily aroused than females, subjects with casual sex experience did not rate themselves more easily aroused than subjects without that experience. Increasing the frequency of casual sex increased the likelihood that an individual perceived him or herself as "different" from others of the same gender. Males and nonreligious subjects were more likely to engage in intercourse without affection than females and religious subjects. Future research should explore the correlates of casual sexual behavior in married and divorced individuals of different ages.

Numerous studies have explored gender differences in the incidence and prevalence of premarital sexual intercourse. Among the findings, men are more likely than women to have a greater number of premarital sexual partners (Simon, 1989), and to report a more permissive sexual standard (Sprecher, 1989).

Personality attributes are related to more frequent intercourse experiences for single college students (e.g., Keller, Elliot, & Gunberg, 1982; Leary & Snell, 1988). However, the relationship context of intercourse experiences was not clearly defined in these studies. In this regard, intercourse may occur in an "emotionally-committed" relationship versus on a "casual basis" without any emotional commitment given or expected.

Prior research suggests a positive association between the frequency of sexual intercourse and the variable "sexual arousability" (Hoon, Hoon, & Wincze, 1976) for *both* males and females (e.g., Harris, Yulis, & Lacoste, 1980); yet, no differentiation was made between the frequency of intercourse occurring within the context of an ongoing "emotionally-committed" relationship versus the frequency of intercourse in noncommitted relationships. Because sexual arousal at least partially influences the decision to engage in intercourse for sexually experienced individuals (Christopher & Cate, 1984), we predicted a positive association between "sexual arousability" and casual sexual behavior. We differentiated individuals on their intercourse experiences by defining "casual sexual intercourse" as "having sexual intercourse with a person on one occasion only." The subject may have known the partner for a period of time or may have just met the partner prior to having sex. "Afterwards, the partner may never have been seen again, or the person may have maintained a platonic relationship with the partner, but sexual intercourse occurred on that one occasion only." An additional focus was to identify variables associated with sexual activity.

METHOD

Subjects

The subjects were 362 (159 males, 203 females) never-married individuals recruited from three undergraduate human sexuality courses at the University of South Carolina. The subjects' mean age was 20, ranging from 18 to 43. Eighty-five percent of the subjects reported having had at least one sexual intercourse experience. The virgins (never experienced sexual intercourse) consisted of

---

Address correspondence to Peter R. Kilmann, Ph.D., M.P.H., Department of Psychology, University of South Carolina, Columbia, SC 29208.

14 males and 33 females. Twenty males and 77 females reported no casual sex experiences and only had sex with affection. Eighty-six males and 76 females reported one to five casual sex experiences. Thirty-nine males and 17 females had six or more casual sex experiences. In essence, 60% of the subjects had experienced casual sex at least once.

Measures

Sexual Behavior and Attitudes Questionnaire. This questionnaire, developed by the experimenters, consisted of 148 multiple choice and Likert scale items designed to elicit information about: (a) demographic variables (e.g., sex, race, marital status, religion); (b) various personality characteristics; (c) descriptions of partners and ratings of satisfaction in casual sex experiences as compared with steady love relationships and partners; and (d) extent of agreement or disagreement with various statements about casual sex.

The Sexual Arousability Inventory (SAI). This inventory was developed by Hoon, Hoon, and Wincze (1976) to measure subjective sexual arousability in women. Anderson, Broffit, Karlson, and Turnquist (1989) found this instrument to be an internally consistent and stable measure of female sexual arousability. The questionnaire items also seem pertinent to the assessment of male sexual arousability.

Procedure

All subjects responded to the measures anonymously. Care was taken to insure privacy by leaving a space of at least one seat between subjects.

RESULTS

A 2 × 2 analysis of variance was conducted on the SAI scores finding that males scored significantly higher than females, $F(1, 336) = 5.40$, $p < .05$. The mean SAI score for males was 90.6 with a standard deviation of 16.9, and for females, 85.4 with a standard deviation of 22.3. Neither the main effect nor the interaction for a single casual intercourse experience were significant. Thus, subjects with casual sex experience did not rate themselves as more easily aroused.

For the statistical analyses, the subjects were divided by gender into four categories: virgins (never had intercourse); sex with affection only (only had intercourse with affection within a steady love relationship); one to five (one to five casual sex experiences); six or more (six or more casual sex experiences).

Descriptive Information

The virgins' beliefs about casual sex were very similar to the subjects who had few such experiences. Most of the "Sex with affection only" subgroup had between one to five sexual partners. These subjects predicted that they would feel empty, guilty, and remorseful if they engaged in casual sex. They predicted that there would be a better chance for a continuing relationship if one got to know the person before having sex. They considered it wrong to have casual sex while married or in an exclusive relationship. Compared to persons having casual sex, they rated themselves as much more mature and responsible, and much higher on the dimensions of attractiveness, intelligence, friendliness, popularity, sexual experience, and knowledge.

Most of the "one to five" subjects were single, white Protestants, and over half reported not actively practicing their religion. These subjects considered themselves to be more attractive, intelligent, friendly, mature, and sexually knowledgeable than their peers. They estimated that more than half of their peers had casual sex. Many reported that casual sex experiences had become less attractive because of a concern about herpes. It is interesting that almost 11% of these subjects had herpes.

Most of the "one to five" subjects reported feeling lonely and/or in need of affection prior to their casual sex experience. Typically, these subjects did not initiate the encounter, and they did not want to continue the relationship. The experience "just happened" and was less emotionally satisfying than sex in a steady relationship. Their last casual sex partner tended to be over 21 years old (50% were 21–30; 30% were over 30). Most would not engage in casual sex again with the same person but perhaps with someone else. Some felt they should not have engaged in casual sex and most preferred steady love relationships. Most believed that casual sex would not lead to a continuing relationship. Most strongly believed it was wrong to have casual sex while involved in marriage or within the context of an exclusive relationship.

The majority of the "six or more" subjects were single, white Protestants who did not actively practice their religion. Nearly all perceived themselves as having average or more sexual experience than their peers. They also estimated that less than half of their peers had experienced even a single casual sex experience. During their most recent experience, these subjects generally were not involved in another relationship, although those who were reported being satisfied with it. Most of these subjects met their casual sex partners either in a bar or in their neighborhood. Like the "one to five" group, they reported that the experience "just happened," and was less emotionally satisfying than sex within the context of a steady relationship. Typically, these subjects went back to their own home to have sexual intercourse within 24 hours of meeting their partner. Casual sex experiences more often than not occurred after drinking, or after smoking "grass," and these participants typically reported that it "just happened" without much planning. Alcohol was usually consumed and frequently involved more than five drinks. Their partners also had been drinking, and about a third of the subjects used drugs. Casual sex was rated as very enjoyable and these subjects were eager to have another experience with the same partner or with a new one.

The "six or more" subjects used more alcohol and drugs, considered casual sex to be more satisfying, and reported less guilt about it than the "one to five" subjects. They generally chose partners who were between 19–21 years old, single, white, less attractive, less interesting, and average or less mature than the subjects' steady lovers. Further, their steady lovers were perceived as having more sexual knowledge and experience than their casual sex partners. The subjects' rated their casual sex experiences as involving less responsibility, satisfaction, and intimacy than sex with a regular partner.

Although most believed that it was wrong to have casual sex while married or in an exclusive relationship, many had experienced casual sex while dating someone exclusively. Casual sex experiences were considered opportunities to test out different partners with varying techniques, and the spontaneity and unpredictability of the situation. Casual sex made them feel attractive and wanted. Although casual sex was intense and passionate, sex in a steady love relationship was seen as more emotionally satisfying.

In comparison with their peers, the "six or more" women perceived themselves as average or less attractive, while the men thought they were more attractive. Both males and females perceived themselves as average or less responsible, but more friendly, mature, and sexually knowledgeable, and as average or more intelligent, popular, sexually active, and sexually experienced.

Discriminant Function Analyses

In order to further contrast the subjects within the four groupings, we performed a discriminant function analysis separately for males and females on a subset of the questionnaire items. Some items were not responded to by the entire sample, (i.e., subjects who had never had casual sexual intercourse according to our definition responded to 100 of the 148 items), and some items were not appropriate for use as discriminators. Eighteen items were chosen for the analysis; three of the items were the demographic variables for age, race, and year in college, while 15 items were the respondents' self-ratings of personal characteristics such as intelligence, friendliness, likeability, and attitudes about sex.

Similar to the procedure used above, the entire sample was partitioned by gender into four subgroups based on their sexual intercourse experience level; Group 1 consisted of virgins; Group 2 of respondents who had not had a casual sex experience but who had had intercourse with affection only, and the subjects in the remaining two groups were classified as either having experienced relatively few (1-5) or many (6 or more) casual sex experiences (Groups 3 and 4 respectively). The Wilks' method, which determines the appropriate number of functions that used the overall multivariate F-ratio for the test of differences of group centroids, was employed in the analyses.

Analysis of Males. Two significant functions of a possible three existed among the data. The first function, which accounted for approximately 53% of the between group variance that was accountable for by all functions ($X(54) = 103.62$, $p < .001$), primarily utilized the self-rating items for conventionality, responsibleness, and age of partner in their last steady relationship. The males in Group 4 (6 or more casual sex experiences) rated themselves as being relatively less responsible, less conventional, and had older partners in their last relationships. To a lesser extent, the self-rating items for extroversion, age, and a belief about what percentage of the population engages in casual sex, were used in the first discriminant function. The subjects in Group 4 made the highest estimation of all the groups as to how many people engage in casual sex. The males in this group were slightly older and perceived themselves as more outgoing than did the males in the other three groups.

The second function, which accounted for 28% of the between group variance overall ($X(34) = 51.70$, $p < .05$), discriminated Groups 1 and 4 from Groups 2 and 3. That is, the male virgins and males who experienced casual sex six or more times were similar along this second dimension, and yet distinct from the males in the sex-with-affection group (Group 2), and also from the males who reported one to five casual sex experiences (Group 3). Groups 1 and 4 were distinct from Groups 2 and 3 in that their last steady partner was either younger (Group 1) or older (Group 4) than the partners of their peers in groups 2 and 3. The male virgins and the males with frequent casual sex experiences considered themselves more attractive than their counterparts and made higher estimations about how many people engage in casual sex. The male virgins and the males with many casual sex experiences similarly reported higher sexual arousability; they also shared the same belief that casual sex involves the woman being seductive and the man being conquering.

Analysis of Females. One significant function (of a possible three) accounted for approximately 60% of the overall between group variance, ($X(54) = 86.63$, $p < .01$). This function discriminated Groups 1 and 2 (the female virgins and sex with affection females) from Groups 3 and 4 (1–5 casual sex experiences, 6 or more casual sex experiences).

The first function primarily used the self-rating items for conventionality and arousability, and the endorsement of two beliefs about casual sex. The females in Groups 1 and 2 considered themselves more conventional than others, strongly believed that casual sex was both an opportunity to test out different partners without making a commitment, and considered casual sex to involve the woman being seductive and the man being conquering. To a lesser extent, the females in Groups 3 and 4 reported a higher sexual arousability than the females in Groups 1 and 2.

DISCUSSION

We did not find a relationship between the frequency of casual sexual intercourse and level of "sexual arousability" for either gender, suggesting that these two variables are unrelated. Garcia, Brenner, DeCarlo, McGlennan, and Tate (1984) found that females rated erotic stories just as arousing as males when the most active character was also female. From an erotic guided imagery procedure, female subjects experienced a commensurate level of sexual arousal to both committed and casual relationship contexts (Harrell & Stolp, 1985). More evidence is needed using varied measures of subjective sexual arousability (e.g., Mosher, Barton-Henry, Green, 1988) to determine whether the factors contributing to sexual arousal are gender-specific.

As the frequency of casual intercourse experiences increased, males perceived themselves as less conventional, less responsible, more outgoing, and had older partners in their last casual intercourse interaction than did males who had not done so. Females who had not had casual intercourse perceived themselves as more conventional, and believed that casual intercourse was just an opportunity to test out different partners without making a commitment.

The "sex with commitment only" group predicted that casual intercourse would leave them feeling empty, guilty, and remorseful. These subjects may have been reluctant to engage in casual intercourse because of these negative emotions and the lack of expected fulfillment (Keller, Elliott, & Gunberg, 1982). Persons with a "sex with commitment only" philosophy may be more likely to find themselves disagreeing with their partners about the level of desired sexual intimacy in dating (Byers & Lewis, 1988).

The subjects in the "one to five group" reported some negative effects from casual intercourse. Their last casual sex experience was less satisfying, more guilt-producing; these subjects indicated that they would not do it again with the same person. They also reported a greater prevalence of herpes, and a greater reluctance to have casual intercourse due to a fear of contacting a sexually transmitted disease. The partners of these subjects also tended to be much older, and were not perceived to offer a future relationship.

Future research should determine whether there are different intrapersonal and interpersonal variables associated with casual sexual contacts, either as separate from an ongoing relationship or as a lifestyle. Future research also should explore the variables associated with the casual intercourse frequency in married and divorced individuals of different ages using more restrictive definitions of "casual intercourse" (e.g., "intercourse occurring only one time with a 'stranger'; i.e., a person who was not known before the encounter nor ever seen again").

REFERENCES

Anderson, B. L., Broffit, B., Karlson, J. A., & Turnquist, D. C. (1989). A psychometric analysis of the Sexual Arousability Index. *Journal of Consulting and Clinical Psychology, 57,* 123–130.

Byers, E. S., & Lewis, K. (1988). Dating couples' disagreements over the desired level of sexual activity. *Journal of Sex Research, 24,* 15–29.

Christopher, F. S., & Cate, R. M. (1984). Factors involved in premarital decision-making. *Journal of Sex Research, 20,* 363–376.

Garcia, L. T., Brenner, K., DeCarlo, M., McGlennan, R., & Tait, S. (1984). Sex differences in sexual arousal to different erotic stories. *Journal of Sex Research, 20,* 391–402.

Harrell, T. H., & Stolp, R. D. (1985). Effects of erotic guided imagery on female sexual arousal and emotional response. *Journal of Sex Research, 21,* 292–304.

Harris, R., Yulis, S., & LaCoste, D. (1980). Relationships among sexual arousability, imagery ability, and introversion-extraversion. *Journal of Sex Research, 16,* 72–86.

Hoon, E. F., Hoon, P. W., & Wincze, J. P. (1976). An inventory for the measurement of female arousability: The SAI. *Archives of Sexual Behavior, 5,* 291–300.

Keller, J. F., Elliott, S. S., & Gunberg, E. (1982). Premarital sexual intercourse among single college students: A discriminant analysis. *Sex Roles, 8,* 21–32.

Leary, M. R., & Snell, W. E., Jr. (1988). The relationship of instrumentality and expressiveness to sexual behavior in males and females. *Sex Roles, 18,* 509–522.

Mosher, D. L., Barton-Henry, M., Green, S. E. (1988). Subjective sexual arousal and involvement: Development of multiple indicators. *Journal of Sex Research, 25,* 412–425.

Simon, A. (1989). Promiscuity as sex difference. *Psychological Reports, 64,* 802.

Sprecher, S. (1989). Premarital sexual standards for different categories of individuals. *The Journal of Sex Research, 26,* 232–248.

Homosexuality, the Bible, and us— a Jewish perspective

Dennis Prager

Of all the issues that tear at our society, few provoke as much emotion, or seem as complex, as the question of homosexuality.

Most homosexuals and their heterosexual supporters argue that homosexuality is an inborn condition, and one, moreover, that is no less valid than heterosexuality. They maintain that to discriminate in any way against a person because of his or her sexual orientation is the moral equivalent of discrimination against a person on the basis of color or religion; that is to say, bigotry plain and simple.

On the other hand there are those who feel, no less passionately, that homosexuality is wrong, that society must cultivate the heterosexual marital ideal, or society's very foundations will be threatened.

In the middle are many who are torn between these two claims. I have been one of them. Generally speaking, I do not concern myself with the actions of consenting adults in the privacy of their homes, and I certainly oppose government involvement with what consenting adults do in private. In addition, both lesbians and homosexual men have been part of my life as friends and relatives.

At the same time, I am a Jew who reveres Judaism. And my religion not only prohibits homosexuality, it unequivocally, unambiguously, and in the strongest language at its disposal, condemns it. Judaism—and Christianity—hold that marital sex must be the ideal to which society aspires. Thus my instinct to tolerate all non-coercive behavior runs counter to the deepest moral claims of my source of values.

This is not all. Adding to the seeming complexity are the questions of choice and psychopathology. Current homosexual doctrine holds that homosexuals are born homosexual, and that homosexuality is in no way a psychological or emotional deviation. Are these claims true? And if they are, what are we to do with Western society's (i.e., Judaism's and Christianity's) opposition to homosexuality? What are we to do with our gut instinct that men and women should make love and marry each other, not their own sex? Have Judaism and Christianity been wrong? Is our instinctive reaction no more than a heterosexual bias? And what about those of us who have two gut instincts— one that favors heterosexual love, and one that believes "live and let live"? These two feelings seem irreconcilable, and they have caused me and millions of others anguish and confusion.

After prolonged immersion in the subject, I continue to have anguish about the subject of homosexuality, but, to my great surprise, much less confusion. I hope that the reader will undergo a similar process, and it is to this end that I devote this article.

THE NATURE OF SEX

Man's nature, undisciplined by values, will allow sex to dominate his life and the life of society. When Judaism first demanded that all sexual activity be channeled into marriage, it changed the world. It is not overstated to say that the Hebrew Bible's prohibition of non-marital sex made the creation of Western civilization possible. Societies that did not place boundaries around sexuality were stymied in their development. The subsequent dominance of the Western world can, to a significant extent, be attributed to the sexual revolution, initiated by Judaism and later carried forward by Christianity.

This revolution consisted of forcing the sexual genie into the marital bottle. It ensured that sex no longer dominated society, it heightened male-female love and sexuality (and thereby almost alone created the possibility of love and eroticism within marriage), and it began the arduous task of elevating the status of women.

It is probably impossible for us who live thousands of years after Judaism began this process to perceive the extent to which sex can dominate, and has dominated, life. Throughout the ancient world, and up to the recent past in many parts of the world, sexuality infused virtually all of society.

Human sexuality, especially male sexuality, is polymorphous, or utterly wild (far more so than animal sexuality). Men have had sex with women and with men; with little girls and young boys; with a single partner and in large groups; with immediate family members; and with a variety of domesticated animals. They have achieved orgasm with inanimate objects such as leather, shoes, and other pieces of clothing; through urinating and defecating on each other (interested readers can see a photograph of the former at select art museums in America exhibiting the works of the gay photographer Robert Mapplethorpe); by dressing in women's garments; by watching other human beings being tortured; by fondling children of either sex; by listening to a man or woman's disembodied voice (e.g., phone sex); and, of course, by looking at pictures of bodies, or parts of bodies. There is little, animate or inanimate, that has not excited some men to orgasm.

Of course, not all of these practices have been condoned by societies—parent-child incest and seducing another man's wife have rarely been countenanced—but many have, and all illustrate what the unchanneled, or in Freudian terms, the "unsublimated," sex drive can lead to.

DESEXUALIZING GOD AND RELIGION

Among the consequences of the unchanneled sex drive is the sexualization of everything—including religion. Unless the sex drive is appropriately harnessed (not squelched, which leads to its own consequences), higher religion cannot develop.

Thus, the first thing the Hebrew Bible did was to desexualize God: "In the beginning God created the heavens and the earth"—by His will, not through any sexual behavior. This was an utterly radical break with all religion, and it alone changed human history. The gods of virtually all civilizations engaged in sexual relations. The gods of Babylon, Canaan, Egypt, Greece, and Rome were, in fact, extremely promiscuous, both with other gods and with mortals.

Given the sexual activity of the gods, it is not surprising that the religions themselves were replete with all forms of sexual activity. In the ancient Near East and elsewhere, virgins were deflowered by priests before marriage, and sacred or ritual prostitution was almost universal.

The Hebrew Bible was the first to place controls on sexual activity. It could no longer dominate religion and social life. It was to be sanctified—which in Hebrew means "separated"—from the world and placed in the home, in the bed of husband and wife. The restriction of sexual behavior by Judaism (and later Christianity) was one of the essential elements that enabled society to progress.

THE UBIQUITY OF HOMOSEXUALITY

The new restrictions were nowhere more radical, more challenging to the prevailing assumptions of mankind, than with regard to homosexuality. Indeed, for all intents and purposes, Judaism may be said to have invented the notion of homosexuality, for in the ancient world sexuality was not divided between heterosexuality and homosexuality. That division was the Bible's doing. Before the Bible, the world divided sexuality between penetrator (active partner) and penetrated (passive partner).

As Martha Nussbaum, professor of philosophy at Brown University, has written, the ancients were no more concerned with people's gender preference than people today are with others' eating preferences:

Ancient categories of sexual experience differed considerably from our own. . . . The central distinction in sexual morality was the distinction between active and passive roles. *The gender of the object . . . is not in itself morally problematic.* Boys and women are very often treated interchangeably as objects of [male] desire. What is socially important is to penetrate rather than to be penetrated. Sex is understood fundamentally not as an interaction, but as a doing of something to someone. . . ."[1] [emphasis added]

Judaism changed this. It rendered the "gender of the object" very morally problematic"; it declared that no one is "interchangeable" sexually; and, as a result, it ensured that sex would in fact be "fundamentally interaction" and not simply "a doing of something to someone." The Hebrew Bible condemned homosexuality in the most powerful and unambiguous language it could: "Thou shalt not lie with mankind, as with womankind; it is an abomination."

To appreciate the extent of the revolution wrought by this prohibition of homosexuality, and the demand that all sexual interaction be male-female, it is first necessary to appreciate just how universally accepted and practiced homosexuality has been throughout the world.

It is biblical sexual values, not homosexuality, that have been deviant. In order to make this point clear, I will cite but a handful of historical examples. Without these examples, this claim would seem unbelievable.

Ancient Near East

Egyptian culture believed that "homosexual intercourse with a god was auspicious," writes New York University sociology professor David Greenberg in *The Construction of Homosexuality*. Having anal intercourse with a god was the sign of a man's mastery over fear of the god. Thus one Egyptian coffin text reads, "Atum [a god] has no power over me, for I copulate between his buttocks."[2] In another coffin text, the deceased person vows, "I will swallow for myself the phallus of [the god] Re."[3] In Mesopotamia, Hammurabi, the author of the famous legal code bearing his name, had male lovers.[4]

Greece

Homosexuality was not only a conspicuous feature of life in ancient Greece, it was exalted. The seduction of young boys by older men was expected and honored. Those who could afford, in time and money, to seduce young boys, did so. Graphic depictions of man-boy sex adorn countless Greek vases.

"Sexual intimacy between men was widespread throughout ancient Greek civilization. . . . What was accepted and practiced among the leading citizens was bisexuality; a man was expected to sire a large number of offspring and to head a family while engaging a male lover."[5]

As Greenberg writes, "The Greeks assumed that ordinarily sexual choices were not mutually exclusive, but rather that people were generally capable of responding erotically to beauty in both sexes. Often they could and did."

"Sparta, too, institutionalized homosexual relations between mature men and adolescent boys." In Sparta, homosexuality "seems to have been universal among male citizens."

Rome

Homosexuality was so common in Rome that Edward Gibbon, in his *History of the Decline and Fall of the Roman Empire,* wrote that "of the first fifteen emperors Claudius was the only one whose taste in love was entirely correct" (i.e., not homosexual).[6]

According to psychiatrist and social historian Norman Sussman, "In contrast to the self-conscious and elaborate efforts of the Greeks to glorify and idealize homosexuality, the Romans simply accepted it as a matter of fact and as an inevitable part of human sexual life. Pederasty was just another sexual activity. Many of the most prominent men in Roman society were bisexual if not homosexual. Julius Caesar was called by his contemporaries every woman's man and every man's woman."[7]

The Arab World

Greenberg notes that "a de facto acceptance of male homosexuality has prevailed in Arab lands down to the modern era." As early as the tenth century, German historians depicted Christian men as preferring martyrdom to submitting to Arab sexual demands.

In the words of one of the world's great scholars of Islam, Marshall G. S. Hodgson, "The sexual relations of a mature man with a subordinate youth were so readily accepted in upper-class circles that there was often little or no effort to conceal their existence."[8]

Edward Westermark observed that "it is a common belief among the Arabic-speaking mountaineers of Northern Morocco that a boy cannot learn the Koran well unless a scribe commits pederasty with him. So also an apprentice is supposed to learn his trade by having intercourse with his master."[9]

Greenberg writes: "In Morocco . . . pederasty has been an 'established custom,' with boys readily available in the towns. . . .

"In nineteenth-century Algeria, 'the streets and public places swarmed with boys of remarkable beauty who more than shared with the women the favor of the wealthier natives.' "[10]

As for non-Arab Islam, "the situation," Greenberg concludes, "has been little different."

And this is only a cursory review. Homosexuality was also prevalent among pre-Columbian Americans; the Celts, Gauls, and pre-Norman English; the Chinese, Japanese, and Thai; and dozens of other nationalities and cultures. Greenberg summarizes the ubiquitous nature of homosexuality in these words: "With only a few exceptions, male homosexuality was not stigmatized or repressed so long as it conformed to norms regarding gender and the relative ages and statuses of the partners. . . . The major exceptions to this acceptance seem to have arisen in two circumstances."

Both of these circumstances were Jewish.

JUDAISM AND HOMOSEXUALITY

The Hebrew Bible, in particular the Torah (the first five books of the Bible), has done more to civilize the world than any other book or idea in history. It is the Hebrew Bible that gave humanity such ideas as a universal, moral, loving God; ethical obligations to this God; the need for history to move forward to moral and spiritual redemption; the belief that history has meaning; and the notion that human freedom and social justice are the divinely desired states for all people. It gave the world the Ten Commandments and ethical monotheism.

Therefore, when this Bible makes strong moral proclamations, I listen with great respect. And regarding male homosexuality—female homosexuality is not mentioned—this Bible speaks in such clear and direct language that one does not have to be a religious fundamentalist in order to be influenced by its views. All that is necessary is to consider oneself a serious Jew or Christian.

Jews or Christians who take the Bible's views on homosexuality seriously are not obligated to prove that they are not fundamentalists or literalists, let alone bigots (though people have used the Bible to defend bigotry). The onus is on those who view homosexuality as compatible with Judaism or Christianity to reconcile this view with their Bible.

Given the unambiguous nature of the biblical attitude toward homosexuality, however, such a reconciliation is not possible. All that is possible is to declare: "I am aware that the Bible condemns homosexuality, and I consider the Bible wrong." That would be an intellectually honest approach.

But this approach leads to another problem. If one chooses which of the Bible's moral values to take seriously (and the Bible states its prohibition of homosexuality not only as a law, but as a value—"it is an abomination"), of what moral use is the Bible?

Advocates of religious acceptance of homosexuality respond that while the Bible is morally advanced in some areas, it is morally regressive in others. Its condemnation of homosexuality is cited as one example, and the Torah's acceptance of slavery as another.

Far from being immoral, however, the Torah's prohibition of homosexuality was a major part of its liberation of the human being from the bonds of unrestrained sexuality and of women from being peripheral to men's lives.

As for slavery, while the Bible declares homosexuality wrong, it never declares slavery good. If it did, I would have to reject the Bible as a document with moral relevance to our times. With its notion of every human being created in God's image and with its central event being liberation from slavery, it was the Torah which first taught humanity that slavery is wrong. The Torah's laws regarding slavery exist not to perpetuate it, but to humanize it. And within Jewish life, these laws worked. Furthermore, the slavery that is discussed in the Torah bears no resemblance to black slavery or other instances with which we are familiar. Such slavery, which includes the kidnapping of utterly innocent people, was prohibited by the Torah.

Another argument advanced by advocates of religious acceptance of homosexuality is that the Bible prescribes the death penalty for a multitude of sins, including such seemingly inconsequential acts as gathering wood on the Sabbath. Since we no longer condemn people who violate the Sabbath, why continue to condemn people who engage in homosexual acts?

2. SEXUAL BIOLOGY, BEHAVIOR, AND ORIENTATION: Sexual Orientation

The answer is that we do not derive our approach toward homosexuality only from the fact that the Torah made it a capital offense. We learn it from the fact that the Bible *makes a moral statement* about homosexuality. It makes no such statement about gathering wood on the Sabbath. The Torah uses its strongest term of disapprobation, "abomination," to describe homosexuality. It is the Bible's moral evaluation of homosexuality that distinguishes homosexuality from other offenses, capital or otherwise. As Professor Greenberg, who betrays no inclination toward religious belief, writes, "When the word *toevah* ("abomination") does appear in the Hebrew Bible, it is sometimes applied to idolatry, cult prostitution, magic, or divination, and is sometimes used more generally. *It always conveys great repugnance"* [emphasis added].

Moreover, it lists homosexuality together with child sacrifice among the "abominations" practiced by the peoples living in the land about to be conquered by the Jews. The two are certainly not morally equatable, but they both characterized the morally primitive world that Judaism opposed. They both characterized a way of life opposite to the one that God demanded of Jews (and even of non-Jews—homosexuality is among the sexual offenses that is covered by one of the "seven laws of the children of Noah" which Judaism holds all people must observe).

Finally, the Bible adds a unique threat to the Jews if they engage in homosexuality and the other offenses of the Canaanites: "You will be vomited out of the land" just as the non-Jews who practice these things were vomited out of the land. Again, as Greenberg notes, this threat "suggests that the offenses were considered serious indeed."

WHY JUDAISM OPPOSES HOMOSEXUALITY

It is impossible for Judaism to make peace with homosexuality, because homosexuality denies many of Judaism's most fundamental values. It denies life; it denies God's expressed desire that men and women cohabit; and it denies the root structure that the Bible prescribes for all mankind, the family.

"Choose life"

If one can speak of Judaism's essence, it is contained in the Torah statement, "I have set before you life and death, the blessing and the curse, and you shall choose life." Judaism affirms whatever enhances life, and it opposes or separates whatever represents death. Thus, meat (death) is separated from milk (life); menstruation (death) is separated from sexual intercourse (life); carnivorous animals (death) are separated from vegetarian, kosher animals (life). This is probably why the Torah juxtaposes child sacrifice with male homosexuality. Though they are not morally analogous, both represent death: One deprives children of life, the other prevents their having life.

Men need women

God's first declaration about man (the human being generally, and the male specifically) is, "It is not good for man to be alone." Now, presumably, in order to solve the problem of man's aloneness, God could have made another man, or even a community of men. However, God solved man's aloneness by creating one other person, a woman—not a man, not a few women, not a community of men and women. Man's solitude was not a function of his not being with other people; it was a function of his being without a woman.

Of course, Judaism also holds that women need men. But both the Torah statement and Jewish law have been more adamant about men marrying than about women marrying. Judaism is worried about what happens to men and to society when men do not channel their drives into marriage. In this regard, the Torah and Judaism were highly prescient: The overwhelming majority of violent crimes are committed by unmarried men.

In order to become fully human, male and female must join. In the words of Genesis, "God created the human . . . male and female He created them." The union of male and female is not merely some lovely ideal; it is the essence of the biblical outlook on becoming human. To deny it is tantamount to denying a primary purpose of life.

The family

Throughout their history, one of the Jews' most distinguishing characteristics has been their commitment to family life. To Judaism, the family—not the nation, and not the individual—is to be the fundamental unit, the building block of society. Thus, when God blesses Abraham, He says, "Through you all the families of the earth will be blessed."

Homosexuality's effect on women

Yet another reason for Judaism's opposition to homosexuality is homosexuality's negative effect on women. There appears to be a direct correlation between the prevalence of male homosexuality and the relegation of women to a low societal role. At the same time, the emancipation of women has been a function of Western civilization, the civilization least tolerant of homosexuality.

In societies where men sought out men for love and sex, women were relegated to society's periphery. Thus, for example, ancient Greece, which elevated homosexuality to an ideal, was characterized, in Sussman's words, by "a misogynistic attitude." Homosexuality in ancient Greece, he writes, "was closely linked to an idealized concept of the man as the focus of intellectual and physical activities."

Classicist Eva Keuls describes Athens at its height of philosophical and artistic greatness as "a society dominated by men who sequester their wives and daughters, denigrate the female role in reproduction, erect monuments to the male genitalia, have sex with the sons of their peers. . . ."

In medieval France, when men stressed male-male love, it "implied a corresponding lack of interest in women. In the *Song of Roland,* a French mini-epic given its final form in the late eleventh or twelfth century, women appear only as shadowy, marginal figures: 'The deepest signs of affection in the poem, as well as in similar ones, appear in the love of man for man...."[11]

The women of Arab society, wherein male homosexuality has been widespread, have a notably low status. In traditional Chinese culture, as well, the low state of women has been linked to widespread homosexuality.[12]

While traditional Judaism is not as egalitarian as many late twentieth century Jews would like, it was Judaism, very much through its insistence on marriage and family and its rejection of infidelity and homosexuality, that initiated the process of elevating the status of women. While other cultures were writing homoerotic poetry, the Jews wrote the *Song of Songs,* one of the most beautiful poems depicting male-female sensual love ever written.

The male homosexual lifestyle

A final reason for opposition to homosexuality is the homosexual lifestyle. While it is possible for male homosexuals to live lives of fidelity comparable to those of heterosexual males, it is usually not the case. While the typical lesbian has had fewer than ten sexual partners, the typical male homosexual in America has had over 500.[13] In general, neither homosexuals nor heterosexuals confront the fact that it is this male homosexual lifestyle, more than the specific homosexual act, that disturbs most people.

This is probably why less attention is paid to female homosexuality. When male sexuality is not controlled, the consequences are considerably more destructive than when female sexuality is not controlled. Men rape. Women do not. Men, not women, engage in fetishes. Men are more frequently consumed by their sex drive, and wander from sex partner to sex partner. Men, not women, are sexually sadistic.

The indiscriminate sex that characterizes much of male homosexual life represents the antithesis of Judaism's goal of elevating human life from the animal-like to the God-like.

THE JEWISH SEXUAL IDEAL

Judaism has a sexual ideal—marital sex. All other forms of sexual behavior, though not equally wrong, deviate from that ideal. The further they deviate, the stronger Judaism's antipathy. Thus there are varying degrees of sexual wrongs. There is, one could say, a continuum of wrong which goes from premarital sex, to adultery, and on to homosexuality, incest, and bestiality.

We can better understand why Judaism rejects homosexuality by understanding its attitudes toward these other unacceptable practices. For example, if a Jew were to argue that never marrying is as equally valid a lifestyle as marrying, normative Judaism would forcefully reject this claim. Judaism states that a life without marrying is a less holy, less complete, and a less Jewish life. Thus, only married men were allowed to be high priests, and only men who had children could sit as judges on the Jewish supreme court, the Sanhedrin.

To put it in modern terms, while an unmarried rabbi can be the spiritual leader of a congregation, he would be dismissed by almost any congregation if he publicly argued that remaining single is as Jewishly valid a way of life as married life.

Despite all this, no Jew could argue that single Jews must be ostracized from Jewish communal life. Single Jews are to be loved and included in Jewish family, social, and religious life.

These attitudes toward not marrying should help clarify Judaism's attitude toward homosexuality. First, it contradicts the Jewish ideal. Second, it cannot be held to be equally valid. Third, those publicly committed to it may not serve as public Jewish role models. But fourth, homosexuals must be included in Jewish communal life and loved as fellow human beings and as Jews.

We cannot open the Jewish door to non-marital sex. For once one argues that any non-marital form of sexual behavior is as valid as marital sex, the door is opened to *all* other forms of sexual expression. If consensual homosexual activity is valid, why not consensual incest between adults? Why is sex between an adult brother and sister more objectionable than sex between two adult men? If a couple agrees, why not allow consensual adultery? Once non-marital sex is validated, how can we draw any line? Why shouldn't gay liberation be followed by incest liberation?

Accepting homosexuality as the social, moral, or religious equivalent of heterosexuality would constitute the first modern assault on the extremely hard-won, millenia-old battle for a family-based, sexually monogamous society. While it is labeled as progress, the acceptance of homosexuality would not be new at all.

IS HOMOSEXUALITY WRONG (EVEN IF HOMOSEXUALS HAVE NO CHOICE)?

To all the previous arguments offered against homosexuality, the most frequent response is: But homosexuals have no choice. To many people this claim is so emotionally powerful that no further reflection seems necessary. How can we oppose actions that people have not chosen?

But upon a moment's reflection, the answer becomes very clear: "Homosexuals have no choice," when true, is a defense of the homosexual, not of his conduct.

It may be necessary to oppose actions even if they are not performed voluntarily. We do it all the time, and in all spheres of life. It is what keeps psychiatrists and the courts so busy.

The issue of whether homosexuals have any choice may be terribly important, but even if we were to conclude that they do not, that conclusion would in no way invalidate any of the objections Judaism raises against homosexuality. *Whether or not homosexuals choose homosexuality is entirely unrelated to the question of whether society ought to regard it as an equally valid way of life.*

If Judaism's arguments against homosexuality are valid, then even if we hold that homosexuals have no choice, we will have to conclude that nature or early nurture has foisted upon some people a tragic burden. But how to deal with a tragic burden is a very different question from whether Judaism, Christianity, and Western civilization should drop their heterosexual marital ideal.

2. SEXUAL BIOLOGY, BEHAVIOR, AND ORIENTATION: Sexual Orientation

In fact, to society at large, gays do not generally argue that a homosexual life is entirely as valid as a heterosexual life. Even if they believe this, few heterosexuals would agree with it. So, gays offer the argument that garners the most heterosexual sympathy—that homosexuals have no choice.

And to those homosexuals who truly have no choice, we do owe sympathy. But sympathy is one thing, and the denial of our value system is quite another. Chosen or not, homosexuality remains opposable. If chosen, we argue against the choice; if not chosen, we offer compassion while retaining our heterosexual marital ideals.

IS HOMOSEXUALITY CHOSEN?

The question of choice, then, is unrelated to the question of homosexuality's rightness or wrongness. But we must still try to resolve the question of whether homosexuality is chosen.

The question is always posed as, "Do homosexuals choose homosexuality?" When phrased this way, the answer usually seems obvious. One hardly imagines an adult sitting down and debating whether to become a homosexual or a heterosexual. But the question is much more instructive when posed in a more specific way: Is homosexuality biologically programmed from birth, or is it socially and psychologically induced?

There is clearly no one answer that accounts for all homosexuals. What can be said for certain is that some homosexuals were started along that path in early childhood, and that most homosexuals, having had sex with both sexes, prefer homosexual sex to heterosexual sex.

We can say "prefer" because the vast majority of gay men have had intercourse with women. As a four-year study of 128 gay men by a UCLA professor of psychology revealed, "More than 92 percent of the gay men had dated a woman at some time, two-thirds had sexual intercourse with a woman."[14]

Moreover, if homosexuality is biologically determined, how are we to account for the vastly differing numbers of homosexuals in different societies? As far as we know, most upper-class men practiced homosexuality in ancient Greece, yet we know that there has been very little homosexuality, for example, among Orthodox Jews.

Wherever homosexuality has been encouraged, far more people have engaged in it. And wherever heterosexuality has been discouraged, homosexuality has similarly flourished, as, for example, in prisons and elsewhere: As Greenberg has written, "High levels of homoeroticism develop in boarding schools, monasteries, isolated rural regions, and on ships with all-male crews."[15]

As for female homosexuality, many lesbian spokeswomen argue passionately that lesbianism is indeed a choice to be made, not a biological inevitability. To cite but two of many such examples, Charlotte Bunch, an editor of *Lesbians and the Women's Movement* (1975), wrote: "Lesbianism is the key to liberation and only women who cut their ties to male privilege can be trusted to remain serious in the struggle against male dominance." And Jill Johnson, in her book, *Lesbian Nation: The Feminist Solution* (1973), wrote: "The continued collusion of any woman with any man is an event that retards the progress of women's supremacy."

Of course, one could argue that homosexuality is biologically determined, but that society, if it suppresses it enough, causes most homosexuals to suppress their homosexuality. Yet, if this argument is true, if society can successfully repress homosexual inclinations, it can lead to either of two conclusions—that society should do so (socially, not legally) for its own sake, or that society should not do so for the individual's sake. Once again we come back to the question of values.

Or, one could argue that people are naturally (i.e., biologically) bisexual (and given the data on human sexuality, this may be true). Ironically, however, if this is true, the argument that homosexuality is chosen is strengthened, not weakened. For if we all have bisexual tendencies, and most of us successfully suppress our homosexual impulses, then obviously homosexuality is frequently both surmountable and chosen. And once again we are brought back to our original question of what sexual ideal society ought to foster—heterosexual marital or homosexual sex.

To sum up:

1) Homosexuality may be biologically induced, but is certainly psychologically ingrained (perhaps indelibly) at a very early age in some cases. Presumably, these individuals always have had sexual desires only for their own sex. Historically, they appear to constitute a minority among homosexuals.

2) In some cases, homosexuality appears not to be indelibly ingrained. These individuals have gravitated toward homosexuality from heterosexual experiences, or have always been bisexual, or live in a society that encourages homosexuality. As Greenberg, who is very sympathetic to gay liberation, writes, "Biologists who view most traits as inherited, and psychologists who think sexual preferences are largely determined in early childhood, may pay little attention to the finding that many gay people have had extensive heterosexual experience."

3) Therefore, the evidence overwhelmingly leads to this conclusion: By and large, it is society, not the individual, that chooses whether homosexuality will be widely practiced. A society's values, much more than individuals' tendencies, determine the extent of homosexuality in that society.

Thus we can have great sympathy for the exclusively homosexual individual while strongly opposing social acceptance of homosexuality. In this way we retain both our hearts and our values.

WHAT ARE WE TO DO?

We could conceivably hold that while heterosexual sex ought to be society's ideal, society should not discriminate against homosexuals. This solution, however, while tempting, is not as tidy as it sounds. For the moment one holds that homosexuality is less socially or morally desirable than heterosexuality, discrimination, in some form, becomes inevitable. For example, it is very difficult to hold that marriage and family must be society's ideal and at the same time advocate homosexual marriage.

12. Homosexuality, the Bible, and Us

More than other issues, homosexuality seems to force one into an extreme position. Either you accept homosexuality completely or you end up supporting some form of discrimination. The moment you hesitate to sanction homosexual marriage, or homosexual men as Big Brothers to young boys, or the ordaining of avowed homosexuals, you have agreed to discrimination against homosexuals. And then the ACLU, gay activists, and others will lump you with the religious right wing.

This is why many liberals find it difficult not to side with *all* the demands of gay activists. They terribly fear being lumped with right-wingers. And they loathe the thought of discriminating against minorities. Gay activists have been quite successful at depicting themselves as another persecuted minority, and this label tugs at the conscience of moral individuals, both liberal and conservative. Of course, in some ways this label is deserved, since gays are a minority, and they certainly have been persecuted. But they are not a persecuted minority in the same way that, let us say, blacks have been. Sexual lifestyle is qualitatively different from skin color.

Since blacks have been discriminated against for what they are and homosexuals have been discriminated against for what they do, a moral distinction between the two types of discrimination can be made in a handful of areas. This in no way exonerates gay-bashing or gay-baiting, let alone such evils as the Nazi or communist incarcerations of gays. But it does mean that a moral distinction between discrimination against behavior and discrimination against color is possible. For example, there is no moral basis to objecting to blacks marrying whites, but there is a moral basis for objecting to homosexual marriage.

That is why gay activists fight against every single vestige of discrimination against homosexuality. They intuit that even one form of discrimination—prohibiting homosexual marriages, for example—means that society differs only in degree from those who declare homosexuality "an abomination."

This is a problem with which I continue to wrestle. I want gays to have the rights that I have. But not everything I am allowed to do is a right. Marriage, for example, is not a universal right (see below), nor, even more so, is religious ordination.

DECRIMINALIZING HOMOSEXUALITY

Before dealing with areas where discriminating on behalf of marital heterosexuality may be proper, let us deal with the areas where discrimination is not morally defensible.

Twenty-three states in the United States continue to have laws against private homosexual relations. I am opposed to these laws. Whatever my misgivings about homosexuality may be, they do not undo my opposition to the state's interference in private consensual relations between adults. Those who wish to retain such laws need to explain where, if ever, they will draw their line. Should we criminalize adultery? After all, adultery is prohibited by the Ten Commandments.

What should be permitted in private, however, does not have to be permitted in all areas of society. Thus, for example, while I am for decriminalizing prostitution, I would not allow the transactions to take place in public or permit prostitutes to advertise on billboards, radio, or television.

To decriminalize an act is not to deem it as socially acceptable as any other act. But social acceptance is precisely what gay liberation aims for—and also where the majority of society disagrees with gay liberation.

I suspect that in this regard most people feel as I do—antipathy to gay-baiting, gay-bashing, and to the criminalizing of private gay behavior, while simultaneously holding that homosexuality is not an equally viable alternative. Given these admittedly somewhat contradictory positions, what are we to do?

I believe that we ought to conduct public policy along two guidelines:

1) We may distinguish between that which grants homosexuals basic rights and that which honors homosexuality as a societally desirable way to live.

2) Therefore, we may discriminate on behalf of the heterosexual marital ideal, but not against the individual homosexual in the private arena—for example, where and how a homosexual lives.

HOMOSEXUAL ORDINATION

The most obvious area wherein the distinction between civil rights and public acceptance of homosexuality manifests itself is religion. It is, after all, Western religion that most fought for confining sexual activity to marriage.

It is therefore not surprising that few Christian or Jewish mainstream denominations, even liberal ones, ordain individuals who publicly declare themselves homosexual.

The issue is only secondarily the individual's sex life. It is primarily one of values. If a candidate for ordination at any of the Jewish seminaries engaged in cross-dressing, a clear violation of a Torah law, or took a personal vow of celibacy, another violation of Jewish law (at least for men), but in neither instance announced it, it would not be the admissions committee's task to inquire about such things. But if a rabbinic student were to announce that he is a transvestite or that remaining single is as desirable to Judaism as being married, he should not be ordained.

In sexual matters, the issue is what is advocated and what is lived publicly far more than what is privately practiced. An organization should be able to choose spokesmen who publicly support its ideals; that sort of "discrimination" is perfectly legitimate.

HOMOSEXUAL MARRIAGE

Gay activists and some liberal groups such as the ACLU argue for the right of homosexuals to marry. Generally, two arguments are advanced—that society should not deny anyone the right to marry, and that if male homosexuals were given the right to marry, they would be considerably less likely to cruise.

The first argument is specious because there is no "right to marry." There is no right to marry more than one partner at a

2. SEXUAL BIOLOGY, BEHAVIOR, AND ORIENTATION: Sexual Orientation

time, or to marry an immediate member of one's family. Society does not allow either practice. Though the ACLU and others believe that society has no rights, only individuals do, most Americans feel otherwise. Whether this will continue to be so, as Judaism and Christianity lose their influence, remains to be seen.

The second argument may have some merit, and insofar as homosexual marriages would decrease promiscuity among gay men, it would be a very positive development for both gays and society. But homosexual marriage would be unlikely to have such an effect. The male propensity to promiscuity would simply overwhelm most homosexual males' marriage vows. It is women who keep most heterosexual men monogamous, or at least far less likely to cruise, but gay men have no such brake on their cruising natures. Male nature, not the inability to marry, compels gay men to wander from man to man. This is proven by the behavior of lesbians, who, though also prevented from marrying each other, are not promiscuous.

HOMOSEXUAL EMPLOYMENT

In general, not hiring a person because he or she is gay is morally indefensible. There are, however, at least two exceptions which necessitate the use of the qualifier, "in general."

In some rare cases in which sexual attraction, or non-attraction, is an absolutely relevant aspect of a job, a case can be made for discrimination against gays in hiring. The armed forces are one possible example. One reason for not admitting gays into combat units is the same reason for not allowing women and men to share army barracks. The sexual tension caused by individuals who may be sexually interested in one another could undermine effectiveness.

Big Brothers provides a second example. Just as heterosexual men are not allowed to serve as Big Brothers to girls, gay men should not be allowed to serve as Big Brothers to boys. The reason is not anti-homosexual any more than not allowing heterosexual men to be Big Brothers to girls is anti-heterosexual; it is common sense. We do not want Big Brothers to be potentially sexually attracted to the young people with whom they are entrusted. It is not because we trust homosexual men less; it is because we do not trust male sexual nature with any minor to whom a male may be sexually attracted.

My own view is that, in general, if employees work responsibly, their off-duty hours are their own business. This is not the view of many liberals and conservatives today, however. Off-hours "womanizing" ended the career of the leading contender for the Democratic Party's presidential nomination of 1988. And it was a major reason for not approving a secretary of defense-designate. Ironically, the voting public often seems far more tolerant of "manizing" than of womanizing. Rep. Barney Frank's male lover ran a male prostitution ring from the congressman's apartment, yet Frank seems to be as popular with his constituency as ever. Such behavior on the part of a heterosexual would doubtless have led to his resignation from Congress.

BEHAVIOR TOWARD HOMOSEXUALS

Violence against homosexuals has claimed numerous lives over the past decade, and too often the law seems to regard it as less of a crime than the murder of heterosexuals. In 1976, when a gay college student was beaten to death by teenagers in front of a Tucson bar, the judge imposed no penalty. In 1984, a Bangor, Maine, judge released to custody of their parents three teenage boys who had beaten and thrown a young gay man into a stream. In 1988, a Texas judge eased the sentence of a man who murdered two homosexuals because, in the judge's words, "I put prostitutes and gays at about the same level, and I'd be hard put to give somebody life for killing a prostitute."[16]

According to a National Gay Task Force study, one-fourth to one-third of gay men have been assaulted or threatened with violence. Even if the figures are exaggerated by a factor of two, they are terrible. And they may actually be understated, since many homosexuals do not wish to report such crimes for fear of embarrassment.

Unfortunately, religious opponents of homosexuality can abet this type of behavior. It should go without saying, but, unfortunately, it needs to be said that the homosexual is created in God's image as much as every other person, and that a homosexual can be as decent a human being as anyone else.

It should also go without saying but, again, it needs to be said that to hurt a homosexual, to be insensitive to a homosexual because of the person's homosexuality, is despicable. Likewise, I believe that when a parent severs relations with a child because of the child's homosexuality, it is a terrible and mutually destructive act.

Gay-bashing, gay-baiting, and jokes that mock (as opposed to poking good-natured fun at) homosexuals have no place in a decent society.

I can confirm from personal experience the truth of the gay activist claim that nearly all of us know or come into regular contact with gay people. From childhood, I was aware that a member of our family circle, one of my mother's cousins, was a gay man; my closest friend during my college year in England was a homosexual; and a proofreader of my journal for two years, one of my closest co-workers, was a lesbian.

I have regarded these people as no less worthy of friendship than my priest friends whose celibacy I do not agree with, or my bachelor friends whose decisions not to marry I disagree with.

"HOMOPHOBIA"

Just as we owe homosexuals humane, decent, and respectful conduct, homosexuals owe the same to the rest of us. Homosexuals' use of the term "homophobic," however, violates this rule as much as heterosexuals' use of the term "faggot" does.

When the term "homophobic" is used to describe anyone who believes that heterosexuality should remain Western society's ideal, it is quite simply a contemporary form of McCarthyism. In fact, it is more insidious than the late senator's use of "communist." For one thing, there was and is such a thing as a communist. But "homophobia" masquerades as a scientific description of a phobia that does not exist in any medical list of phobias.

Yet the insidiousness of the term really lies elsewhere. It abuses psychology in order to dismiss a human being whose values the name-caller does not like. It dismisses a person's views as being the product of unconscious pathological fears. It is not only demeaning, it is unanswerable. Indeed, the more one denies it, the more the label sticks.

Whenever I hear the term, unless it is used to describe thugs who beat innocent homosexuals, I know that the user of the term has no argument, only McCarthy-like demagoguery, with which to rebut others. To hold that heterosexual marital sex is preferable to all other expressions of sexuality is no more "homophobic" than it is "incest-phobic" to oppose incest, or "beast-phobic" to want humans to make love only to their own species.

Finally, those who throw around the term "homophobic" ought to recognize the principle of "that which goes around comes around." We can easily descend into name-calling. Shall we start by labeling male homosexuals "women-phobic" and "vagina-phobic," and lesbians "men-phobic" and "penis-phobic"? It makes as much sense, and it is just as filthy a tactic.

Good people can differ about the desirability of alternate modes of sexual expression. There are many good people who care for homosexuals, and yet fear the chiseling away of the West's family-centered sex-in-marriage ideal. They merit debate, not the label "homophobic." And there are good homosexuals who argue otherwise. They, too, merit debate, not the label "faggot."

WHAT IS AT STAKE

The creation of Western civilization has been a terribly difficult and unique thing. It took a constant delaying of gratification, and a rechanneling of natural instincts; and these disciplines have not always been well received. There have been numerous attempts to undo Judeo-Christian civilization, not infrequently by Jews (through radical politics) and Christians (through antisemitism).

And the bedrock of this civilization, and of Jewish life, of course, has been the centrality and purity of family life. But the family is not a natural unit so much as it is a *value* that must be cultivated and protected. The Greeks assaulted the family in the name of beauty and Eros. The Marxists assaulted the family in the name of progress. And, today, gay liberation assaults it in the name of compassion and equality. I understand why gays would do this. Life has been miserable for many of them. What I have not understood is why Jews and Christians would join the assault.

I do now. They do not know what is at stake. At stake is our civilization. It is very easy to forget what Judaism has wrought and what Christians have created in the West. But those who loathe this civilization never forget. The radical Stanford University faculty and students who chanted, "Hey, hey, ho, ho, Western civ has got to go," were referring to much more than their university's syllabus.

And no one is chanting that song more forcefully than those who believe and advocate that sexual behavior doesn't play a role in building or eroding a civilization.

NOTES

1. Martha Nussbaum, "The Bondage and Freedom of Eros," *Times Literary Supplement,* June 1–7, 1990.
2. Terence J. Deakin, "Evidence of Homosexuality in Ancient Egypt," *International Journal of Greek Love.* Cited in David E. Greenberg, *The Construction of Homosexuality* (Chicago: University of Chicago Press, 1988).
3. Raymond O. Faulkner, *The Ancient Egyptian Coffin Texts* (Aris and Phillips, 1973). Cited in Greenberg.
4. W. L. Moran, "New Evidence from Mari on the History of Prophecy," *Biblica: 50, 1969.* Cited in Greenberg.
5. Norman Sussman, "Sex and Sexuality in History," *The Sexual Experience,* eds. Sadock, Kaplan and Freedman (Baltimore: Williams & Wilkins, 1976).
6. Edward Gibbon, *History of the Decline and Fall of the Roman Empire,* Vol. 1, London, 1898. Cited in John Boswell, *Christianity, Social Tolerance, and Homosexuality* (Chicago: University of Chicago Press, 1980).
7. Sussman.
8. Marshall G. S. Hodgson, *The Venture of Islam,* Vol. 2 (Chicago: University of Chicago Press, 1974).
9. Edward Westermark, *Ritual and Belief in Morocco,* Vol. 1 (London: Macmillan, 1926). Cited in Greenberg.
10. Greenberg.
11. Greenberg.
12. Cited in Arno Karlen, *Sexuality and Homosexuality* (New York: Norton, 1971).
13. Alan Bell and Martin Weinberg, *Homosexualities,* Alfred Kinsey Institute for Sex Research (New York: Simon and Schuster, 1978).
14. Letitia Anne Peplau, "What Homosexuals Want," *Psychology Today,* March 1981.
15. Greenberg.
16. "Texas Judge Eases Sentence for Killer of 2 Homosexuals," *New York Times,* December 17, 1988.

An introduction to a muddled and sometimes contentious world of scientific research—one whose findings, now as tentative as they are suggestive, may someday shed light on the sexual orientation of everyone

HOMOSEXUALITY AND BIOLOGY

Chandler Burr

Chandler Burr holds a masters degree in international economics and Japan studies, and has reported from Manila for The Christian Science Monitor; *Burr's article in this issue will form the basis of a book on the same subject.*

HE ISSUE OF HOMOSEXUALITY HAS ARRIVED at the forefront of America's political consciousness. The nation is embroiled in debate over the acceptance of openly gay soldiers in the U.S. military. It confronts a growing number of cases in the courts over the legal rights of gay people with respect to marriage, adoption, insurance, and inheritance. It has seen referenda opposing gay rights reach the ballot in two states and become enacted in one of them—Colorado, where local ordinances banning discrimination against homosexuals were repealed. The issue of homosexuality has always been volatile, and it is sure to continue to inflame political passions.

It is timely and appropriate that at this juncture a scientific discipline, biology, has begun to ask the fundamental question What *is* homosexuality? And it has begun to provide glimmers of answers that may in turn not only enhance our self-knowledge as human beings but also have some influence, however indirect, on our politics.

What makes the science in this case so problematic, quite apart from the usual technical difficulties inherent in biological research—particularly neurobiological research, which accounts for much of the present investigation—is the ineffable nature of our psychosexual selves. This encompasses a vast universe of stimulation and response, of aesthetic and erotic sensibilities. There are those who see an element of hubris in the quest to explain such things in biological terms. Others see not so much hubris as hype: certain well-publicized findings, they fear, could turn out to be milestones on the road to an intellectual dead end.

It is undeniably true that neurobiological research is often pursued in a context of great ignorance. The brain remains an organ of mystery even in general, not to mention with regard to specific functions. "We don't know" may be the most frequently used words in neurobiology, and they seem to be used with special frequency when the subject of sexual orientation comes up. Once, I mentioned to a researcher how often I heard these words on the lips of her colleagues, and she replied, "Good—then they're saying the right thing." In this context, and also considering that the subject matter is politically charged, professional rivalries are inevitable and occasionally bitter. Some of those involved in the research are motivated not only by scientific but also by personal concerns. Many of the scientists who have been studying homosexuality are gay, as am I.

Homosexuality's invitation to biology has been standing for years. Homosexuals have long maintained that sexual orientation, far from being a personal choice or lifestyle (as it is often called), is something neither chosen nor changeable; heterosexuals who have made their peace with homosexuals have often done so by accepting that premise. The very term "sexual orientation," which in the 1980s replaced "sexual preference," asserts the deeply rooted nature of sexual desire and love. It implies biology.

Researchers can look back on two histories: a century-long, highly problematic psychological investigation of homosexuality, and a short but extremely complex history of biological research that started out as an examination of ovulation in rats. Three distinct but interrelated biological fields are involved in the recent work on sexual orientation: neuroanatomy, psychoendocrinology, and genetics.

13. Homosexuality and Biology

The Background

IOLOGISTS EMBARKED UPON RESEARCH INTO homosexuality in response to an intellectual vacuum created by the failure of other sciences to solve the riddle of sexual orientation. "Other sciences" mostly means psychiatry. As Michael Bailey and Richard Pillard, the authors of one of the most important genetic inquiries into homosexuality, have observed, decades of psychiatric research into possible environmental causes of homosexuality—that is to say, social and cultural causes—show "small effect size and are causally ambiguous."

As a distinct concept, homosexuality is relatively recent. David Halperin points out in *One Hundred Years of Homosexuality* that the term itself first appeared in German (*Homosexualität*) in a pamphlet published in Leipzig in 1869; it entered the English language two decades later. That some human beings engage in sexual activity with others of the same sex has, of course, been noted since antiquity. Historically, however, the focus was on the acts themselves rather than on the actors. The historian John Boswell, of Yale, has noted that during the Middle Ages "same-sex sex" was regarded as a sin, but those who committed that sin were not defined as constituting a type of people different from others. Between the sixteenth and the eighteenth century same-sex sex became a crime as well as a sin, but again, those who committed such crimes were not categorized as a class of human being. This changed in the nineteenth century, when modern medicine and particularly the science of psychiatry came to view homosexuality as a form of mental illness. By the 1940s homosexuality was discussed as an aspect of psychopathic, paranoid, and schizoid personality disorders.

Having defined homosexuality as a pathology, psychiatrists and other doctors made bold to "treat" it. James Harrison, a psychologist who produced the 1992 documentary film *Changing Our Minds*, notes that the medical profession viewed homosexuality with such abhorrence that virtually any proposed treatment seemed defensible. Lesbians were forced to submit to hysterectomies and estrogen injections, although it became clear that neither of these had any effect on their sexual orientation. Gay men were subjected to similar abuses. *Changing Our Minds* incorporates a film clip from the late 1940s, now slightly muddy, of a young gay man undergoing a transorbital lobotomy. We see a small device like an ice pick inserted through the eye socket, above the eyeball and into the brain. The pick is moved back and forth, reducing the prefrontal lobe to a hemorrhaging pulp. Harrison's documentary also includes a grainy black-and-white clip from a 1950s educational film produced by the U.S. Navy. A gay man lies in a hospital bed. Doctors strap him down and attach electrodes to his head. "We're going to help you get better," says a male voice in the background. When the power is turned on, the body of the gay man jerks violently, and he begins to scream. Doctors also tried castration and various kinds of aversion therapy. None of these could be shown to change the sexual orientation of the people involved.

Among those who looked into the matter was the sex researcher Alfred Kinsey, whose 1948 report *Sexual Behavior in the Human Male* showed homosexuality to be surprisingly common across lines of family, class, and educational and geographic background. In his book *Being Homosexual*, the psychoanalyst Richard Isay writes,

> Kinsey and his co-workers for many years attempted to find patients who had been converted from homosexuality to heterosexuality during therapy, and were surprised that they could not find one whose sexual orientation had been changed. When they interviewed persons who claimed they had been homosexuals but were now functioning heterosexually, they found that all these men were simply suppressing homosexual behavior... and that they used homosexual fantasies to maintain potency when they attempted intercourse. One man claimed that, although he had once been actively homosexual, he had now "cut out all of that and don't even think of men—except when I masturbate."

HOMOSEXUALITY'S INVITATION TO BIOLOGY HAS BEEN STANDING FOR YEARS. HOMOSEXUALS HAVE LONG MAINTAINED THAT SEXUAL ORIENTATION, FAR FROM BEING A PERSONAL CHOICE, IS SOMETHING NEITHER CHOSEN NOR CHANGEABLE.

Psychiatry not only consistently failed to show that homosexuality was a preference, a malleable thing, susceptible to reversal; it also consistently failed to show that homosexuality was a pathology. In 1956, in Chicago, a young psychologist named Evelyn Hooker presented a study to a meeting of the American Psychological Association. Hooker had during her training been routinely instructed in the theory of homosexuality as a pathology. A group of young gay men with whom she had become friendly seemed, however, to be quite healthy and well adjusted. One of them, a former student of hers, sat her down one day and, as she recalls in *Changing Our Minds,* said, "Now, Evelyn, it is your scientific duty to study men like me." She demurred. It was only when a fellow scientist remarked to her, "He's right—we know nothing about them," that Hooker sought and received a study grant from the National Institute of Mental Health. She chose a group of thirty gay men as the objects of her research and thirty straight men as controls; none of the sixty had ever sought or undergone psychiatric treatment. "It was the first time [homosexuals] had been studied outside a medical setting or prison," she says. "I was prepared, if I was so convinced, to say that these men were not as well adjusted as they seemed on the surface."

Hooker administered psychological tests to her sixty subjects, including the Rorschach ink-blot test, producing sixty psychological profiles. She removed all identifying marks, including those indicating sexual orientation, and, to eliminate her own biases, gave them for interpretation to three eminent psychologists. One of these was Bruno Klopfer, who believed that he would be able to distinguish homosexuals from heterosexuals by means of the Rorschach test. As it turned out, none of the three could tell the homosexuals and heterosexuals apart. In side-by-side comparisons of matched profiles, the heterosexuals and homosexuals were indistinguishable, demonstrating an equal distribution of pathology and mental health. Reviewing Hooker's results from a test in which the subject creates pictures with cutout figures, one of the interpreters, a psychologist named Edwin Shneidman, stumbled onto a particular subject's orientation only when he came across a cutout scene depicting two men in a bedroom. Shneidman remembers, "I said to Evelyn, 'Gee, I wish I could say that I see it all now, that this is the profile of a person with a homosexual orientation, but I can't see it at all.'"

Hooker's research throughout her long career was driven by the belief that for psychiatry to be minimally scientific, pathology must be defined in a way that is objective and empirically observable. Her study was the first of many showing that homosexuality could not be so defined as pathology. In 1973 the American Psychiatric Association removed homosexuality from its official *Diagnostic and Statistical Manual,* signifying the end of homosexuality's official status as a disease. Today's psychiatrists and psychologists, with very few exceptions, do not try to change sexual orientation, and those aspiring to work in the fields of psychiatry and psychology are now trained not to regard homosexuality as a disease.

Anatomy Lessons

ITH HOMOSEXUALITY MOVED FROM THE realm of psychiatric pathology into the realm of normal variants on human sexual behavior, research efforts took a new turn. Psychiatry had succeeded in defining what homosexuality is *not*—not in explaining what it is. Questions of etiology, in this as in other psychiatric matters, thus became by default questions for neurobiology. Are homosexuals and heterosexuals biologically different? In thinking about this question, biologists have been greatly influenced by findings that involve what may be a related question: Just how, neurologically, do men differ from women?

In 1959, at the University of California at Los Angeles, the neuroendocrinologist Charles Barraclough found that if a female rat was injected shortly before or after birth with testosterone, a male sex hormone, the abnormal amount of this hormone would make the rat permanently sterile, unable to ovulate. "Ovulation" as used here is in part a technical term: it refers both to what a lay person would think of as ovulation—the movement of an egg from the ovary into the fallopian tube—and to the series of hormonal interactions that cause that event.

Rats have short estrous cycles. Every four days various glands in the rat's body start pumping estrogens, or female sex hormones, into the bloodstream, setting in motion a series of chemical events. Estrogen levels reach a certain concentration and stimulate part of the hypothalamus, the small portion of the brain that regulates (among other things) body temperature, hunger, thirst, and sexual drive. The hypothalamus in turn stimulates the pituitary gland; the pituitary then releases a burst of something called luteinizing hormone, which causes the ovary to release an egg. Barraclough discovered that in female rats even a single perinatal exposure to testosterone will prevent this entire process from ever occurring.

If that discovery was intriguing, a subsequent one was even more so: the discovery that male rats can ovulate—at least in the sense of going through the hormonal preliminaries. In 1965 Geoffrey Harris, a neuroendocrinologist at Oxford University, castrated a group of newborn male rats, depriving them of the testosterone from their testes. He found that if estrogen was injected into the bodies of these rats after they reached adulthood, it stimulated the hypothalamus, which initiated the sequence of hormone releases described above. The male rats obviously had no ovaries or wombs, but they went through the biochemical motions of ovulation. If one grafted an ovary onto a male rat, he would ovulate perfectly.

Further tests revealed a strange asymmetry. Whereas newborn male rats deprived of testosterone will, as Harris found, experience female-like ovulation, newborn female rats deprived of estrogen will continue to develop as females. In adulthood they will not seem somehow male. Although the rats' ovaries have been removed, their brains will still produce the stimulus to ovulate. Scientists realized that without testosterone the genetic blueprint for masculinity was essentially worthless. Indeed, they learned, for a male rat's brain to become truly organized as male, the rat must be exposed to testosterone within the first five days of life. After the fifth day the masculinizing window of opportunity is closed, and the genetic male will grow up with a "female" brain. In contrast, the brain of a female needs no estrogen for organization; left alone, it will become female.

Thus it came to be understood that what one might think of as the "default brain" for both sexes of the rat is feminine, and that testosterone is as necessary in the creation of a masculine brain as it is in the creation of masculine genitals. This concept, which is the basis of one approach to the neurobiological search for the origins of sexual orientation, is known as the "sexual differentiation of the brain."

Roger Gorski, a neurobiologist at the University of California at Los Angeles who has long been involved in research on sexual differentiation, looked back recently on the development of his field: "We spent much of our professional careers trying to understand this process of sexual differentiation, and what functions happen within it—male sex behavior, female sex behavior, control of ovulation, control of food intake, body weight, aggressive behavior, some aspects of maternal behavior. You know why male dogs lift their legs when they pee? Because the brain has changed. So this is really a fundamental concept, that the brain is inherently female and to develop as male it must be exposed to masculinizing hormones."

Several years after Harris's experiment other researchers at Oxford University succeeded in confirming anatomically what the principle of the sexual differentiation of the brain had strongly implied: that an observable difference exists between the brains of male rats and those of female rats. In 1971 the anatomists Geoffrey Raisman and Pauline Field published a paper that compared the synapses, or connections between brain cells, in the hypothalamuses of male and female rats. The prevailing view at the time was that all structures of male and female brains were alike. Raisman and Field found that female and male rat brains differed in the number of synaptic connections between brain cells in the hypothalamus: females had more. Rat brains, which varied by sex in terms of function, also varied in terms of structural shape—were "sexually dimorphic." In 1977 a team of neurobiologists led by Roger Gorski located a second sexual dimorphism, again in the rat hypothalamus: a small nucleus, or cluster of cells, five times larger in volume in the male rat than in the female. Gorski found that with the naked eye he could sex rats' brains with almost 100 per-

---

WHAT MAKES THE SCIENCE IN THIS CASE SO PROBLEMATIC IS THE INEFFABLE NATURE OF OUR PSYCHOSEXUAL SELVES. THIS ENCOMPASSES A VAST UNIVERSE OF STIMULATION AND RESPONSE, OF AESTHETIC AND EROTIC SENSIBILITIES.

---

cent accuracy. Gorski's team named the nucleus, logically, the sexually dimorphic nucleus. Its function is not known.

The groundwork had been laid in rodents. The next step was to see if sexual dimorphism of some kind could be found in the brains of human beings. In 1982 the cell biologist Christine de Lacoste-Utamsing and the physical anthropologist Ralph Holloway published in *Science* an examination of a structure in the human brain called the corpus callosum. The corpus callosum, which is made up of nerve fibers known as axons, is a long, narrow structure that connects and transmits information between the brain's right and left hemispheres. It is one of the largest and most clearly identifiable portions of the brain, and has for years figured prominently in brain research. De Lacoste-Utamsing and Holloway found that the shape of a portion of the corpus callosum called the splenium differed so dramatically be-

tween the sexes, with the splenium being larger in women than in men, that impartial observers were able to sex brains easily by looking at this single feature. The De Lacoste-Utamsing and Holloway study is well known and frequently cited, despite the failure of many of the attempts to replicate it. Whether the dimorphism found by De Lacoste-Utamsing and Holloway truly exists remains a matter of considerable debate.

In 1985, three years after the publication of the De Lacoste-Utamsing and Holloway article, Dick Swaab, a researcher at the Netherlands Institute for Brain Research, in Amsterdam, reported that he, too, had found evidence of sexual dimorphism in human brains—in the form of a human homologue of the sexually dimorphic nucleus that Gorski had found in rats.

Swaab announced an even more remarkable discovery five years later, in 1990. He had found, he wrote in an article in the journal *Brain Research*, that a cluster of cells in the human brain called the suprachiasmatic nucleus was dimorphic—but dimorphic according to sexual orientation rather than sex. Swaab said that the suprachiasmatic nucleus was nearly twice as large in homosexual men as it was in heterosexual men.

If true, this was something wholly new: an anatomical difference between homosexuals and heterosexuals.

SIMON LEVAY IS A YOUNG NEUROBIOLOGIST WHO AT the time of Swaab's second discovery was conducting research at the Salk Institute, in La Jolla, California. LeVay would soon become the author of what is surely the most publicized neurobiological article on homosexuality that has appeared to date. I spoke with him one day recently in his West Hollywood apartment. LeVay is a wiry, muscular man, remarkably intense. Perhaps the most striking thing about him is the way he talks. In a crisp British accent he zeroes in on each point and then moves on with an air of impatience.

"You shouldn't draw such a distinction between biological and psychological mechanisms," he chided me at one point during our conversation. "What people are really getting at is the difference between innately determined mechanisms and culturally determined mechanisms, but people screw that up and say that's the difference between biology and psychology. It isn't. It's two different approaches for looking at the same thing: the mind. Biologists look at it from the bottom up, from the level of synapses and molecules, and psychologists are looking at it from the top down, at behavior and such."

LeVay had been intrigued by Swaab's research, but he was troubled by the fact that the portion of the brain examined by Swaab seemed to have nothing to do with the regulation of sexual behavior, at least not in animals. The suprachiasmatic nucleus governs the body's daily rhythms; dimorphism there according to sexual orientation might be provocative, certainly, but it would seem to constitute an effect, not a cause. Why not check out the hypothalamus, a region that is intimately involved with sexual behavior?

Laura Allen, a postdoctoral assistant in Gorski's laboratory, had identified four small groups of neurons in the anterior portion of the hypothalamus, naming them the interstitial nuclei of the anterior hypothalamus (INAH) 1, 2, 3, and 4. Allen's research had shown that INAH 2 and INAH 3 were sexually dimorphic in human beings—significantly larger in men than in women. Was it possible that these nuclei were dimorphic according to sexual orientation as well? That was the focus of LeVay's research, and he presented his conclusions in a short paper titled "A Difference in Hypothalamic Structure Between Heterosexual and Homosexual Men." It was published in *Science* in August of 1991. In the introduction LeVay defined sexual orientation as "the direction of sexual feelings or behavior toward members of one's own or the opposite sex" and hypothesized that Allen's INAH nuclei were involved in the generation of "male-typical sexual behavior." He went on,

> I tested the idea that one or both of these nuclei exhibit a size dimorphism, not with sex, but with sexual orientation. Specifically, I hypothesized that INAH 2 or INAH 3 is large in individuals sexually oriented toward women (heterosexual men and homosexual women) and small in individuals sexually oriented toward men (heterosexual women and homosexual men).

LeVay dissected brain tissue obtained from routine autopsies of forty-one people who had died at hospitals in New York and California. There were nineteen homosexual men, all of whom had died of AIDS; sixteen presumed heterosexual men, six of whom had been intravenous drug abusers and had died of AIDS; and six presumed heterosexual women. No brain tissue from lesbians was available. LeVay's conclusions included the following:

> INAH 3 did exhibit dimorphism. . . . [T]he volume of this nucleus was more than twice as large in the heterosexual men . . . as in the homosexual men. . . . There was a similar difference between the heterosexual men and the women. . . . These data support the hypothesis that INAH 3 is dimorphic not with sex but with sexual orientation, at least in men.

The results were sufficiently clear to LeVay to allow him to state, "The discovery that a nucleus differs in size between heterosexual and homosexual men illustrates that sexual orientation in humans is amenable to study at the biological level."

The study, as LeVay himself readily admits, has several problems: a small sample group, great variation in individual nucleus size, and possibly skewed results because all the gay men had AIDS (although LeVay found "no significant difference in the volume of INAH 3 between the heterosexual men who died of AIDS and those who died of other causes"). As of this writing, LeVay's findings have yet to be replicated by other researchers. LeVay himself has extended his search for dimorphism according to sexual orientation to the corpus callosum, which he is studying by means of magnetic-resonance imaging. Until his original findings are confirmed, the notion that homosexuals and heterosexuals are in some way anatomically distinct must hold the status of tantalizing supposition.

It needs also to be remembered that, as noted earlier, the issue of dimorphism of any kind in the brain is hotly contested. The idea that the brains of heterosexuals and homosexuals may be different morphologically is derived from the idea that the brains of men and women are different morphologically—recall the corpus callosum study by De Lacoste-Utamsing and Holloway. But that study is itself problematic, efforts to replicate it having turned up inconsistent results. Anne Fausto-Sterling is a developmental geneticist at Brown University. She, along with William Byne, a neurobiologist and psychiatrist at Columbia University, has been among the chief critics of neurobiological investigations of homosexuality. Fausto-Sterling during an interview not long ago itemized some of the results from a long line of attempts to replicate sexual dimorphism: "1985: no sex differences in shape, width, or area. 1988: three independent observers unable to distinguish male from female. 1989: women had smaller callosal areas but larger percent of area in splenium, more-slender CCs, and more-bulbous splenium." A new corpus callosum study by Laura Allen, conducted in 1991, *did* find sexual dimorphism—and the debate continues. Part of the difficulty is methodological, involving whose brains are being compared, and how. Dead people or living people? Old or young or mixed? Healthy or sick? By means of brain sections or magnetic-resonance imaging? LeVay calls studies of the corpus callosum "the longest-running soap opera in neurobiology." And, of course, he himself is now part of the cast.

Even if LeVay's hypothalamus study stands up to scrutiny, it will not justify drawing extravagant conclusions. Establishing a distinction is not the same thing as finding a cause. Anatomy is not etiology, but it may offer a starting point for a journey backward in search of the ultimate origins of sexual orientation. That journey takes us into the realm of hormones and genetics.

The Puzzles of Chemistry

N A LARGE ROOM AT THE UCLA DEPARTMENT of anatomy, Roger Gorski and I recently stood facing a dozen black-topped lab tables, each below a ceiling-mounted video monitor. We were about to watch a tape of rats having sex. Gorski, an eternally cheerful, almost elfin man of fifty-seven, was energetically describing the tape. "There are six couples," he explained, though at the moment I saw only one uninterested-looking white rat. "That's an unaltered female," he said. "They're going to put in another female that has been injected with testosterone." Sure enough,

13. Homosexuality and Biology

someone's hand reached down into the screen and a second rat landed in the cage. The rats at first edged around each other, but in just a few seconds on the dozen monitors I saw the testosterone-injected female begin to sniff the other female rat and then mount her aggressively. At the lab tables a handful of medical students went on with their work, paying no attention. After a few moments the tape cut to two males, one perinatally castrated and injected with estrogen, one unaltered. After some initial maneuvering the castrated male responded to the advances of the unaltered male by bending his back and offering himself in what was to me indistinguishable from female-rat lordosis—behavior indicating receptivity to sex, pictures of which Gorski had shown me in his office. The altered rat submitted as the other male mounted him. The tape continued with similar scenes. It was quite dramatic.

Such research in animals has led to hypotheses that hormones are, in some way, a cause of homosexuality in human beings. No one, of course, suggests that the sexuality of rats and that of human beings are strictly comparable; some critics of neurobiological research on homosexuality question the utility of animal models entirely. Nonetheless, it was investigations involving animals that got researchers thinking.

Of the scientists who have concentrated on hormonal or psychoendocrinological studies of homosexuality, Günter Dörner, of Germany, is one of the best known. In the 1970s Dörner classified homosexuality as a "central nervous pseudohermaphroditism," meaning that he considered male homosexuals to have brains with the mating centers of women but, of course, the bodies of men. For decades endocrinologists had speculated that because male sex hormones are known to be responsible in human beings for masculine body characteristics and in animals for certain aspects of male sexual behavior, it follows that adult homosexual men should have lower levels of testosterone, or else higher levels of estrogen, in the bloodstream than adult heterosexual men, and that homosexual and heterosexual women should display the opposite pattern. This is known as the "adult hormonal theory" of sexual orientation, and Dörner claimed that some initial studies bore it out.

In 1984 Heino Meyer-Bahlburg, a neurobiologist at Columbia University, analyzed the results of twenty-seven studies undertaken to test the theory. According to Meyer-Bahlburg, a score of the studies in fact showed no difference between the testosterone or estrogen levels of homosexual and heterosexual men. Three studies did show that homosexuals had significantly lower levels of testosterone, but Meyer-Bahlburg believed that two of them were methodologically unsound and that the third was tainted by psychotropic drug use on the part of its subjects. Two studies actually reported higher levels of testosterone in homosexual men than in heterosexual men, and one unhelpfully showed the levels to be higher in bisexuals than in either heterosexuals or homosexuals.

As it came to be widely accepted that adult hormone levels were not a factor in sexual orientation, scientists shifted their attention to prenatal hormone exposure. Many of the glands in a human being's hormone system are busily functioning even before birth—tiny hormone factories that produce the chemicals that help to mold the person who will eventually emerge. Perhaps, it was thought, different levels of prenatal hormones produce different sexual orientations. For obvious reasons, the sometimes brutal hormonal experiments done on monkeys and rats cannot be done on human beings, but nature at times provides a narrow window onto the mysteries of prenatal hormonal effects in ourselves.

Congenital adrenal hyperplasia (CAH) has been called

THE RATS AT FIRST EDGED AROUND EACH OTHER, BUT IN JUST A FEW SECONDS ON THE DOZEN MONITORS I SAW THE TESTOSTERONE-INJECTED FEMALE BEGIN TO SNIFF THE OTHER FEMALE RAT AND THEN MOUNT HER AGGRESSIVELY.

---

by Meyer-Bahlburg a "model endocrine syndrome" for examining the effects of abnormal amounts of prenatal sex hormones. CAH, which can affect both males and females, is caused by a simple problem: an enzyme defect makes it impossible for a fetus's adrenal gland to produce cortisol, an important hormone. In a normal fetus, as the adrenal gland produces cortisol, the brain stands by patiently, waiting for the signals that the cortisol level is appropriately high and production can be shut off. But in CAH fetuses, which lack the enzyme to create cortisol, the brain doesn't get those signals, and so it orders the adrenal gland to continue production. The adrenal gland continues pumping out what it thinks is cortisol, but it is unknowingly producing masculinizing androgens. It dumps these into the fetus's system, thereby overexposing it to male hormones.

The consequences are most dramatic in females. Once, in his office, Roger Gorski dug into a desk drawer and grabbed a few photographs. "What sex is it?" he asked. I squinted at close-ups of a child's genitals and saw a penis, plain as day. "It's a boy," I said confidently, Gorski's eyebrows shot up. "Where are the testicles?" he asked. I looked closer. Oops.

This was a CAH baby. In this case, Gorski told me, the doctors had decided at the time of birth that the child was a boy with undescended testicles, a relatively common and minor condition. But in fact I was looking at a genetic female.

With surgery a CAH female's external genitals can be made to look feminine, as her internal apparatus already fully is, and she will be raised as a girl. But hormones may have already had their effect in an area that plastic surgery cannot touch: the brain. Or at least so proponents of the prenatal-hormone theory of sexual orientation would argue. The sexual orientation of CAH females tends to bear them out. A 1984 study by the Johns Hopkins University sex researcher John Money found that 37 percent of CAH women identified themselves as lesbian or bisexual; the current estimate of the proportion of lesbians in the general female population is from two to four percent.

One possible clue as to whether the prenatal-hormone theory of sexual orientation is a profitable line of inquiry involves something called luteinizing-hormone (LH) feedback. The brain releases several hormones, including LH, which initiate the development of an egg in a woman's ovary. As the egg develops, the ovary releases increasing amounts of estrogen, stimulating the brain to produce more LH, which in turn promotes the production of still more estrogen. The process is called positive feedback. In men, estrogen usually acts to suppress the production of luteinizing hormone—it results in *negative* feedback. These differences in LH feedback in human beings, together with the discovery that male rats hormonally altered after birth will display both positive LH feedback and same-sex sexual behavior, led some researchers to a hypothesis. They speculated that gay men, their brains presumably not organized prenatally by testicular hormones, just as women's are not, would show a positive LH feedback, like that of a heterosexual female, rather than the negative feedback of the typical heterosexual male. If such feedback were to be found consistently in homosexual men—by means of chemical analysis of the blood after injection with estrogen—could this not be taken as evidence that some decisive prenatal hormonal event, with important bearing on subsequent sexual orientation, had indeed occurred?

This line of inquiry has given rise to an active field of study that as yet has little to show for itself. The uncertainties are of two kinds. The first one involves the following question: Do LH feedback patterns of the sort being sought in fact exist in human beings? The second comes down to this: Even if LH feedback patterns of the sort being sought do exist, will they really tell us anything about events that occurred before birth? Unfortunately, neuroscientists lack unequivocal answers to both questions, despite considerable efforts. Different studies have yielded conflicting data. No one has yet come up with what one neurobiologist facetiously terms a "gay blood test."

In an article published in 1990 in the *Journal of Child and Adolescent Psychopharmacology*, Heino Meyer-Bahlberg surveyed the work done so far on hormonal research in general and concluded: "The evidence available to date is inconsistent, most studies are methodologically unsatisfactory, and alternative interpretations of the results cannot be ruled out." On the other hand, Meyer-Bahlberg went on, "not all potential avenues to a psychoendocrine explanation of homosexuality have been exhausted."

Among the unexhausted avenues is one being explored by Richard Pillard.

A PSYCHIATRIST AT THE BOSTON UNIVERSITY SCHOOL of Medicine, Richard Pillard is a tall, pleasant man in his fifties with a neatly trimmed moustache and a relaxed manner. Even when talking seriously, he remains good-natured. When we spoke one afternoon in his Boston townhouse, he joked that he is uniquely equipped to investigate whether homosexuality has a biological basis: he, his brother, and his sister are gay, and Pillard believes that his father may have been gay. One of Pillard's three daughters from a marriage early in life is bisexual. This family history seems to invite a biological explanation, and it made Pillard start thinking about the origins of sexual orientation.

Pillard says that it had long puzzled him why transsexuals—men or women who wish to live in bodies of the opposite sex—are so different from gay people: "You'd think they'd be on the far end of the spectrum, the 'gayest of the gay.'" And yet transsexuals are not in fact gay. Whereas gay men, quite comfortably and unalterably, see themselves as men, male transsexuals see themselves as women trapped in men's bodies. Pillard and a colleague, James Weinrich, a psychobiologist at the University of California at San Diego, began to theorize that gay men are men who in the womb went through only a partial form of sexual and psychosexual differentiation. More precisely, Pillard and Weinrich theorized that although gay men do undergo masculinization—they are, after all, fully male physically—they go incompletely if at all through another part of the process: defeminization.

As fetuses, Pillard points out, human beings of both sexes start out with complete female and male "anlages," or precursors of the basic interior sexual equipment—vagina, uterus, and fallopian tubes for women, and vas deferens, seminal vesicles, and ejaculatory ducts for men. These packages are called the Müllerian (female) and Wolffian (male) ducts, and are tubes of tissue located in the lower abdomen. How do the sexual organs develop? It happens differently in men and women.

At the moment of conception an embryo is given its chromosomal sex, which determines whether it will develop testes or ovaries. In female human beings (as in female rats) the female structures will simply develop, without any help from hormones; the Wolffian duct will shrivel up. The process of becoming male, however, is more complex. Where women need none, men need two kinds of hormones: androgens from the testes to prompt the Wolffian duct into development, and a second substance, called Müllerian inhibiting hormone, to suppress the Müllerian duct and defeminize the male fetus.

Pillard speculates that Müllerian inhibiting hormone, or a substance analogous to it, may have brain-organizing effects. Its absence or failure to kick in sufficiently may prevent the brain from defeminizing, thereby creating what Pillard calls "psychosexual androgyny." In this view, gay men are basically masculine males with female aspects, including perhaps certain cognitive abilities and emotional sensibilies. Lesbian women could be understood as women who have some biologically induced masculine aspects.

An experimental basis is provided by research by the psychiatrist Richard Green, of the University of California at Los Angeles, which shows that children who manifest aspects of gender-atypical play are often gay. Green has concluded that an inclination toward gender-atypical play in prepubescent boys—for example, dressing in women's clothes, playing with dolls, or taking the role of the mother when playing house—indicates a homosexual orientation 75 percent of the time. If that is true, it is important, because it would be an example of a trait linked to sexual orientation which does not involve sexual behavior—suggesting how deeply rooted sexual orientation is. Discussing this line of research, Simon LeVay told me, "It's well known from animal work that sex-typical play behavior is under hormonal control. Robert Goy [at the University of Wisconsin at Madison] has done many studies over the years showing that you can reverse the sex-typical play behavior of infant monkeys by hormonal manipulations in prenatal life. [Play] is an example of a sex-reversed trait in gay people that's not directly related to sex. It's not sex, it's play. When you get to adulthood, these things become blurred. It's easier to tell a gay kid than a gay adult—kids are much of a muchness. Most gay men, even those who are very macho as adults, recall at least some gender-atypical behavior as children."

> PILLARD AND WEINRICH THEORIZED THAT ALTHOUGH GAY MEN DO UNDERGO MASCULINIZATION—THEY ARE, AFTER ALL, FULLY MALE PHYSICALLY—THEY GO INCOMPLETELY IF AT ALL THROUGH ANOTHER PART OF THE PROCESS: DEFEMINIZATION.

The Pillard-Weinrich theory also accords with what Green refers to as male "vulnerability" during the process of sexual differentiation. A considerably larger number of male embryos come into existence than female embryos, and yet males and females come into the world in about the same numbers. Therefore, phenomena linked to sex must reduce the number of males who survive to term. Many disorders are, in fact, more common in men than women, and some of these could result from problems originating in masculine differentiation. Although good statistics do not exist, it appears that there may be two gay men for every gay woman, which would be consistent with the vulnerability theory.

It is important to remember that although homosexuals and heterosexuals may be "sex-reversed" in some ways, in other ways they are not. For example, neither gay nor straight men tend to be confused on the subject of what sex they are: male. LeVay says, "It's not just that you look down and see you have a penis and you say, 'Oh, I'm a boy. Great.' I think there must be some internal representation of what sex you are, independent of these external signals like the appearance of your body. I think most gay men are aware of some degree of femininity in themselves, yet there is no reversal of gender identity." Gay men and straight men also seem to display an identical strong drive for multiple sexual partners; lesbians and straight women seem to be alike in favoring fewer sexual partners.

The evidence from hormonal research may circumstantially implicate biology in sexual orientation, but it is far from conclusive. William Byne raises a warning flag: "If the prenatal-hormone hypothesis were correct, then one might expect to see in a large proportion of homosexuals evidence of prenatal endocrine disturbance, such as genital or gonadal abnormalities. But we simply don't find this." Moreover, the hormonal research does not answer the question of ultimate cause. If hormones help to influence sexual orientation, what is influencing the hormones?

The Genetic Quest

N 1963 KULBIR GILL, A VISITING SCIENTIST from India working at Yale University, was conducting research into genetic causes of female sterility. His experiments involved exposing the fruit fly *Drosophila melanogaster*, that workhorse of genetic research, to X-rays, and observing the behavior of the resulting offspring. Gill noticed that a certain group of mutant male flies were courting other males, following each other and vibrating their wings to make characteristic courtship "songs." Gill published his findings in a short note in the publication *Drosophila Information Service* and then returned to the question of female sterility.

A decade later Jeffrey Hall, a biologist at Brandeis University, followed up on Gill's odd discovery. Every discovered *Drosophilia* gene mutation is given a name, and Gill had called his mutation "fruity." Hall, considering this name to be denigrating, redubbed it, still somewhat tongue-in-cheek, "fruitless." Hall explains that the fruitless mutation produces two distinct behaviors. First, fruitless-bearing male flies, unlike nonmutant male flies, actively court other males as well as females, although for reasons that remain poorly understood, they are unable actually to achieve intercourse with members of either sex. Second, fruitless-bearing males elicit and are receptive to courtship *from* other males, which nonmutant males reject.

Fruit flies can live for two or three months, and this "bisexual" fly strain has existed behaviorally unchanged through hundreds of generations. Some gene mutations are lethal to flies; fruitless is not one of these, nor does it cause illness. It is, Hall says, a nonpathological genetic mutation that causes a consistent, complex behavior. And fruitless displays an anatomical sexual dimorphism, bringing LeVay's study to mind. In the abdomen of male *Drosophila* flies there is a muscle, the so-called muscle of

2. SEXUAL BIOLOGY, BEHAVIOR, AND ORIENTATION: Sexual Orientation

Lawrence, whose function is unknown; female fruit flies don't have it, and neither do fruitless males.

Although fruitless flies don't mate, the perpetuation of the fruitless trait is made possible by the fact that it is recessive—a full pair of the mutations is needed for fruitless behavior to be expressed. When males that carry a single fruitless gene mate with a fruitless-carrying female, a percentage of their offspring will carry the full pair and display typical fruitless behavior. If a genetic component of homosexuality in human beings exists, it could possibly operate by means of a comparable mechanism.

Angela Pattatucci, a geneticist at the National Institutes of Health, gave me a demonstration a few months ago in her lab. She took a small glass container of tiny *Drosophila* flies, popped off the top, and plugged an ether-soaked cotton ball into the mouth. Within a few seconds the flies were lying stunned on the glass floor. Using a plastic stick Pattatucci separated out a few of the flies into a larger glass jar. I looked at a group of males and females through a microscope, their bodies vibrating, red eyes bulging. Pattatucci showed me how to differentiate the genitalia at the end of the abdomen—smooth and light-colored for females, furry and dark for males.

Pattatucci said that researchers are relatively close to finding the actual fruitless gene. It is already known that fruitless is located physically on the right arm of the third chromosome. After establishing the precise location (or locations) of the mutation, researchers can determine the sequence of biochemical information in fruitless's genetic code—the order of thousands of units of the basic genetic components adenine, thymine, guanine, and cytosine. Once the combination is known, the search can begin for a similar combination—a fruitless analogue—in human beings.

In the jar the males, separated out, eventually came back to awareness. "Watch that one," Pattatucci said, pointing to a fly that had come up behind another fly, vibrating his wings in courtship. He then climbed on top of the male he was courting. I watched the two flies, one atop the other, the one on the bottom wandering around as if a bit bored. As noted, for a fruitless fly that is as far as things can go.

I once asked Jeffrey Hall if courtship alone could be satisfying for a fly. "Could be," he said. "Maybe it's delicious, maybe he's frustrated. But this becomes ludicrous. How do you know when a fruit fly is frustrated?" It is an important point: the danger of anthropomorphizing insect behavior is great, and I found myself doing it almost by reflex when watching Pattatucci's flies. How can we equate fly behavior with a vast something that in human beings generates aesthetic and intellectual perceptions—with something that encompasses emotional need and love and the pain of love? So Hall is careful to describe fruitless as "a mutation that leads to a mimic of bisexuality." He is skeptical that finding a fruitless analogue will lead to a full explanation of human homosexuality. DNA analogues for all sorts of fruit-fly genes do exist in human beings, and the process of looking for them is relatively straightforward. But, as Hall points out, "it is very unlikely that the genetics of homosexuality will ever devolve to a single factor in humans with such major effects as it has in *Drosophila*."

WHEN BIOLOGISTS ARE INTERESTED IN ESTABLISHING whether genetics is involved in the appearance of certain characteristics or conditions, one obvious place to look is among people who are closely related to one another. In "A Genetic Study of Male Sexual Orientation," a study that has now achieved almost as much renown as LeVay's, the Northwestern University psychologist Michael Bailey and Boston University's Richard Pillard compared fifty-six "monozygotic" twins (identical twins, from the same zygote, or fertilized egg), fifty-four "dizygotic" (fraternal) twins, and fifty-seven genetically unrelated adopted brothers. Identical twins are important in sexual-orientation research because, of course, they have identical genomes, including the sex-chromosome pair. If homosexuality is largely genetic in origin, then the more closely related that people are, the greater should be the concordance of their sexual orientation.

That is, in fact, what the study found. Bailey and Pillard reported a gay-gay concordance rate of 11 percent for the adoptive brothers, 22 percent for the dizygotic twins, and 52 percent for the monozygotic twins. The findings suggest that homosexuality is highly attributable to genetics—by some measures up to 70 percent attributable, according to Pillard. This figure is based on something geneticists call "heritability," a painstakingly calculated indicator of how much genes have to do with a given variation among people. If heritability is less than 100 percent, then the characteristic being studied is by definition "multifactorial." Eye color is 100 percent dependent on genetics. Height, on the other hand, though about 90 percent genetic, is also affected by nutrition, and thus is multifactorial.

If a large contribution to homosexuality comes from genes, where does the rest of it come from? The range of environmental and biological inputs a developing child receives is both enormous and enormously complex. "Whatever the other variables are," Pillard says, "they must be present early in life. I think this because the gender-atypical behavior that so strongly prefigures an adult homosexual orientation can be observed early in development." And he goes on: "There certainly could be different paths to the same outcome. With individual cases, there are doubtless some that are mostly or all genes, and others that might be all environment. Our analysis [of twins] doesn't say anything about the individual." Jeffrey Hall can be so underwhelmed by the prospect of finding a human analogue of the fruitless mutation because, as he points out, if we do find it, we still will not have fully accounted for the etiology of homosexuality even in identical twins. "You will effectively know nothing from this genetic knowledge," Hall says. A behavior as simple as jumping, he notes, is quite complex genetically, having to do with all kinds of genes and other, unknown factors. He says, "We are not about to create a genetic surgical procedure which makes you Michael Jordan." LeVay made the same point in the course of our conversation: "It's one thing to say that genes are involved, as they almost certainly are. It's a whole other thing to actually identify those genes, because homosexuality may be polygenic, with each gene having a small effect."

Whatever the uncertainties ahead, though, the important point is that the genetic work is already fairly compelling. A new Bailey and Pillard genetic study of lesbian twins, to be published soon in the *Archives of General Psychiatry,* echoes the researchers' original male-twin findings with strongly similar results. "We're getting a lot of consistency where we should be getting it," Bailey says.

The most interesting question is perhaps becoming not *whether* genetics plays a role in homosexuality but *how*. Why does nature preserve genes that influence sexual behavior and yet do not facilitate reproduction? Does less than 100 percent heritability mean that the Bailey and Pillard study is incompatible with a bipolar model of sexual orientation? In his study LeVay defined homosexuality in terms of the sex of a person's sexual-object choice: *either* men or women, *either* homosexual or heterosexual. Pillard and Bailey's multifactorial model suggests a shaded continuum of sexual orientations, and of origins and causes, more complex and subtle than a simple either-or model can accommodate, and closer to what may be the quirks and ambiguities of our real lives.

The Ramifications of Science

HAT DOES IT ALL MEAN? AS WE HAVE SEEN, scientists must sift for their conclusions through ambiguous results from a disparate group of studies that are excruciatingly difficult to interpret. Yet even at this relatively early date, out of the web of complexities it is becoming ever clearer that biological factors play a role in determining human sexual orientation. Richard Green said to me, "I suspect that at least in your lifetime we will find a gene that contributes substantially to sexual orientation." Michael Bailey says, "I would—and have—bet my career on homosexuality's being biologically determined." The pace of neurobiological and genetic research is only increasing.

The search is not without its opponents. Some, recalling earlier psychiatric "treatments" for homosexuality, discern in the biological quest the seeds of genocide. They conjure

> IF A LARGE CONTRIBUTION TO HOMOSEXUALITY COMES FROM GENES, WHERE DOES THE REST OF IT COME FROM? THE RANGE OF ENVIRONMENTAL AND BIOLOGICAL INPUTS A DEVELOPING CHILD RECEIVES IS BOTH ENORMOUS AND ENORMOUSLY COMPLEX.

up the specter of the surgical or chemical "rewiring" of gay people, or of abortions of fetal homosexuals who have been hunted down in the womb. "I think all of us working in this field," Pattatucci says, "have delusions of grandeur in thinking we can control the way this knowledge will be used." Certainly the potential for abuse is there, but that is true of much biomedical knowledge. It is no reason to forswear knowledge of ourselves, particularly when the potential benefits are great.

Some of the benefits could be indirect. Laura Allen points out, for example, that there are many now-mysterious disease—autism, dyslexia, schizophrenia—that affect men and women differently, hiding inside parts of the human mind and body that we cannot penetrate. Neurobiological research into sexual differentiation may help us to understand and cure these diseases, as well as to unlock other mysteries—the mysteries of sexuality.

And then there is the question with which we began—that of the acceptance of gay people in American society. The challenge posed by homosexuality is one of inclusion, and, as Evelyn Hooker would say, the facts must be allowed to speak. Five decades of psychiatric evidence demonstrates that homosexuality is immutable, and nonpathological, and a growing body of more recent evidence implicates biology in the development of sexual orientation.

Some would ask: How can one justify discriminating against people on the basis of such a characteristic? And many would answer: One cannot. Yet it would be wise to acknowledge that science can be a rickety platform on which to erect an edifice of rights. Science can enlighten, can instruct, can expose the mythologies we sometimes live by. It can make objective distinctions—as, for example, between sexual pathology on the one hand and sexual orientation on the other. But we cannot rely on science to supply full answers to fundamental questions involving human rights, human freedom, and human tolerance. The issue of gay people in American life did not arise in the laboratory. The principles needed to resolve it will not arise there either.

THE GAY DEBATE

Is homosexuality a matter of choice or chance?

Meredith F. Small

Meredith F. Small, an associate professor of anthropology at Cornell University, has been researching the sexual behavior of primates for 15 years, concentrating on macaque monkeys. The culmination of her work, Female Choices: Sexual Behavior of Female Primates, *is published by Cornell University Press.*

As never before, the issue of homosexuality has become a political football. As a candidate, President Bill Clinton campaigned on a promise to lift the ban on homosexuals in the armed forces. The Republican Party platform opposed "efforts by the Democratic Party to include sexual preference as a protected minority," a theme reiterated by various speakers at the Republican national convention last August in Houston. In the general campaign, Vice President Dan Quayle declared that homosexuality is a personal choice, and not a very good one at that. And last November, a measure to limit gay rights was defeated in Oregon while a similar bill passed in Colorado, prompting entertainer Barbara Streisand and other Hollywood notables to call for a boycott of the state.

Choice or chance? Nature or nurture? Biology or environment? What *are* the origins of homosexuality? While many Americans share Quayle's view that homosexuality is essentially a lifestyle choice, others are convinced it's an inborn tendency beyond the influence of social or psychological forces. In fact, however, recent research—focusing almost exclusively on male homosexuality (see "Lesbians: Less Is Known")—suggests that the answer is more complicated than advocates on either side of the debate would have us believe.

In 1991 neuroscientist Simon LeVay of the Salk Institute in La Jolla, Calif., made international headlines with a report that a certain area of the brain tended to be smaller in homosexual men than in heterosexual men. Although Dr. LeVay has been cautious about interpreting his results, he has suggested that since this particular area of the brain may be closely connected with sexual behavior, it could well affect sexual orientation.

Meanwhile, other researchers have also reported structural differences between homosexual and heterosexual men in regions of the brain having nothing to do with sex or reproduction. This flurry of research signals a fundamental shift in the way scientists think about sexual orientation, with the emphasis now strongly highlighting the biological—including the possibility that sexual orientation might be at least partially inherited. There is even work in progress to discover what gene or genes might be associated with sexual orientation. So far there have been only a handful of biological studies, few of which have been successfully repeated. Still, many scientists are looking to biology for answers other approaches have failed to provide.

The current focus on biology contrasts sharply with the medical community's traditional portrait of the gay man, which was based on the psychoanalytic concept of the Oedipus conflict. According to this theory, a male child who reaches the stage of development where he separates psychologically from his mother and identifies with his father will become heterosexual. But if a boy grows up with a domineering mother who prevents his detaching from her or with a distant or hostile father who discourages identification, he will turn out homosexual.

Freud described this scenario as one of arrested development, though not a sickness in itself. Other practitioners disagreed, however, and when in 1952 the American Psychiatric Association produced a manual for categorizing mental illness, the Diagnostic and Statistical Manual of Mental Disorders (DSM), homosexuality was classified as a sociopathic personality disorder. Because this manual was (and still is) used by mental health professionals as a standard, homosexuality was officially regarded as an illness. Psychoanalyst Judd Marmor, a professor emeritus of psychiatry at UCLA, says the major flaw with the homosexuality-as-sickness description is that it came from the people psychiatrists saw in their offices. "If we based our understanding of heterosexuality on our patients," says Dr. Marmor, "we'd have an odd picture of that too. Overprotective mothers and withdrawing fathers create neurotic kids, not homosexuals." Marmor was a leader in the often bitter battle to have homosexuality removed from the DSM list of pathologies, which happened in 1973. "Not all homosexuals are alike," he says, "and it's a mistake to throw them all into one basic category. There are multiple factors that contribute to homosexuality—genetic, hormonal and environmental."

In looking for biological factors in human behavior, researchers often turn to the animal world for clues. But homosexuality—at least as practiced by humans—appears to be uniquely human. In scientific terms, there is no good animal model for male homosexuality. Though in most species of nonhuman primates, our closest living relatives, males will frequently mount other males, this behavior appears to have less to do with sexual pleasure than social status. It's a ploy by the dominant male, or done in play among young animals. Penetration rarely, if ever, occurs. Male baboons sometimes grab the testicles of other males, but this too is believed to be more a social interaction—a way for males to cement friendships or calm each other down—than a sexual one.

Only in one species of primate, the bonobo (a kind of chimpanzee), do males regularly take part in what we might consider real sex with each other: Little bonobo males engage in what appears to be fellatio (oral sex) with male playmates. But whether the act is, in fact, sexual or merely social is unknown, and the behavior disappears as the youngsters mature and concentrate on females. What's more, bonobos as a species are highly sexual primates—more so than humans—and it seems reasonable that young bonobo males experiment in all sorts of ways to prepare them for a life of intense sexuality.

Though virtually unknown among most primates, male homosexual orientation appears consistently across human societies, a fact that is sometimes cited in support of its having a biological basis. "There hasn't been a society known to anthropologists where some kind of emotional/erotic bonding between males doesn't exist," says Dr. William Leap, an associate professor of anthropology at American University in Washington. In primitive and peasant cultures, the role of the homosexual is usually very clearly defined. Most Native American tribes, for example, historically had "berdaches," men who dressed as women and performed women's tasks, including, in some cases, sexual relations with men. These men were often revered as shamans or priests. Among some New Guinea tribes, anal and oral sex between males is part of the initiation of boys into full manhood. In most tribal societies, according to Leap, homosexuality is approved or tolerated, at worst the subject of mild ridicule or teasing. "It would be most peculiar to find committed antigay sentiments in these groups," he says.

Closer to home, sex research pioneer Alfred Kinsey and his colleagues were the first to systematically study the prevalence of homosexual behavior in our own society. In their landmark 1948 book *Sexual Behavior in the Human*

LESBIANS: LESS IS KNOWN

Until quite recently research on the biological basis of homosexuality has focused almost exclusively on male homosexuality. To some degree this fact reflects the lesser attention science has given to women and to women's issues in general. It may also reflect the lower profile of homosexual women in our society.

According to Dr. June Reinisch, director of the Kinsey Institute, 2% to 3% of American women are predominantly homosexual—about half the accepted figure for male homosexuality. This smaller percentage, however, may not reflect the totality of women who are erotically oriented toward members of their own sex. Researchers believe many women with homosexual tendencies choose not to act on them in response to stronger societal pressure to fulfill the traditional roles of wife and mother.

And even more often than in research on male homosexuality, studies of lesbianism are clouded by lack of clear definitions. For example, it is believed that a greater percentage of women who call themselves lesbians have significant heterosexual histories than men who call themselves homosexual or gay. It is also thought that lesbians tend to acknowledge their sexual orientation later in life than homosexual men. There may even be some number of lesbians who have chosen a homosexual lifestyle as a political statement, though in so doing they are likely expressing a facet of their existing sexuality rather than switching orientations.

"At the height of the feminist movement in the 1970s," says Dr. Lillian Faderman, author of *Odd Girls and Twilight Lovers: A History of Lesbian Life in Twentieth Century America*, "many women did indeed come out as lesbians through the movement, which empowered them to express themselves homosexually."

Although the brain anatomy of lesbians has yet to be investigated, even to the limited extent of that of gay men, at least one study has raised the possibility of a biological influence on lesbianism. Dr. Michael Bailey, an assistant professor of psychiatry at Northwestern University, and Dr. Richard Pillard, a professor of psychiatry at Boston University School of Medicine, authors of a recent study on gay twins (see main story), recruited 115 lesbians who had identical twin sisters. As in their study of gay men, the researchers reported that almost half of the identical twins of lesbians were themselves either lesbian or bisexual. For non-identical twins the rate was only about half that, and only one in six adopted sisters of lesbians had a similar sexual orientation. Pillard and Bailey interpret their findings to suggest that lesbianism might be moderately heritable, though other researchers caution that these studies, like those of gay male twins, are not definitive.

What explains the paucity of research on lesbianism? "Women's sexuality in general has seldom been taken seriously by the research community," says Dr. Anne Fausto-Sterling, a professor of medical science in the Brown University Division of Biology and Medicine. "And because lesbianism is not seen as a threat to the social order to the same extent that male homosexuality is, neither is it such a source of interest to researchers and those who fund their work."

2. SEXUAL BIOLOGY, BEHAVIOR, AND ORIENTATION: Sexual Orientation

Male, they reported that 37% of American men had at least one homosexual encounter during their lifetimes. As Kinsey pointed out, "This is more than one male in three of the persons one may meet as he passes along a city street." The sharpest rise in the incidence of homosexual experiences reported to Kinsey was during adolescence. As men got older and married, incidence declined. When it came to a more enduring sexual orientation, Kinsey found that between 8% and 13% of the male population could be rated as predominantly homosexual for at least three years of their lives between the ages of 16 and 55. About 4% of the men Kinsey studied were exclusively homosexual throughout their lives, and today Dr. June Reinisch, director of the Kinsey Institute, estimates that 4% to 8% of American men are predominantly homosexual. Kinsey took care to point out that there is a wide spectrum of sexuality. "Males do not represent two discrete populations, heterosexual and homosexual," he wrote. "The living world is a continuum in each and every one of its aspects. The sooner we learn this concerning human sexual behavior the sooner we shall reach a sound understanding of the realities of sex."

IF HOMOSEXUALITY does have a biological basis, is it genetically determined? Can a man actually inherit homosexual tendencies? The classic scientific approach to such a question is to compare the behavior of identical twins, who develop from a single egg fertilized by a single sperm and thus share the same genetic instruction set. Such studies are designed to determine how much of a behavior is heritable, or can be attributed to genetics, and how much is caused by the environment.

Dr. Michael Bailey, an assistant professor of psychiatry at Northwestern University, and Dr. Richard Pillard, a professor of psychiatry at Boston University School of Medicine, have used twin data to look at the possible role of genetics in homosexuality. They recruited male homosexuals who had twin brothers and then classified the twin pairs as either identical or fraternal (developed from two eggs, fertilized by different sperm, and thus no more alike than any pair of siblings). Of the 56 gay men with an identical twin, 52% had twin brothers who were also homosexual. Of the gay men with fraternal brothers, only 22% of the brothers were also homosexual. The fact that identical twins are about twice as likely as fraternal twins to both be homosexual suggests that genes play an important role in determining sexual orientation.

"Our study suggests that homosexuality is moderately heritable," concludes Bailey, "but I wouldn't trust any specific figure. It doesn't follow a clear inheritance pattern, but then people haven't been able to follow homosexuality through generations as they have with other possible genetically related conditions."

DR. WILLIAM BYNE, a resident in psychiatry at Columbia-Presbyterian Medical Center, points out that the fact that almost half of the homosexual identical twins in Bailey and Pillard's study had heterosexual twin brothers suggests that factors other than genes must also guide sexual orientation. Those other influences could be biological, social or both. "We don't know yet," concedes Bailey, "but I suspect it is some sort of subtle biological factor, such as hormonal differences that affect one twin but not the other."

In other words, even if sexual orientation isn't determined solely by genes, it might be a result of a biological event that occurs early in life. Which brings us back to the brain—and to animals. Though studies of animal behavior have shed little light on the behavior of human homosexuals, research on animal brain structure and development has been a beacon for new work on sexual orientation and the human brain. In several mammalian species, for example, there are consistent, well-documented differences between the size and structure of male and female brains. The portion of the brain Simon LeVay reported to be smaller in gay men, known as the third interstitial nucleus of the anterior hypothalamus (INAH 3), is closer in size to the corresponding area in female brains.

And when Dr. Dick Swaab, director of the Netherlands Institute for Brain Research at Amsterdam University, investigated a section of the hypothalamus that helps regulate the daily rhythms of the body, he found that this area was larger and contained more cells in the brains of homosexual males than in either females or other males who were presumably heterosexual. Unlike LeVay's work, Swaab's studies suggest that while the brains of homosexual men may indeed differ structurally from those of heterosexual men, homosexuals' brains are not uniformly "feminized."

Most recently, Dr. Laura Allen, a postdoctoral scholar, and Dr. Roger Gorski, a professor of anatomy and cell biology, both at UCLA's School of Medicine, reported that an area called the anterior commissure, which connects the two sides of the brain, is generally larger in females than in males—and can be even larger in homosexual males.

Like the twin studies, this brain research has been challenged on several grounds. Three of the homosexual men in LeVay's study, for instance, had INAH 3 areas as large as those of heterosexual men, as did two of the presumably heterosexual women. Another problem was that all the homosexual men and some of the heterosexual men in the study had died of AIDS, and no one really knows the effects of AIDS and its complications on the size and shape of the dying brain. Nor has anyone demonstrated a relationship between INAH 3 and sexual behavior in humans. What's more, only the male AIDS patients in LeVay's study—and not the presumed heterosexuals who had died of other causes—had been asked their sexual orientation before they died. And in the research on sex differences in the anterior commissure, the only other study contradicted Allen and Gorski's findings.

HOW DOES A BRAIN get to be masculine or feminine in the first place? In male animals and humans, there are critical periods during gestation and early postnatal development when the brain becomes awash in male hormones that independently masculinize and defeminize it. Scientists hypothesize that if something out of the ordinary takes place during these critical periods, the brain of a male could be incompletely masculinized or defeminized. Some researchers have suggested that maternal stress during pregnancy might alter the fetal environment and result in a male whose brain is essentially female and thus "wired" for female sexual behavior.

Evidence to support this hypothesis is scant, however. One intriguing clue: When male rats are castrated soon after birth, eliminating the male hormone testosterone at a critical period in the masculinizing process, as adults the animals regularly display the female mating position, called lordosis. But oddly enough, these same rats continue to act like males as well—they also mount other rats in the typical male manner.

The mating behavior of rats, however, is under strict hormonal control, whereas human sexuality can't be manipulated quite so easily. And while some of these studies may support the notion that the brains of homosexuals are organized differently from those of heterosexuals, they still fall short of presenting a coherent picture of what constitutes a homosexual brain. As Dr. John Bancroft, a psychiatrist in the medical research unit of the Royal Edinburgh Hospital in Scotland, puts it: "To say a small part of the hypothalamus is involved in the sexual drive is only a slightly more refined statement than saying the brain is involved in the sex drive."

WHERE THEN DOES all of this leave the environmental influences—such as family structure—on homosexuality? According to William Byne, who has just begun a study designed to replicate LeVay's brain anatomy research, the best predictor of future homosexual orientation at present is something researchers call "gender nonconformity" in childhood. Byne says that studies show that about 85% of adult male homosexuals in this country report some degree of childhood gender nonconformity. The fact that it occurs early in development leads some experts to suggest that it might be innate. Byne points out, however, that not all homosexuals recall gender nonconformity as children, and up to 55% of heterosexual men also recall some degree of it. He also notes that early occurrence does not necessarily suggest it is innate, since boys and girls are treated differently from the moment of birth.

Byne doubts that homosexuality itself is inborn; rather, he suggests that genes and hormones are more likely to influence certain personality traits than sexual orientation. These traits, he speculates, might then predispose individuals to homosexual development in certain environments. For example, boys who by temperament do not enjoy rough-and-tumble activities such as sports may disappoint parents who expect their children to adhere to rigid sex roles. In such families—particularly if their sons' aversion to these activities interferes with their relationships with other boys—these children may develop low masculine self-regard. If their first close relationships with other boys happen during adolescence, when their bodies are awakening sexually, their feelings of affiliation may become eroticized.

"Certainly this doesn't mean that every kid who doesn't like baseball is going to turn out homosexual," says Byne. "If his father can say,'Okay, let's find something else we can enjoy together,' the boy will probably emerge feeling fine about his masculinity."

Dr. John Money, director of the psychohormonal research unit of the Johns Hopkins University School of Medicine, also sees homosexuality as developing from an interplay between biology and environment: "I use a language analogy," he says. "You don't choose your native language, even though you are born without it. You assimilate it into a brain prenatally prepared to receive a native language. Once assimilated, a native language becomes locked in—just as if it were preprogrammed genetically by hormonal influences or brain chemistry. Sexual status is essentially the same."

If scientists ever do decide that biology plays a major role in why a man prefers other men sexually, what then? Will those who now see homosexuality as a choice, and who condemn those who choose it, become more accepting? Some may. But as Money notes, "skin color is biologically determined, and that hasn't deterred racial oppression."

Interpersonal Relationships

- **Establishing Sexual Relationships (Articles 15–16)**
- **Responsible Quality Relationships (Articles 17–19)**

Most people are familiar with the term "sexual relationship." It denotes an important dimension of sexuality: interpersonal sexuality, or sexual interactions occurring between two (and sometimes more) individuals. This unit focuses attention on these types of relationships.

No woman is an island. No man is an island. Interpersonal contact forms the basis for self-esteem and meaningful living; conversely, isolation results in loneliness and depression for most human beings. People seek and cultivate friendships for the warmth, affection, supportiveness, and sense of trust and loyalty that such relationships can provide.

Long-term friendships may develop into intimate relationships. The qualifying word in the previous sentence is "may." Today many people, single as well as married, yearn for close or intimate interpersonal relationships but

Unit 3

fail to find them. Discovering how and where to find potential friends, partners, lovers, and soul mates is reported to be more difficult today than in times past. Fear of rejection causes some to avoid interpersonal relationships, others to present a false front or illusory self that they think is more acceptable or socially desirable. This sets the stage for a game of intimacy that is counterproductive to genuine intimacy. For others a major dilemma may exist—the problem of balancing closeness with the preservation of individual identity in a manner that at once satisfies the need for personal and interpersonal growth and integrity. In either case, partners in a relationship should be advised that the development of interpersonal awareness (the mutual recognition and knowledge of others as they really are) rests upon trust and self-disclosure—letting the other person know who you really are and how you truly feel. In American society this has never been easy, and it is especially difficult in the area of sexuality.

The above considerations in regard to interpersonal relationships apply equally well to achieving meaningful and satisfying sexual relationships. Three basic ingredients lay the foundation for quality sexual interaction. These are self-awareness, acceptance of the partner's needs, and desires. Without these, misunderstandings may arise, bringing anxiety, frustration, dissatisfaction, and/or resentment into the relationship, as well as a heightened risk of contracting AIDS or another STD, experiencing an unplanned pregnancy, or sexual dysfunction by one or both partners. These basic ingredients, taken together, contribute to sexual responsibility. Clearly, no one person is completely responsible for quality sexual relations.

As might already be apparent, there is much more to quality sexual relationships than our popular culture recognizes. Such relationships are not established on penis size, beautiful figures, or correct sexual techniques. Rather, it is the quality of the interaction that makes sex a celebration of our sexuality. A person-oriented (as opposed to genitally oriented) sexual awareness coupled with a leisurely, whole-body/mind sensuality and an open attitude toward exploration make for quality sexuality.

The subsection *Establishing Sexual Relationships* opens with "The Mating Game," an article suggesting that characteristics and behaviors from attraction to the interaction involved when two people decide they are interested sexually in one another have changed little since the Stone Age. The second article, "Ways of Loving," proposes a three-style model of loving and couple interaction and allows readers to assess their "attachment" feelings, fears, and behaviors, including their cuddling, romantic, trusting, and tuned-into-your partner levels.

In the subsection *Responsible Quality Relationships*, the examination of sexual and intimate relationships continues through a blending of research, experience, and advice-giving articles. The first two, "What Is Love?" and "The Lessons of Love," focus on what has only recently become a subject of scientific research but has been a topic of human pondering individually and through literature, music, and drama throughout the ages: love. "Forecast for Couples" looks at intimate relationships through a systems approach involving "recurring cycles of promise and betrayal." Each allows readers to explore their own expectations and feelings about the kind of relationships they experience and seek, while having a backdrop of others' experiences with which to compare their own.

Looking Ahead: Challenge Questions

What do you see as the greatest barriers to satisfying intimate relationships? Are some people destined to fail at establishing and/or maintaining them? If so, who and why?

How do you (or how have you heard others) assess a potential sexual/intimate partner? Is there a kind of system or flow chart? How does the pursuit or mating game unfold?

How important are hugging and cuddling to you? Are they important in and of themselves, and not just a prelude to lovemaking? Why?

List 5 to 10 answers to the question "What is love?" Are these things you've heard from others? Believe yourself? Experienced yourself?

What are some "lessons about love" you could try (because we often do not listen) to pass on to others that you believe would help them find more satisfying intimate relationships?

Do you feel that disappointment and conflict are part of the couple development process? If so, how can they be handled and resolved so as to increase the bond in the relationship?

THE Mating GAME

The sophisticated sexual strategies of modern men and women are shaped by a powerful Stone Age psychology.

It's a dance as old as the human race. At cocktail lounges and church socials, during office coffee breaks and dinner parties—and most blatantly, perhaps, in the personal ads in newspapers and magazines—men and women perform the elaborate ritual of advertisement and assessment that precedes an essential part of nearly every life: mating. More than 90 percent of the world's people marry at some point in their lives, and it is estimated that a similarly large number of people engage in affairs, liaisons, flings or one-night stands. The who, what, when and where of love, sex and romance are a cultural obsession that is reflected in everything from Shakespeare to soap operas and from Tristram and Isolde to 2 Live Crew, fueling archetypes like the coy ingénue, the rakish cad, the trophy bride, Mrs. Robinson, Casanova and lovers both star-crossed and blessed.

It all may seem very modern, but a new group of researchers argues that love, American style, is in fact part of a universal human behavior with roots stretching back to the dawn of humankind. These scientists contend that, in stark contrast to the old image of brute cavemen dragging their mates by the hair to their dens, our ancient ancestors—men and women alike—engaged in a sophisticated mating dance of sexual intrigue, shrewd strategizing and savvy negotiating that has left its stamp on human psychology. People may live in a thoroughly modern world, these researchers say, but within the human skull is a Stone Age mind that was shaped by the mating concerns of our ancient ancestors and continues to have a profound influence on behavior today. Indeed, this ancient psychological legacy influences everything from sexual attraction to infidelity and jealousy—and, as remarkable new research reveals, even extends its reach all the way down to the microscopic level of egg and sperm.

These new researchers call themselves evolutionary psychologists. In a host of recent scientific papers and at a major conference last month at the London School of Economics, they are arguing that the key to understanding modern sexual behavior lies not solely in culture, as some anthropologists contend, nor purely in the genes, as some sociobiologists believe. Rather, they argue, understanding human nature is possible only if scientists begin to understand the evolution of the human mind. Just as humans have evolved

---How we choose---

Women are more concerned about whether mates will invest time and resources in a relationship; men care more about a woman's physical attractiveness, which in ancient times reflected her fertility and health.

specialized biological organs to deal with the intricacies of sex, they say, the mind, too, has evolved customized mental mechanisms for coping with this most fundamental aspect of human existence.

Gender and mind. When it comes to sexuality and mating, evolutionary psychologists say, men and women often are as different psychologically as they are physically. Scientists have long known that people typically choose mates who closely resemble themselves in terms of weight, height, intelligence and even earlobe length. But a survey of more than 10,000 people in 37 cultures on six continents, conducted by University of Michigan psychologist David Buss, reveals that men consistently value physical attractiveness and youth in a mate more than women do; women, equally as consistently, are more concerned than men with a prospective mate's ambition, status and resources. If such preferences were merely arbitrary products of culture, says Buss, one might expect to find at least one society somewhere where men's and women's mating preferences were reversed; the fact that attitudes are uniform across cultures suggests they are a fundamental part of human psychology.

Evolutionary psychologists think many of these mating preferences evolved in response to the different biological challenges faced by men and women in producing children—the definition of success in evolutionary terms. In a seminal paper, evolutionary biologist Robert Trivers of the University of California at Santa Cruz points out that in most mammals, females invest far

15. Mating Game

more time and energy in reproduction and child rearing than do males. Not only must females go through a long gestation and weaning of their offspring, but childbirth itself is relatively dangerous. Males, on the other hand, potentially can get away with a very small biological investment in a child.

Human infants require the greatest amount of care and nurturing of any animal on Earth, and so over the eons women have evolved a psychology that is particularly concerned with a father's ability to help out with this enormous task—with his clout, protection and access to resources. So powerful is this psychological legacy that nowadays women size up a man's finances even when, as a practical matter, they may not have to. A recent study of the mating preferences of a group of medical students, for instance, found that these women, though anticipating financial success, were nevertheless most interested in men whose earning capacity was equal to or greater than their own.

Healthy genes. For men, on the other hand, reproductive success is ultimately dependent on the fertility of their mates. Thus males have evolved a mind-set that homes in on signs of a woman's health and youth, signs that, in the absence of medical records and birth certificates long ago, were primarily visual. Modern man's sense of feminine beauty—clear skin, bright eyes and youthful appearance—is, in effect, the legacy of eons spent diagnosing the health and fertility of potential mates.

This concern with women's reproductive health also helps explain why men value curvaceous figures. An upcoming paper by Devendra Singh of the University of Texas at Austin reveals that people consistently judge a woman's figure not by whether she is slim or fat but by the ratio of waist to hips. The ideal proportion—the hips roughly a third larger than the waist—reflects a hormonal balance that results in women's preferentially storing fat on their hips as opposed to their waists, a condition that correlates with higher fertility and resistance to disease. Western society's modern-day obsession with being slim has not changed this equation. Singh found, for instance, that while the winning Miss America has become 30 percent thinner over the past several decades, her waist-to-hip ratio has remained close to this ancient ideal.

Women also appreciate a fair face and figure, of course. And what they look for in a male's physique can also be explained as an evolved mentality that links good looks with good genes. A number of studies have shown that both men and women rate as most attractive

WHOM WE MARRY

More than 90 percent of all people marry and, they typically choose mates who closely resemble themselves, from weight and height, to intelligence and values, to nose breadth and even earlobe length.

faces that are near the average; this is true in societies as diverse as those of Brazil, Russia and several hunting and gathering tribes. The average face tends to be more symmetrical, and, according to psychologist Steven Gangestad and biologist Randy Thornhill, both of the University of New Mexico, this symmetry may reflect a person's genetic resistance to disease.

People have two versions of each of their genes—one from each parent—within every cell. Sometimes the copies are slightly different, though typically each version works just as effectively. The advantage to having two slightly different copies of the same gene, the researchers argue, is that it is harder for a disease to knock out the function of both copies, and this biological redundancy is reflected in the symmetry in people's bodies, including their faces. Further evidence for a psychological mechanism that links attractiveness with health comes from Buss's worldwide study of mating preferences: In those parts of the world where the incidence of parasites and infectious diseases is highest, both men and women place a greater value on attractive mates.

Some feminists reject the notion that women should alter physical appearance to gain advantage in the mating game. But archaeological finds suggest that the "beauty myth" has been very much a part of the human mating psychology since the times of our ancient ancestors—and that it applies equally to men. Some of the very first signs of human artistry are carved body ornaments that date back more than 30,000 years, and findings of worn nubs of ochre suggest that ancient humans may have used the red and black chalklike substance as makeup. These artifacts probably served as social signs that, like lipstick or a Rolex watch today, advertised a person's physical appearance and status. In one grave dating back some 20,000 years, a male skeleton was found bedecked with a tunic made from thousands of tiny ivory beads—the Stone Age equivalent of an Armani suit.

Far from being immutable, biological mandates, these evolved mating mechanisms in the mind are flexible, culturally influenced aspects of human psychology that are similar to people's tastes for certain kinds of food. The human sweet tooth is a legacy from a time when the only sweet things in the environment were nutritious ripe fruit and honey, says Buss, whose book "The Evolution of Desire" is due out next year. Today, this ancient taste for sweets is susceptible to modern-day temptation by candy bars and such, though people have the free will to refrain from indulging it. Likewise, the mind's mating mechanisms can be strongly swayed by cultural influences such as religious and moral beliefs.

Playing the field. Both men and women display different mating psychologies when they are just playing around as opposed to searching for a lifelong partner, and these mental mechanisms are also a legacy from ancient times. A new survey by Buss and his colleague David Schmitt found that when women are looking for "short term" mates, their preference for attractive men increases substantially. In a study released last month, Doug Kenrick and Gary Groth of Arizona State University found that while men, too, desire attractive mates when they're playing the field, they will actually settle for a lot less.

Men's diminished concern about beauty in short-term mates reflects the fact that throughout human evolution, men have often pursued a dual mating strategy. The most successful strategy for most men was to find a healthy, fertile, long-term mate. But it also didn't hurt to take advantage of any low-risk opportunity to sire as many kin as possible outside the relationship, just to hedge the evolutionary bet. The result is an evolved psychology that allows a man to be sexually excited by a wide variety of women even while committed to a partner. This predilection shows up in studies of men's and women's sexual fantasies today. A study by Don Symons of the University of California at Santa Barbara and Bruce Ellis of the University of Michigan found that while both men and women actively engage in sexual fantasy, men typically have more fantasies about anonymous partners.

Surveys in the United States show

3. INTERPERSONAL RELATIONSHIPS: Establishing Sexual Relationships

that at least 30 percent of married women have extramarital affairs, suggesting that, like men, women also harbor a drive for short-term mating. But they have different evolutionary reasons for doing so. Throughout human existence, short-term flings have offered women an opportunity to exchange sex for resources. In Buss and Schmitt's study, women value an "extravagant lifestyle" three times more highly when they are searching for a brief affair than when they are seeking a long-term mate. Women who are secure in a relationship with a committed male might still seek out attractive men to secure healthier genes for their offspring. Outside affairs also allow women to shop for better partners.

Sperm warfare. A woman may engage the sexual interest of several men simultaneously in order to foster a microscopic battle known as sperm competition. Sperm can survive in a woman's reproductive tract for nearly a week, note biologists Robin Baker and Mark Bellis of the University of Manchester, and by mating with more than one man within a short period of time, a woman sets the stage for their sperm to compete to sire a child—passing this winning trait on to her male offspring as well. In a confidential survey tracking the sexual behavior and menstrual cycles of more than 2,000 women who said they had steady mates, Baker and Bellis found that while there was no pattern to when women had sex with their steady partners, having sex on the side peaked at the height of the women's monthly fertility cycles.

Since in ancient times a man paid a dear evolutionary price for being cuckolded, the male psychology produces a physiological counterstrategy for dealing with a woman's infidelity. Studying the sexual behavior of a group of couples, Baker and Bellis found that the more time a couple spend apart, the more sperm the man ejaculates upon their sexual reunion—as much as three times higher than average.

This increase in sperm count is unrelated to when the man last ejaculated through nocturnal emission or masturbation, and Baker and Bellis argue that it is a result of a man's evolved psychological mechanism that bolsters his chances in sperm competition in the event that his mate has been unfaithful during their separation. As was no doubt the case in the times of our ancient ancestors, these concerns are not unfounded: Studies of blood typings show that as many as 1 of every 10 babies born to couples in North America is not the offspring of the mother's husband.

Despite men's efforts at sexual subterfuge, women still have the last word on the fate of a man's sperm in her reproductive tract—thanks to the physiological effects of the female orgasm. In a new study, Baker and Bellis reveal that if a woman experiences an orgasm soon after her mate's, the amount of sperm retained in her reproductive tract is far higher than if she has an earlier orgasm or none at all. Apparently a woman's arousal, fueled by her feelings as well as her mate's solicitous attentions, results in an evolutionary payoff for both.

Cads and dads. Whether people pursue committed relationships or one-night stands depends on their perceptions of what kind of mates are in the surrounding sexual environment. Anthropologist Elizabeth Cashdan of the University of Utah surveyed hundreds of men and women on whether they thought the members of their "pool" of potential mates were in general trustworthy, honest and capable of commitment. She also asked them what kinds of tactics they used to attract mates. Cashdan found that the less committed people thought their potential mates would be, the more they themselves pursued short-term mating tactics. For example, if women considered their world to be full of "cads," they tended to dress more provocatively and to be more promiscuous; if they thought that the world was populated by potential "dads"—that is, committed and nurturing men—they tended to emphasize their chastity and fidelity. Similarly, "cads" tended to emphasize their sexuality and "dads" said they relied more on advertising their resources and desire for long-term commitment.

These perceptions of what to expect from the opposite sex may be influenced by the kind of home life an individual knew as a child. Social scientists have long known that children from homes where the father is chronically absent or abusive tend to mature faster physically and to have sexual relations earlier in life. Psychologist Jay Belsky of Pennsylvania State University argues that this behavior is an evolved psychological mechanism, triggered by early childhood experiences, that enables a child to come of age earlier and leave the distressing situation. This psychological mechanism may also lead to a mating strategy that focuses on short-term affairs.

The green monster. Whether in modern or ancient times, infidelities can breed anger and hurt, and new research suggests subtle differences in male and female jealousy with roots in the ancient past. In one study, for example, Buss asked males and females to imagine that their mates were having sex with someone else or that their mates were engaged in a deep emotional commitment with another person. Monitoring his subjects' heart rates, frowning and stress responses, he found that the stereotypical double standard cuts both ways. Men reacted far more strongly than

---

JEALOUS PSYCHE

Men are most disturbed by sexual infidelity in their mates, a result of uncertainty about paternity. Women are more disturbed by emotional infidelity, because they risk losing their mate's time and resources.

BEAUTY QUEST

The most attractive men and women are in fact those whose faces are most average, a signal that they are near the genetic average of the population and are perhaps more resistant to disease.

15. Mating Game

EVOLVED FANTASIES
Eroticism and gender

For insights into the subtle differences between men's and women's mating psychologies, one need look no further than the local bookstore. On one rack may be magazines featuring scantily clad women in poses of sexual invitation—a testimony to the ancient legacy of a male psychology that is acutely attuned to visual stimulus and easily aroused by the prospect of anonymous sex. Around the corner is likely to be a staple of women's erotic fantasy: romance novels.

Harlequin Enterprises Ltd., the leading publisher in the field, sells more than 200 million books annually and produces about 70 titles a month. Dedicated romance fans may read several books a week. "Our books give women everything," says Harlequin's Kathleen Abels, "a loving relationship, commitment and having sex with someone they care about." Some romance novels contain scenes steamy enough to make a sailor blush, and studies show that women who read romances typically have more sexual fantasies and engage in sexual intercourse more frequently than nonreaders do.

Sexual caricature. Since sexual fantasy frees people of the complications of love and mating in the real world, argue psychologists Bruce Ellis and Don Symons, it is perhaps not surprising that in erotic materials for both men and women, sexual partners are typically caricatures of the consumer's own evolved mating psychology. In male-oriented erotica, for instance, women are depicted as being lust driven, ever willing and unencumbered by the need for emotional attachment. In romance novels, the male lead in the book is typically tender, emotional and consumed by passion for the heroine, thus ensuring his lifelong fidelity and dependence. In other words, say Ellis and Symons, the romance novel is "an erotic, utopian, female counterfantasy" to male erotica.

Of course, most men also enjoy stories of passion and romance, and women can be as easily aroused as men by sexually explicit films. Indeed, several new entertainment ventures, including the magazine *Future Sex* and a video company, Femme Productions, are creating erotic materials using realistic models in more sensual settings in an attempt to appeal to both sexes. Still, the new research into evolutionary psychology suggests that men and women derive subtly different pleasures from sexual fantasy—something that even writing under a ghost name can't hide. According to Abels, a Harlequin romance is occasionally penned by a man using a female pseudonym, but "our avid readers can always tell."

---

women to the idea that their mates were having sex with other men. But women reacted far more strongly to the thought that their mates were developing strong emotional attachments to someone else.

As with our evolved mating preferences, these triggers for jealousy ultimately stem from men's and women's biology, says Buss. A woman, of course, has no doubt that she is the mother of her children. For a man, however, paternity is never more than conjecture, and so men have evolved psychologies with a heightened concern about a mate's sexual infidelity. Since women make the greater biological investment in offspring, their psychologies are more concerned about a mate's reneging on his commitment, and, therefore, they are more attentive to signs that their mates might be attaching themselves emotionally to other women.

Sexual monopoly. The male preoccupation with monopolizing a woman's sexual reproduction has led to the oppression and abuse of women worldwide, including, at its extremes, confinement, domestic violence and ritual mutilation such as clitoridectomy. Yet the new research into the mating game also reveals that throughout human evolution, women have not passively acquiesced to men's sexual wishes. Rath-

---DUELING SPERM---

If a couple has been apart for some time, the man's sperm count goes up during sex at their reunion—an ancient, evolved strategy against a female's possible infidelities while away.

---

er, they have long employed a host of behavioral and biological tactics to follow their own sexual agenda—behaviors that have a huge impact on men's behavior as well. As Buss points out, if all women suddenly began preferring to have sex with men who walked on their hands, in a very short time half the human race would be upside down.

With its emphasis on how both men and women are active players in the mating game, evolutionary psychology holds out the promise of helping negotiate a truce of sorts in the battle of the sexes—not by declaring a winner but by pointing out that the essence of the mating game is compromise, not victory. The exhortations of radical feminists, dyed-in-the-wool chauvinists and everyone in between are all spices for a sexual stew that has been on a slow boil for millions of years. It is no accident that consistently, the top two mating preferences in Buss's survey—expressed equally by males and females worldwide—were not great looks, fame, youth, wealth or status, but *kindness* and *intelligence*. In the rough-and-tumble of the human mating game, they are love's greatest allies.

WILLIAM F. ALLMAN

Ways of Loving

A Matter of Style

Diane Swanbrow

Diane Swanbrow used to be anxious and is now avoidant.

For most of his 50 years, Alan has lived by what might be called the Napoleonic code: "In love, victory goes to the man who runs away." Alan escapes by watching, reading about or playing sports. Partly as a result of therapy suggested by his second wife, he has begun to see his intense interest in sports as an emotional shield. "It's lucky for me that she's so patient," he says.

Detachment certainly isn't Jane's style of romance. Her theme song might be "He Don't Love Me Like I Love Him." When a boyfriend called to cancel a date because he had a last-minute project to finish, Jane stormed over to his office to confront him. "We'd been seeing each other for over a year, but I was never sure where I stood with him," she says. When he lost his temper, she plunged into despair.

Stephanie's relationships are different. "My husband and I are willing to talk about anything and everything," she says. "We have very different interests, but similar values and a great deal of mutual respect. When we get upset with each other, we know we're gonna come out on the other side. We just feel *safe*."

Recent psychological studies reveal that like Alan, Jane and Stephanie, each of us has an "attachment style," a set of expectations about intimacy that was formed in our first years. This template holds important clues about whom and how we'll love. The new research also confirms the age-old observation that when we're in love, a lot of us behave like babies, reverting to a pattern of interaction that we established years before with our opposite-sex parent.

Knowing your attachment style won't help you find Ms. Right or Mr. Right, but it may shed light on the psychological framework that affects your relationships—and may even suggest approaches for gaining a new emotional equilibrium.

Like Alan, about 30% of us are "emotionally avoidant," according to research done by Cindy Hazan, a Cornell University psychologist, in collaboration with Phillip R. Shaver, a psychologist at the State University of New York at Buffalo. Their studies show that avoiders prefer to do almost anything with their partners rather than talk about their feelings. Even when a relationship ends, they often show few outward signs of distress.

Avoiders are of both sexes. "Many men care about relationships just as much as women, and some women are as likely as men to act emotionally detached," says psychologist Patricia Noller of the University of Queensland in Australia. Her studies of attachment behavior in young people and newlyweds show that avoidant types *are* likelier to get divorced and their relationships may not last as long. "But," says Hazan, "they're no less likely than other people to be dating or married."

For the fifth of the population who, like Jane, have an anxious attachment style, love is a variation on "I Got It Bad (and That Ain't Good)." Being in love makes anxious types even more jittery. They feel that if they could only merge completely with the beloved, they'd finally relax. But because that goal is inevitably frustrated, they have jealous tantrums, demand attention at inappropriate times and, ultimately, feel unloved.

That there's a link between our experiences as babies and the way we act as adults in love is age-old news. But the modern scientific study of the phenomenon only began in the 1950s, with the work of British psychoanalyst John Bowlby, who believed that attachment style is powerfully shaped by a mother's responses to her baby's needs. A consistently warm, attentive mother helps a baby feel secure. An ambivalent mother who treats her infant inconsistently contributes to an anxious style, and a cold, rejecting mother encourages avoidance.

A study of 43 newly married couples done by Linda Cooke at Claremont Graduate School and Lynn Miller, a University of Southern California researcher, supports the homespun advice given to generations of young women: If you want a preview of what kind of husband a man will make, look at his relationship with his mother. "If the man in your life had a cold, rejecting mother, or one who ran hot and cold on him, watch out," says Miller. "The influence of a woman's relationship with her father isn't quite as strong, although it's still significant. In general, those who had good, secure relationships with both parents are likeliest to be satisfied with their marriages."

Some attachment styles are more compatible than others. Keith Davis, a psychol-

16. Ways of Loving

> ## What's Your Style?
>
> **Which of these three patterns comes closest to describing the way you or your romantic partner usually feels in close relationships?**
>
> **1.** I am comfortable with closeness and find it relatively easy to trust and depend on others. I don't worry about being hurt by those I'm close to. (*secure*)
>
> **2.** I am comfortable without a lot of closeness. It's important to me to be independent and self-reliant. I'd rather not depend on others or have others depend on me. (*avoidant*)
>
> **3.** I want closeness, but I find that others are reluctant to get as close as I'd like. I worry that others won't care about me as much as I care for them. (*anxious*)
>
> *Adapted from Hazan and Shaver, "Romantic Love Conceptualized as an Attachment Process,"* Journal of Personality and Social Psychology, *1987.*

ogist at the University of South Carolina in Columbia, and Lee Kirkpatrick, a University of Massachusetts psychologist, have found that secure people seem to marry each other. No surprise there. But after studying more than 200 couples, Davis identified another pattern: Anxious types pick avoidant sorts, and vice versa.

Joanna, a homemaker and mother of three, is married to Jack, whose two jobs keep him out from morning until midnight. "I wait and wait for him to come home, just like I remember waiting for my dad to come home on my eighth birthday," she says. "I know Jack loves the kids, and I think he loves me, but sometimes I wonder." Curiously, Davis has found that a match between an avoidant man and an anxious woman is not only one of the commonest, but also among the stablest.

Rest assured, attachment styles can be changed. Of the hundreds of people Hazan has surveyed, a surprising 20% have altered their approach to relationships. Secure types, of course, are the least likely to switch. As Hazan says, "An early sense of security seems to inoculate people against trouble in the wake of one or two romantic disasters."

That important sense of security sometimes comes from outside the family. "One of the most striking findings is that anxious or avoidant people who say that adults other than parents were kind to them during childhood are the likeliest to become secure as grown-ups," Hazan says. "A friend's parent, a kindly teacher or a neighbor somehow provided them with a positive model of how love works. It doesn't matter who the person was—just that someone gave a different message from the family's."

Changing your attachment style, though possible, isn't easy or painless. "Attachment is partly a set of expectations built on experience," says Hazan. "Those expectations, especially if based on early events, are resistant to change."

Facing the past is an important step toward change, according to psychologist Carl Hindy, coauthor of *If This Is Love, Why Do I Feel So Insecure?* (Fawcett Crest, 1990, $4.95). "Understanding the childhood origins of an insecure love pattern is really important," he says. "Being aware of the deep roots of your feelings means you're less inclined to attribute relationship problems solely to your partner.

If the problem *is* your partner, though, there are a few things to try before seeing a marriage counselor—or lawyer. "For example," Hindy says, "if your mate insists on working late most nights, develop your own social life instead of pining at home. By acting powerless and passive, you're sending the message that change isn't possible. And sometimes you can entice a partner to change by pointing out that no matter how differently you show it, your ultimate desires are the same—to have a loving relationship that helps both of you feel secure."

What Is LOVE?

After centuries of ignoring the subject as too vague and mushy, science has undergone a change of heart about the tender passion

By PAUL GRAY

What is this thing called love? What? Is this thing called love? What is this thing called? Love.

However punctuated, Cole Porter's simple question begs an answer. Love's symptoms are familiar enough: a drifting mooniness in thought and behavior, the mad conceit that the entire universe has rolled itself up into the person of the beloved, a conviction that no one on earth has ever felt so torrentially about a fellow creature before. Love is ecstasy and torment, freedom and slavery. Poets and songwriters would be in a fine mess without it. Plus, it makes the world go round.

Until recently, scientists wanted no part of it.

The reason for this avoidance, this reluctance to study what is probably life's most intense emotion, is not difficult to track down. Love is mushy; science is hard. Anger and fear, feelings that have been considerably researched in the field and the lab, can be quantified through measurements: pulse and breathing rates, muscle contractions, a whole spider web of involuntary responses. Love does not register as definitively on the instruments; it leaves a blurred fingerprint that could be mistaken for anything from indigestion to a manic attack. Anger and fear have direct roles—fighting or running—in the survival of the species. Since it is possible (a cynic would say commonplace) for humans to mate and reproduce without love, all the attendant sighing and swooning and sonnet writing have struck many pragmatic investigators as beside the evolutionary point.

So biologists and anthropologists assumed that it would be fruitless, even frivolous, to study love's evolutionary origins, the way it was encoded in our genes or imprinted in our brains. Serious scientists simply assumed that love—and especially Romantic Love—was really all in the head, put there five or six centuries ago when civilized societies first found enough spare time to indulge in flowery prose. The task of writing the book of love was ceded to playwrights, poets and pulp novelists.

But during the past decade, scientists across a broad range of disciplines have had a change of heart about love. The amount of research expended on the tender passion has never been more intense. Explanations for this rise in interest vary. Some cite the spreading threat of AIDS; with casual sex carrying mortal risks, it seems important to know more about a force that binds couples faithfully together. Others point to the growing number of women scientists and suggest that they may be more willing than their male colleagues to take love seriously. Says Elaine Hatfield, the author of *Love, Sex, and Intimacy: Their Psychology, Biology, and History:* "When I was back at Stanford in the 1960s, they said studying love and human relationships was a quick way to ruin my career. Why not go where

MEXICO
Shawls in the Marketplace
In the highlands of Chiapas, weaving skills are treasured, and a colorful, well-made shawl advises potential husbands of its wearer's dexterity.

CHINA
Courtship on Horseback
On the plains of Xinjiang, mounted Kazakh suitors play Catch the Maiden. He chases her in pursuit of a kiss. If he succeeds, she goes after him with a riding crop.

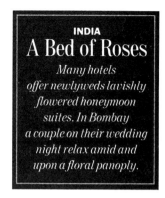

INDIA
A Bed of Roses
Many hotels offer newlyweds lavishly flowered honeymoon suites. In Bombay a couple on their wedding night relax amid and upon a floral panoply.

the real work was being done: on how fast rats could run?" Whatever the reasons, science seems to have come around to a view that nearly everyone else has always taken for granted: romance is real. It is not merely a conceit; it is bred into our biology.

Getting to this point logically is harder than it sounds. The love-as-cultural-delusion argument has long seemed unassailable. What actually accounts for the emotion, according to this scenario, is that people long ago made the mistake of taking fanciful literary tropes seriously. Ovid's *Ars Amatoria* is often cited as a major source of misreadings, its instructions followed, its ironies ignored. Other prime suspects include the 12th century troubadours in Provence who more or less invented the Art of Courtly Love, an elaborate, etiolated ritual for idle noblewomen and aspiring swains that would have been broken to bits by any hint of physical consummation.

Ever since then, the injunction to love and to be loved has hummed nonstop through popular culture; it is a dominant theme in music, films, novels, magazines and nearly everything shown on TV. Love is a formidable and thoroughly proved commercial engine; people will buy and do almost anything that promises them a chance at the bliss of romance.

But does all this mean that love is merely a phony emotion that we picked up because our culture celebrates it? Psychologist Lawrence Casler, author of *Is Marriage Necessary?*, forcefully thinks so, at least at first: "I don't believe love is part of human nature, not for a minute. There are social pressures at work." Then falls a shadow over this certainty. "Even if it is a part of human nature, like crime or violence, it's not necessarily desirable."

Well, love either is or is not intrinsic to our species; having it both ways leads nowhere. And the contention that romance is an entirely acquired trait—overly imaginative troubadours' revenge on muddled literalists—has always rested on some teetery premises.

For one thing, there is the chicken/egg dilemma. Which came first, sex or love? If the reproductive imperative was as dominant as Darwinians maintain, sex probably led the way. But why was love hatched in the process, since it was presumably unnecessary to get things started in the first place? Furthermore, what has sustained romance—that odd collection of tics and impulses—over the centuries? Most mass hallucinations, such as the 17th century tulip mania in Holland, flame out fairly rapidly when people realize the absurdity of what they have been doing and, as the common saying goes, come to their senses. When people in love come to their senses, they tend to orbit with added energy around each other and look more helplessly loopy and self-besotted. If romance were purely a figment, unsupported by any rational or sensible evidence, then surely most folks would be immune to it by now. Look around. It hasn't happened. Love is still in the air.

And it may be far more widespread than even romantics imagined. Those who argue that love is a cultural fantasy have tended to do so from a Eurocentric and class-driven point of view. Romance, they say, arose thanks to amenities peculiar to the West: leisure time, a modicum of creature comforts, a certain level of refinement in the arts and letters. When these trappings are absent, so is romance. Peasants mated; aristocrats fell in love.

But last year a study conducted by anthropologists William Jankowiak of the University of Nevada–Las Vegas and Edward Fischer of Tulane University found evidence of romantic love in at least 147 of the 166 cultures they studied. This discovery, if borne out, should pretty well wipe out the idea that love is an invention of the Western mind rather than a biological fact. Says Jankowiak: "It is, instead, a universal phenomenon, a panhuman characteristic that stretches across cultures. Societies like ours have the resources to show love through candy and flowers, but that does not mean that the lack of resources in other cultures indicates the absence of love."

Some scientists are not startled by this contention. One of them is anthropologist Helen Fisher, a research associate at the American Museum of Natural History and the author of *Anatomy of Love: The Natural History of Monogamy, Adultery and Divorce,* a recent book that is making waves among scientists and the general reading public. Says Fisher: "I've never *not* thought that love was a very primitive, basic human emotion, as basic as fear, anger or joy. It is so evident. I guess anthropologists have just been busy doing other things."

Among the things anthropologists—often knobby-kneed gents in safari shorts—tended to do in the past was ask questions about courtship and marriage rituals. This now seems a classic example, as the old song has it, of looking for love in all the wrong places. In many cultures, love and marriage do not go together. Weddings can have all the romance of corporate mergers, signed and sealed for family or territorial interests. This does not mean, Jankowiak insists, that love does not exist in such cultures; it erupts in clandestine forms, "a phenomenon to be dealt with."

Somewhere about this point, the specter of determinism begins once again to flap and cackle. If science is going to probe and prod and then announce that we are all scientifically fated to love—and to love preprogrammed types—by our genes and chemicals, then a lot of people would just as soon not know. If there truly is a biological predisposition to love, as more and more scientists are coming to believe, what follows is a recognition of the amazing diversity in the ways humans have chosen to express the feeling. The cartoon images of cavemen bopping cavewomen over the head and dragging them home by their hair? Love. Helen of Troy, subjecting her adopted city to 10 years of ruinous siege? Love. Romeo and Juliet? Ditto. Joe in Accounting making a fool of himself around the water cooler over Susan in Sales? Love. Like the universe, the more we learn about love, the more preposterous and mysterious it is likely to appear. —*Reported by Hannah Bloch/New York and Sally B. Donnelly/Los Angeles*

THE LESSONS OF LOVE

Yes, we've learned a few things. We now know that it is the insecure rather than the confident who fall in love most readily. And men fall faster than women. And who ever said sex had anything to do with it?

Beth Livermore

As winter thaws, so too do icicles on cold hearts. For with spring, the sap rises—and resistance to love wanes. And though the flame will burn more of us than it warms, we will return to the fire—over and over again.

Indeed, love holds central in everybody's everyday. We spend years, sometimes lifetimes pursuing it, preparing for it, longing for it. Some of us even die for love. Still, only poets and songwriters, philosophers and playwrights have traditionally been granted license to sift this hallowed preserve. Until recently. Over the last decade and a half, scientists have finally taken on this most elusive entity. They have begun to parse out the intangibles, the *je ne sais quoi* of love. The word so far is—little we were sure of is proving to be true.

OUT OF THE LAB, INTO THE FIRE

True early greats, like Sigmund Freud and Carl Rogers, acknowledged love as important to the human experience. But not till the 1970s did anyone attempt to define it—and only now is it considered a respectable topic of study.

One reason for this hesitation has been public resistance. "Some people are afraid that if they look too close they will lose the mask," says Arthur Aron, Ph.D., professor of psychology at the University of California, Santa Cruz. "Others believe we know all that we need to know." But mostly, to systematically study love has been thought impossible, and therefore a waste of time and money.

No one did more to propagate this false notion than former United States Senator William Proxmire of Wisconsin, who in 1974 launched a very public campaign against the study of love. As a member of the Senate Finance Committee, he took it upon himself to ferret out waste in government spending. One of the first places he looked was the National Science Foundation, a federal body that both funds research and promotes scientific progress.

Upon inspection, Proxmire found that Ellen Berscheid, Ph.D., a psychologist at the University of Minnesota who had already broken new ground scrutinizing the social power of physical attractiveness, had secured an $84,000 federal grant to study relationships. The proposal mentioned romantic love. Proxmire loudly denounced such work as frivolous—tax dollars ill spent.

The publicity that was given Proxmire's pronouncements not only cast a pall over all behavioral science research, it set off an international firestorm around Berscheid that lasted the next two years. Colleagues were fired. Her office was swamped with hate mail. She even received death threats. But in the long run, the strategy backfired, much to Proxmire's chagrin. It generated increased scientific interest in the study of love, propelling it forward, and identified Berscheid as the keeper of the flame. Scholars and individuals from Alaska to then-darkest Cold War Albania sent her requests for information, along with letters of support.

Berscheid jettisoned her plans for very early retirement, buttoned up the country house, and, as she says, "became a clearinghouse" for North American love research. "It became eminently clear that there were people who really did want to learn more about love. And I had tenure."

PUTTING THE SOCIAL INTO PSYCHOLOGY

This incident was perfectly timed. For during the early 1970s, the field of social psychology was undergoing a revolution of sorts—a revolution that made the study of love newly possible.

For decades behaviorism, the school of psychology founded by John B. Watson, dominated the field. Watson argued that only overt actions capable of direct observation and measurement were worthy of study. However, by the early seventies, dissenters were openly calling this approach far too narrow. It excluded unobservable mental events such as ideas and emotions. Thus rose cognitive science, the study of the mind, or perception, thought, and memory.

Now psychologists were encouraged to ask human subjects what they thought and how they felt about things. Self-report questionnaires emerged as a legitimate research tool. Psychologists were encouraged to escape laboratory confines—to study real people in the real world. Once out there, they discovered that there was plenty to mine.

Throughout the seventies, soaring divorce rates, loneliness, and isolation began to dominate the emotional landscape of America. By the end of that decade, love had become a pathology. No longer was the question "What is love?" thought to be trivial. "People in our culture dissolve unions when love disappears, which has a lasting effect on society," says Berscheid. Besides, "we already understood the mating habits of the stickleback fish." It was time to turn to a new species.

Today there are hundreds of research papers on love. Topics range from romantic ideals to attachment styles of the young and unmarried. "There were maybe a half dozen when I wrote my dissertation on romantic attraction in 1969," reports Aron. These days, a national association and an international society bring "close relationship" researchers close together annually. Together or apart they are busy producing and sharing new theories, new questionnaires to use as research instruments, and new findings. Their unabashed aim: to improve the human condition by helping us to understand, to repair, and to perfect our love relationships.

SO WHAT *IS* LOVE?

"If there is anything that we have learned about love it is its variegated nature," says Clyde Hendrick, Ph.D., of Texas Tech University in Lubbock. "No one volume or theory or research program can capture love and transform it into a controlled bit of knowledge."

Instead, scholars are tackling specific questions about love in the hopes of nailing down a few facets at a time. The expectation is that every finding will be a building block in the base of knowledge, elevating understanding.

Elaine Hatfield, Ph.D., now of the University of Hawaii, has carved out the territory of passionate love. Along with Berscheid, Hatfield was at the University of Minnesota in 1964 when Stanley Schacter, formerly a professor there and still a great presence, proposed a new theory of emotion. It said that any emotional state requires two conditions: both physiological arousal and relevant situational cues. Already studying close relationships, Hatfield and Berscheid were intrigued. Could the theory help to explain the turbulent, all-consuming experience of passionate love?

Hatfield has spent a good chunk of her professional life examining passionate love, "a state of intense longing for union with another." In 1986, along with sociologist Susan Sprecher, she devised the Passionate Love Scale (PLS), a questionnaire that measures thoughts and feelings she previously identified as distinctive of this "emotional" state.

Lovers rate the applicability of a variety of descriptive statements. To be passionately in love is to be preoccupied with thoughts of your partner much of the time. Also, you likely idealize your partner. So those of you who are passionately in love would, for example, give "I yearn to know all about—" a score somewhere between "moderately true" and "definitely true" on the PLS.

True erotic love is intense and involves taking risks. It seems to demand a strong sense of self.

The quiz also asks subjects if they find themselves trying to determine the other's feelings, trying to please their lover, or making up excuses to be close to him or her—all hallmarks of passionate, erotic love. It canvasses for both positive and negative feelings. "Passionate lovers," explains Hatfield, "experience a roller coaster of feelings: euphoria, happiness, calm, tranquility, vulnerability, anxiety, panic, despair."

For a full 10 percent of lovers, previous romantic relationships proved so painful that they hope they will never love again.

Passionate love, she maintains, is kindled by "a sprinkle of hope and a large dollop of loneliness, mourning, jealousy, and terror." It is, in other words, fueled by a juxtaposition of pain and pleasure. According to psychologist Dorothy Tennov, who interviewed some 500 lovers, most of them expect their romantic experiences to be bittersweet. For a full 10 percent of them, previous romantic relationships proved so painful that they hope never to love again.

Contrary to myths that hold women responsible for romance, Hatfield finds that both males and females love with equal passion. But men fall in love faster. They are, thus, more romantic. Women are more apt to mix pragmatic concerns with their passion.

And people of all ages, even four-year-old children, are capable of "falling passionately in love." So are people of any ethnic group and socioeconomic stratum capable of passionate love.

Hatfield's most recent study, of love in three very different cultures, shows that romantic love is not simply a product of the Western mind. It exists among diverse cultures worldwide.

Taken together, Hatfield's findings support the idea that passionate love is an evolutionary adaptation. In this scheme, passionate love works as a bonding mechanism, a necessary kind of interpersonal glue that has existed since the start of the human race. It assures that procreation will take place, that the human species will be perpetuated.

UP FROM THE SWAMP

Recent anthropological work also supports this notion. In 1991, William Jankowiak, Ph.D., of the University of Nevada in Las Vegas, and Edward Fischer, Ph.D., of Tulane University published the first study systematically comparing romantic love across 166 cultures.

They looked at folklore, indigenous advice about love, tales about lovers, love potion recipes—anything related. They found "clear evidence" that romantic love is known in 147, or 89 percent, of cultures. Further, Jankowiak suspects that the lack of proof in the remaining 19 cultures is due more to field workers' oversights than to the absence of romance.

Unless prompted, few anthropologists recognize romantic love in the populations that they study, explains Jankowiak. Mostly because romance takes different shapes in different cultures, they do not know what to look for. They tend to recognize romance only in the form it takes in American culture—a progressive phenomenon leading from flirtation to marriage. Elsewhere, it may be a more fleeting fancy. Still, reports Jankowiak, "when I ask them specific questions about behavior, like 'Did couples run away from camp together?', almost all of them have a positive response."

For all that, there is a sizable claque of scholars who insist that romantic love is a cultural invention of the last 200 years or so. They point out that few cultures outside the West embrace romantic love with the vigor that we do. Fewer still build marriage, traditionally a social and economic institution, on the individualistic pillar of romance.

Romantic love, this thinking holds, consists of a learned set of behaviors; the phenomenon is culturally transmitted from one generation to the next by example, stories, imitation, and direct instruc-

3. INTERPERSONAL RELATIONSHIPS: Responsible Quality Relationships

LOVE ME TENDER

How To Make Love to a Man
(what men like, in order of importance)

taking walks together
kissing
candle-lit dinners
cuddling
hugging
flowers
holding hands
making love
love letters
sitting by the fireplace

How To Make Love to a Woman
(what women like, in order of importance)

taking walks together
flowers
kissing
candle-lit dinners
cuddling
declaring "I love you"
love letters
slow dancing
hugging
giving surprise gifts

tion. Therefore, it did not rise from the swamps with us, but rather evolved with culture.

THE ANXIOUS ARE ITS PREY

Regardless whether passionate, romantic love is universal or unique to us, there is considerable evidence that what renders people particularly vulnerable to it is anxiety. It whips up the wherewithal to love. And anxiety is not alone; in fact, there are a number of predictable precursors to love.

To test the idea that emotions such as fear, which produces anxiety, can amplify attraction, Santa Cruz's Arthur Aron recorded the responses of two sets of men to an attractive woman. But one group first had to cross a narrow 450-foot-long bridge that swayed in the wind over a 230-foot drop—a pure prescription for anxiety. The other group tromped confidently across a seemingly safe bridge. Both groups encountered Miss Lovely, a decoy, as they stepped back onto terra firm.

Aron's attractive confederate stopped each young man to explain that she was doing a class project and asked if he would complete a questionnaire. Once he finished, she handed him her telephone number, saying that she would be happy to explain her project in greater detail.

Who called? Nine of the 33 men on the suspension bridge telephoned, while only two of the men on the safe bridge called. It is not impossible that the callers simply wanted details on the project, but Aron suspects instead that a combustible mix of excitement and anxiety prompted the men to become interested in their attractive interviewee.

Along similar if less treacherous lines, Aron has most recently looked at eleven possible precursors to love. He compiled the list by conducting a comprehensive literature search for candidate items. If you have a lot in common with or live and work close to someone you find attractive, your chances of falling in love are good, the literature suggests.

Other general factors proposed at one time or another as good predictors include being liked by the other, a partner's positive social status, a partner's ability to fill your needs, your readiness for entering a relationship, your isolation from others, mystery, and exciting surroundings or circumstances. Then there are specific cues, like hair color, eye expression, and face shape.

Love depends as much on the perception of being liked as on the presence of a desirable partner. Love isn't possible without it.

To test the viability and relative importance of these eleven putative factors, Aron asked three different groups of people to give real-life accounts of falling in love. Predictably, desirable characteristics, such as good looks and personality, made the top of the list. But proximity, readiness to develop a relationship, and exciting surroundings and circumstances ranked close behind.

The big surprise: reciprocity. Love is at heart a two-way event. The perception of being liked ranked just as high as the presence of desirable characteristics in the partner. "The combination of the two appears to be very important," says Aron. In fact, love just may not be possible without it.

Sprecher and his colleagues got much the same results in a very recent cross-cultural survey. They and their colleagues interviewed 1,667 men and women in the U.S., Russia, and Japan. They asked the people to think about the last time they had fallen in love or been infatuated. Then they asked about the circumstance that surrounded the love experience.

Surprisingly, the rank ordering of the factors was quite similar in all three cultures. In all three, men and women consider reciprocal liking, personality, and physical appearance to be especially important. A partner's social status and the approval of family and friends are way down the list. The cross-cultural validation of predisposing influences suggests that reciprocal liking, desirable personality and physical features may be universal elements of love, among the *sine qua non* of love, part of its heart and soul.

FRIENDSHIP OVER PASSION

Another tack to the intangible of love is the "prototype" approach. This is the study of our conceptions of love, what we "think" love is.

In 1988, Beverly Fehr, Ph.D., of the University of Winnipeg in Canada conducted a series of six studies designed to determine what "love" and "commitment" have in common. Assorted theories suggested they could be anything from mutually inclusive to completely separate. Fehr asked subjects to list characteristics of love and to list features of commitment. Then she asked them to determine which qualities were central and which more peripheral to each.

People's concepts of the two were to some degree overlapping. Such elements as trust, caring, respect, honesty, devotion, sacrifice, and contentment were deemed attributes of both love and commitment. But such other factors as intimacy, happiness, and a desire to be with the other proved unique to love (while commitment alone demanded perseverance, mutual agreement, obligation, and even a feeling of being trapped).

The findings of Fehr's set of studies, as well as others', defy many expectations. Most subjects said they consider caring, trust, respect, and honesty central to love—while passion-related events like touching, sexual passion, and physical attraction are only peripheral. "They are not very central to our concept of love," Fehr shrugs.

Recently, Fehr explored gender differences in views of love—and found remarkably few. Both men and women put forth friendship as primary to love. Only in a second study, which asked subjects to match their personal ideal of love to various descriptions, did any differences show up. More so than women, men tended to rate erotic, romantic love closer to their personal conception of love.

Both men and women deem romance and passion far less important than support and warm fuzzies . . .

Still, Fehr is fair. On the whole, she says, "the essence, the core meaning of love differs little." Both genders deem romance and passion far less important than support and warm fuzzies. As even Nadine Crenshaw, creator of steamy romance novels, has remarked, "love gets you to the bathroom when you're sick."

LOVE ME TENDER

Since the intangible essence of love cannot be measured directly, many researchers settle for its reflection in what people do. They examine the behavior of lovers.

Clifford Swensen, Ph.D., professor of psychology at Purdue University, pioneered this approach by developing a scale with which to measure lovers' behavior. He produced it from statements people made when asked what they did for, said to, or felt about people they loved . . . and how these people behaved towards them.

Being supportive and providing encouragement are important behaviors to all love relationships—whether with a friend or mate, Swensen and colleagues found. Subjects also gave high ratings to self-disclosure, or talking about personal matters, and a sense of agreement on important topics.

But two categories of behaviors stood out as unique to romantic relationships. Lovers said that they expressed feelings of love verbally; they talked about how they enjoyed being together, how they missed one another when apart, and other such murmurings. They also showed their affection through physical acts like hugging and kissing.

Elaborating on the verbal and physical demonstrations of love, psychologist Raymond Tucker, Ph.D., of Bowling Green State University in Ohio probed 149 women and 48 men to determine "What constitutes a romantic act?" He asked subjects, average age of 21, to name common examples. There was little disagreement between the genders.

Both men and women most often cited "taking walks" together. For women, "sending or receiving flowers" and "kissing" followed close on its heels, then "candle-lit dinners" and "cuddling." Outright declarations of "I love you came in a distant sixth. (Advisory to men: The florists were right all along. Say it with flowers instead.)

. . . as one romance novelist confides, "love gets you to the bathroom when you're sick."

For men, kissing and "candle-lit dinners" came in second and third. If women preferred demonstrations of love to outright declarations of it, men did even more so; "hearing and saying 'I love you didn't even show up among their top ten preferences. Nor did "slow dancing or giving or receiving surprise gifts," although all three were on the women's top-ten list. Men likewise listed three kinds of activity women didn't even mention: "holding hands," "making love"—and "sitting by the fireplace." For both sexes, love is more tender than most of us imagined.

All in all, says Tucker, lovers consistently engage in a specific array of actions. "I see these items show up over and over and over again." They may very well be the bedrock behaviors of romantic love.

SIX COLORS OF LOVE

That is not to say that once in love we all behave alike. We do not. Each of us has a set of attitudes toward love that colors what we do. While yours need not match your mate's, you best understand your partner's approach. It underlies how your partner is likely to treat you.

There are six basic orientations toward love, Canadian sociologist John Allen Lee first suggested in 1973. They emerged from a series of studies in which subjects matched story cards, which contain statements projecting attitudes, to their own personal relationships. In 1990 Texas Tech's Clyde Hendrick, along with wife/colleague Susan Hendrick, Ph.D., produced a Love Attitude Scale to measure all six styles. You may embody more than one of these styles. You are also likely to change style with time and circumstance.

Both men and women prefer demonstrations of love to outright declarations of it.

You may, for example, have spent your freewheeling college years as an Eros lover, passionate and quick to get involved, setting store on physical attraction and sexual satisfaction. Yet today you may find yourself happy as a Storge lover, valuing friendship-based love, preferring a secure, trusting relationship with a partner of like values.

There are Ludus lovers, game-players who like to have several partners at one time. Their partners may be very different from one another, as Ludus does not act on romantic ideals. Mania-type lovers, by contrast, experience great emotional highs and lows. They are very possessive—and often jealous. They spend a lot of their time doubting their partner's sincerity.

Pragma lovers are, well, pragmatic. They get involved only with the "right" guy or gal—someone who fills their needs or meets other specifications. This group is happy to trade drama and excitement for a partner they can build a life with. In contrast, Agape, or altruistic, lovers form relationships because of what they may be able to give to their partner. Even sex is not an urgent concern of theirs. "Agape functions on a more spiritual level," Hendrick says.

The Hendricks have found some gender difference among love styles. In general, men are more ludic, or game-playing. Women tend to be more storgic,

3. INTERPERSONAL RELATIONSHIPS: Responsible Quality Relationships

THE COLORS OF LOVE

How do I love thee? At least six are the ways.

There is no one type of love; there are many equally valid ways of loving. Researchers have consistently identified six attitudes or styles of love that, to one degree or another, encompass our conceptions of love and color our romantic relationships. They reflect both fixed personality traits and more malleable attitudes. Your relative standing on these dimensions may vary over time—being in love NOW will intensify your responses in some dimensions. Nevertheless, studies show that for most people, one dimension of love predominates.

Answering the questions below will help you identify your own love style, one of several important factors contributing to the satisfaction you feel in relationships. You may wish to rate yourself on a separate sheet of paper. There are no right or wrong answers, nor is there any scoring system. The test is designed to help you examine your own feelings and to help you understand your own romantic experiences.

After you take the test, if you are currently in a relationship, you may want to ask your partner to take the test and then compare your responses. Better yet, try to predict your partner's love attitudes before giving the test to him or her.

Studies show that most partners are well-correlated in the areas of love passion and intensity (Eros), companionate or friendship love (Storge), dependency (Mania), and all-giving or selfless love (Agape). If you and your partner aren't a perfect match, don't worry. Knowing your styles can help you manage your relationship.

Directions: Listed below are several statements that reflect different attitudes about love. For each statement, fill in the response on an answer sheet that indicates how much you agree or disagree with that statement. The items refer to a specific love relationship. Whenever possible, answer the questions with your current partner in mind. If you are not currently dating anyone, answer the questions with your most recent partner in mind. If you have never been in love, answer in terms of what you think your responses would most likely be.

FOR EACH STATEMENT:
A = Strongly agree with the statement
B = Moderately agree with the statement
C = Neutral, neither agree nor disagree
D = Moderately disagree with the statement
E = Strongly disagree with the statement

Eros
Measures passionate love as well as intimacy and commitment. It is directly and strongly correlated with satisfaction in a relationship, a major ingredient in relationship success. Eros gives fully, intensely, and takes risks in love; it requires substantial ego strength. Probably reflects secure attachment style.

1. My partner and I were attracted to each other immediately after we first met.
2. My partner and I have the right physical "chemistry" between us.
3. Our lovemaking is very intense and satisfying.
4. I feel that my partner and I were meant for each other.
5. My partner and I became emotionally involved rather quickly.
6. My partner and I really understand each other.
7. My partner fits my ideal standards of physical beauty/handsomeness.

Ludus
Measures love as an interaction game to be played out with diverse partners. Relationships do not have great depth of feeling. Ludus is wary of emotional intensity from others, and has a manipulative or cynical quality to it. Ludus is negatively related to satisfaction in relationships. May reflect avoidant attachment style.

8. I try to keep my partner a little uncertain about my commitment to him/her.
9. I believe that what my partner doesn't know about me won't hurt him/her.
10. I have sometimes had to keep my partner from finding out about other partners.
11. I could get over my affair with my partner pretty easily and quickly.
12. My partner would get upset if he/she knew of some of the things I've done with other people.
13. When my partner gets too dependent on me, I want to back off a little.
14. I enjoy playing the "game of love" with my partner and a number of other partners.

Storge
Reflects an inclination to merge love and friendship. Storgic love is solid, down to earth, presumably enduring. It is evolutionary, not revolutionary, and may take time to develop. It is related to satisfaction in long-term relationships.

15. It is hard for me to say exactly when our friendship turned to love.
16. To be genuine, our love first required caring for a while.
17. I expect to always be friends with my partner.
18. Our love is the best kind because it grew out of a long friendship.
19. Our friendship merged gradually into love over time.
20. Our love is really a deep friendship, not a mysterious, mystical emotion.
21. Our love relationship is the most satisfying because it developed from a good friendship.

Pragma
Reflects logical, "shopping list" love, rational calculation with a focus on desired attributes of a lover. Suited to computer-matched dating. Related to satisfaction in long-term relationships.

22. I considered what my partner was going to become in life before I committed myself to him/her.
23. I tried to plan my life carefully before choosing my partner.
24. In choosing my partner, I believed it was best to love someone with a similar background.
25. A main consideration in choosing my partner was how he/she would reflect on my family.
26. An important factor in choosing my partner was whether or not he/she would be a good parent.
27. One consideration in choosing my partner was how he/she would reflect on my career.
28. Before getting very involved with my partner, I tried to figure out how compatible his/her hereditary background would be with mine in case we ever had children.

Mania
Measures possessive, dependent love. Associated with high emotional expressiveness and disclosure, but low self-esteem; reflects uncertainty of self in the relationship. Negatively associated with relationship satisfaction. May reflect anxious/ambivalent attachment style.

29. When things aren't right with my partner and me, my stomach gets upset.
30. If my partner and I break up, I would get so depressed that I would even think of suicide.
31. Sometimes I get so excited about being in love with my partner that I can't sleep.
32. When my partner doesn't pay attention to me, I feel sick all over.
33. Since I've been in love with my partner, I've had trouble concentrating on anything else.
34. I cannot relax if I suspect that my partner is with someone else.
35. If my partner ignores me for a while, I sometimes do stupid things to try to get his/her attention back.

Agape
Reflects all-giving, selfless, nondemanding love. Associated with altruistic, committed, sexually idealistic love. Like Eros, tends to flare up with "being in love now."

36. I try to always help my partner through difficult times.
37. I would rather suffer myself than let my partner suffer.
38. I cannot be happy unless I place my partner's happiness before my own.
39. I am usually willing to sacrifice my own wishes to let my partner achieve his/hers.
40. Whatever I won is my partner's to use as he/she chooses.
41. When my partner gets angry with me, I still love him/her fully and unconditionally.
42. I would endure all things for the sake of my partner.

Adapted from Hendrick, Love Attitudes Scale

18. Lessons of Love

Research has shown that men fall in love faster than women; women are more apt to mix pragmatic concerns with their passion.

more pragmatic—and more manic. However, men and women seem to be equally passionate and altruistic in their relationships. On the whole, say the Hendricks, the sexes are more similar than different in style.

Personality traits, at least one personality trait, is strongly correlated to love style, the Hendricks have discovered. People with high self-esteem are more apt to endorse eros, but less likely to endorse mania than other groups. "This finding fits with the image of a secure, confident eros lover who moves intensely but with mutuality into a new relationship," they maintain.

When they turned their attention to ongoing relationships, the Hendricks' found that couples who stayed together over the course of their months-long study were more passionate and less game-playing than couples who broke up. "A substantial amount of passionate love" and "a low dose of game-playing" love are key to the development of satisfying relationships—at least among the college kids studied.

YOUR MOTHER MADE YOU DO IT

The love style you embrace, how you treat your partner, may reflect the very first human relationship you ever had—probably with Mom. There is growing evidence supporting "attachment theory," which holds that the rhythms of response by a child's primary care giver affect the development of personality and influence later attachment processes, including adult love relationships.

First put forth by British psychiatrist John Bowlby in the 1960s and elaborated by American psychologist Mary Ainsworth, attachment theory is the culmination of years of painstaking observation of infants and their adult caregivers—and those separated from them—in both natural and experimental situations. Essentially it suggests that there are three major patterns of attachment; they develop within the first year of life and stick with us, all the while reflecting the responsiveness of the caregiver to our needs as helpless infants.

Those whose mothers, or caregivers, were unavailable or unresponsive may grow up to be detached and nonresponsive to others. Their behavior is Avoidant in relationships. A second group takes a more Anxious-Ambivalent approach to relationships, a response set in motion by having mothers they may not have been able to count on—sometimes responsive, other times not. The lucky among us are Secure in attachment, trusting and stable in relationships, probably the result of having had consistently responsive care.

While attachment theory is now driving a great deal of research on children's social, emotional, and cognitive development, University of Denver psychologists Cindy Hazan and Philip Shaver set out not long ago to investigate the possible effect of childhood relationships on adult attachments. First, they developed descriptive statements that reflect each of the three attachment styles. Then they asked people in their community, along with college kids, which statements best describe how they relate to others. They asked, for example, about trust and jeal-

3. INTERPERSONAL RELATIONSHIPS: Responsible Quality Relationships

ousy, about closeness and desire for reciprocation, about emotional extremes.

The distribution of the three attachment styles has proved to be about the same in grown-ups as in infants, the same among collegians as the fully fledged. More than half of adult respondents call themselves Secure; the rest are split between Avoidant and Ambivalent. Further, their adult attachment patterns predictably reflect the relationship they report with their parents. Secure people generally describe their parents as having been warm and supportive. What's more, these adults predictably differ in success at romantic love. Secure people reported happy, long-lasting relationships. Avoidants rarely found love.

Secure adults are more trusting of their romantic partners and more confident of a partner's love, report Australian psychologists Judith Feeney and Patricia Noller of the University of Queensland. The two surveyed nearly 400 college undergraduates with a questionnaire on family background and love relationships, along with items designed to reveal their personality and related traits.

In contrast to the Secure, Avoidants indicated an aversion to intimacy. The Anxious-Ambivalent participants were characterized by dependency and what Feeney and Noller describe as "a hunger" for commitment. Their approach resembles the Mania style of love. Each of the three groups reported differences in early childhood experience that could account for their adult approach to relationships. Avoidants, for example, were most likely to tell of separations from their mother.

It may be, Hazan and Shaver suggest, that the world's greatest love affairs are conducted by the Ambitious-Ambivalents—people desperately searching for a kind of security they never had.

THE MAGIC NEVER DIES

Not quite two decades into the look at love, it appears as though love will not always mystify us. For already we are beginning to define what we think about it, how it makes us feel, and what we do when we are in love. We now know that it is the insecure, rather than the confident, who fall in love more readily. We know that outside stimuli that alter our emotional state can affect our susceptibility to romance; it is not just the person. We now know that to a certain extent your love style is set by the parenting you received. And, oh yes, men are more quickly romantic than women.

The best news may well be that when it comes to love, men and women are more similar than different. In the face of continuing gender wars, it is comforting to think that men and women share an important, and peaceful, spot of turf. It is also clear that no matter how hard we look at love, we will always be amazed and mesmerized by it.

FORECAST FOR Couples

Painful and confusing as they may be, intimate relationships today actually follow particular dynamic patterns; they evolve through recurring cycles of promise and betrayal. Herewith, a map of the territory.

Barry Dym and Michael Glenn, M.D.

For both men and women, intimate relationships have come to assume an importance that is perhaps unprecedented. And while the sexes are having a devilishly hard time getting together in these days of rapid role change, they are clearly struggling to make things work in a way that satisfies both partners. What is so surprising is that the struggles have been taking place almost entirely in the absence of a general cultural understanding about the nature of relationships. In occasional articles over the past 10 issues, PSYCHOLOGY TODAY has sought to relay what family therapists know—that relationships have rules of their own. Each is a system, something larger and different from each partner individually. In "The Reinvention of Marriage [Jan/Feb 1992]," for example, PT introduced a developmental perspective to relationships, explaining how they naturally develop over time. In this article, two noted family therapists add another dimension: Relationships progress not in a straight line, but through endless cycles of advance and retreat.—The Editors

Not so long ago, a simple story stated that a couple began when a man and a woman fell in love. They would then marry and form a family. The woman would take care of the home and children; the man would support them by toiling in the heartless world. They would both sacrifice their individual goals to the greater good of the family. Their romance would gradually melt into affection and partnership. The man would be the acknowledged leader, following law and custom, but the woman would rule in domestic matters.

Not every couple followed this prescription—far from it. Forms of coupling varied from couple to couple and from community to community. But each couple, whatever they did, had to contend with this story, this cultural narrative. Some adopted it with relative ease; some twisted and changed themselves in order to accommodate; others were defiant, but their very defiance proved the story's continued vitality. Anyone could invoke it as an authority against a partner who failed to play the assigned role. The same is true today, albeit in response to a different cultural narrative.

The contemporary couple is changing rapidly, responding to shifts in where and how people live, in the economics of employment, in the different kinds of power women and men wield, in beliefs about how things are supposed to be between the sexes, and in the nature of the family. As couples change, so does the cultural narrative about them. One result of the rapid changes is that both men and women tend to overestimate the power the *other* sex wields in intimate relations today. Both feel like victims in the war between the sexes.

Fascination with couples fills today's media and shapes our popular imagination. The romantically engaged couple is

3. INTERPERSONAL RELATIONSHIPS: Responsible Quality Relationships

the icon of our time, a major focus of movies, television, books, and music. Most people devote tremendous energy to trying to find the perfect partner. And yet the couple is an isolated and fragile form, caught between great expectations and declining resources. It is supposed to be the cure for all that ails you. In fact, our commitment to the inner life of relationships has grown as our commitment to the larger society recedes. But the couple falls apart almost as easily as it comes together: half of all marriages end in divorce; early love often fades into domestic boredom. Contemporary couples must develop in the shadow of their potential demise.

There have always been many different kinds of couples: "just living together" couples, gay and lesbian couples, childless couples, interracial couples, post-divorce

> **Contemporary couples must develop in the shadow of their potential demise. As they struggle to avoid breaking up, they often distort the very relationships they are trying to preserve.**

couples, couples of vastly different age, and so on. The life course of real couples varies widely; few march in a straight line past every predictable milepost, from the first romantic attachment to the birth of children to the empty-nest syndrome, and finally into retirement together. But *certain* stories regularly prevail. Against them, social diversity continues to build, often in unexpected ways (as by the impact of new immigrant families).

People—psychotherapists included—often participate in, theorize about, and try to fix ailing couples without a clear sense of what a couple is or an understanding of how it has gotten that way. This is like trying to treat the heart without knowing something about its normal functioning. Intimate couple relationships, painful and confusing as they may be, follow particular patterns; yet couples today have only the most rudimentary map of the territory through which life takes them. They are in a psychological and moral wilderness. Self-help books and psychotherapists try to help but often fail. What is needed is the creation of a living narrative, new language, new concepts, and new metaphors—a map of couples in our time.

As family therapists, we began our thinking with a simple observation: so many people seem disappointed in their relationships. What is the disappointment all about? Psychotherapists look for the roots of disappointment in unresolved childhood conflicts; philosophers and psychologists note its origins in our attachments to specific goals and material comfort. But the more we thought about it, the more a simpler answer emerged: relationships are disappointing because they do not seem to fulfill their early promise.

Our culture asks so much of couple relationships—romance and passion, partnership, friendship, and nurturance—that disappointment is inevitable. The expansive promise of new beginnings often comes to seem like a youthful illusion at best, a cruel hoax at worst. The implicit contracts people make with each other—which are based more on potential than on past performance—come tumbling down. Partners break promises; individuals break their own resolutions. Husbands and wives are forever noting, "This is not the person I thought I had married" and "If I had known then what I know today, I never would have married her." These statements are not simply sour grapes or the distorted complaints of dissatisfied individuals. They reflect the truth of broken promises.

CONTRADICTIONS

This is a revolutionary time in male and female relationships and therefore in the lives of couples. In times of change, contradictions sharpen. This process marks the lives of contemporary couples, making both partners tense and excited. We can point out three basic conflicts with which couples today must cope.

1. The clash between great expectations and limited resources.

According to our cultural narrative, the romantically engaged couple is an answer for everything. We want more from our partners, but we're less and less able to give of ourselves. Our partners must be passionate lovers as well as loyal confidantes, willing to join us intensely when we want, but leaving us alone when we need "private space." We ask for romance in our quiet moments, but want a sturdy partner to help raise children, maintain a household, and coordinate schedules. These activities interfere with one another, and our expectations don't mix.

The couple is supposed to be a stable haven in a cool, hostile, unpredictable world. In the past, women had the role of maintaining domestic relationships, but now that two incomes are required to get by, more and more couples are made up of two working partners. Many couples, even those without children, return home each day exhausted. No one stands at the threshold to welcome them and soothe their return.

> **The couple is supposed to be a stable haven in a cool, hostile unpredictable world.**

At the same time, couples are more than ever isolated from the resources that used to sustain them, such as extended families and communities. We all have friends, but fewer of us live close to our families. Who can we depend on, no questions asked, to take care of the kids when we are in a pinch? Who will support us and offer us wisdom through the hard times? Most couples are jammed for time, for emotional energy, and for patience. "I just need a minute to myself" has become our modern litany. Our partner's company sometimes drains us more than it enhances us. We probably do more for one another these days, but we expect so much that we're still often disappointed.

2. The clash between the individual and the couple.

We always marvel at those selfless individuals who place others' needs and comforts first. In an age such as ours, individual pleasures, development, and fulfillment often come first. The contemporary concern with self intensifies the basic tension between our allegiance to the relationship and allegiance to ourselves.

In couple relationships this tension is often polarized by gender: women have tended to stand for the relationship, connection, and mutual dependence; men for individualism and independence. Such polarization, where it exists, exaggerates and distorts and leads to dramatic confrontations such as those in which women feel abandoned while men feel controlled. This is probably the most common dilemma presented to couple therapists today, and can be seen as the archetypal struggle of the modern couple.

But there is a growing trend to dissolve this simple division by gender. Women are also concerned with their own development, with being independent, respected partners, capable of pursuing their goals outside of the relationship. The question then arises: just *who* in the couple is committed to the relationship?

In other eras, romantic love centered on the partner. "What can I do to win you?" was a burning question. These days we look for partners who can bring out the best in ourselves. "What can you do for *me?*" we ask. The ideal partner today is a cross between a psychotherapist and a good parent. Even generosity, we are told, proceeds best from self-fulfillment: only if we feel good about ourselves will we be good to our partners. But when we feel bad about ourselves, and our partners are not filling our needs, we may soon lose our commitment to the relationship. We and our partner then become two islands in an unfriendly sea.

3. The clash between staying together and splitting up, marriage and divorce.

Many relationships last a short time. We discard our partners—or they discard us—and we move on. Even longer relationships have a way of fizzling out after a year or so: they just don't seem right any more; nasty arguments turn us sour; our involvement fades away. Even those relationships that lead to marriage have trouble holding fast. And yet we keep starting relationships again, hoping each time we'll find the right partner—or at least take a more realistic attitude towards them.

We seem less angry, less disillusioned with relationships than with ourselves or our current partner. As difficulties in a relationship mount, we often persist because we have so much "invested" in it; but eventually we wonder if it makes sense to put any more into such a losing relationship.

Most of us become less willing to accept a stale relationship. As breakup and divorce have become easier, so has our dream of the good partner. We imagine anew that someone out there will save us from loneliness, redeem us as individuals, and help us avoid the problems that destroyed our last relationship.

We're vividly aware that breakup and divorce are possible. Such awareness can take the edge off our own commitment: it is an escape clause, a skepticism built into contemporary relationships. We react to this skepticism by nervously maintaining a safer distance, withholding a part of ourselves, and trying to let go of some of our romantic intensity. As we struggle to avoid breaking up, we often distort the very relationships we are trying to preserve.

In trying to understand the disappointment of couples, we began asking couples about their original promises. What was it they had originally pledged to one another, what contract had they tacitly made? And how did this contract affect the dissolution or reconciliation that followed their sense of betrayal? Further, how did couples move *beyond* their outrage? How did the resolution of their disappointment affect how they subsequently thought of themselves—both as individuals and as a couple? Out of the answers and our observations of hundreds of couples arose our notion that couples continually move through a three-stage cycle of promise, betrayal, and resolution. It was surprisingly simple. But the more we turned it over and measured it against our experience, the more it seemed to fit.

Our basic idea was that couples initially pass through three recognizable stages: Expansion and Promise; Contraction and Betrayal; and Resolution. The early expansiveness of relationships expresses our desire for romance, our yearnings to burst through the walls of our isolation and alienation to connect with another person, and our longings to be more than insignificant beings on this "little" planet.

Later in relationships we contract and pull back into our skin. This contraction demonstrates our pessimism, our cynicism, our capacity to see ourselves as victims, and our lack of vision and enduring discipline. It expresses our belief that men and women are not natural allies but naturally at war, and our conviction that we were fools for believing in romance.

When in our lives we bring these two opposing currents together, when we struggle past our pessimism with a sense of perspective and compromise, there is a period of resolution, a time of apparent stability. But new challenges, like the birth of a child or one partner's press towards self-fulfillment and growth, often threaten and topple these stable places. No couple can stay at a point of resolution forever; they must always adjust. The character of couples is thus constantly evolving.

COUPLE DEVELOPMENT

In order for a couple to endure, the partners must resolve the problems that emerge in their relationship. No couple does this by moving in a straight line.

19. Forecast for Couples

Instead all pass through series after series of endlessly spiraling three-stage cycles of Expansion and Promise, Contraction and Betrayal, and Resolution.

Couples first move through times of positive hopes and experiences, then through times of trouble and disappointment—perhaps the positive experiences were not deep enough, perhaps they did not last long enough. Then they move into some middle ground between the two opposing conditions. Each cycle reflects their effort to recognize and reconcile a conflict: the freedom and the promise of the early relationship versus the crushing defeat that invariably follows.

Initially, two people come together enough to form a lasting relationship. This is the task of the first Expansive Stage. According to today's cultural narrative, couples should begin in a burst of romance, exploration, and sexual attraction. But not every couple, and not every partner, falls in love. Instead, couples commonly begin with a shared experience of expansiveness and promise, which may include romantic love, but may also arise from a warm and respectful friendship.

In this stage, individuals feel somehow larger, more witty and charming, stronger yet more vulnerable—in short, closer to their ideal selves than ever before. The developmental trajectories of men and women converge for a moment, so that men take time to talk and understand, while women appear more independent. Each partner's appreciation spurs the other to expand his or her capacities. Early relationships lack the constricting patterns that eventually emerge. They are spacious instead, encouraging both exploration and experimentation.

The Expansive Stage is one of the few times when we tell our whole story to another person, who bears witness to it and helps shape it further. The two individual narratives are then woven into a couple narrative, which takes on a life, an identity, of its own. People will say, "This is how we do things" and "That is just how we are." Individual identity becomes inextricably bound to the character of the couple.

But couples must also find a way to include the fears and insecurities, the ineptness and even the cruelty that figures prominently in their lives. Introducing this material into the relationship is the task of the Contraction and Betrayal Stage.

This second stage begins when one partner pulls back to routine ways. The withdrawal may be neutral, not angry; but the person who is left feels abandoned and

3. INTERPERSONAL RELATIONSHIPS: Responsible Quality Relationships

betrayed. When she (it is almost always the woman who stays connected longer) objects, he may feel controlled and withdraw further; she may then be both frightened and furious, insistently asking that the person she had gotten to know reemerge. In response, he may build his shell thicker, and so the sequence grows.

This nightmarish cycle makes caricatures of the two partners. The great potential of the Expansive Stage, when men and women shared "male" and "female" attributes, dissolves into cruel stereotypes. Each partner feels trapped and betrayed—not only by the other but also by himself or herself. More than anything, people wish to remain the person they were in the Expansive Stage, the person they had striven to be through years of dreaming and preparing. Now they feel immensely let down by their own failures. They blame both self and other, and a mood of accusation permeates the relationship.

Just as the Expansive Stage brings us closer to our ego ideal, so the Contraction Stage confronts us with our greatest fears and our poorest self-image. During this stage, distinctive, repetitive struggles form and consolidate. They seem to define the whole relationship. The struggles are so distressing that the couple may draw someone, like a child or parent, or something, like alcohol or excessive work, into the relationship to buffer the conflict. These patterns become integral parts of the couple's moments together—and recur throughout the life of the couple. They become as familiar and distinctive as the implicit promises of expansion. Couples grow very accustomed to the predictable experiences of contraction.

Even though it is a difficult stage, contraction is essential. Unless partners can bring their wounds and uncertainties into the relationship, they will feel neither real nor whole, and the vigilance required to protect themselves will make them guarded and superficial. In contraction, critical themes from the partners' past enter the couple's experience, further deepening their character. Contraction, then, is not a "negative" stage; it is as necessary as the others. We confront ourselves honestly in contraction's harsh light, telling the truth about our limitations and those of our partner. The insights must be folded into the relationship. Couples who endure contraction will look back on it as a time when they were tested and triumphed.

Resolution

To survive, couples must climb out of the Stage of Contraction without entirely excluding its messages. They must at least partially reconcile the first two stages. This is the task of the third stage, the Stage of Resolution.

This is a stage of compromise, negotiation, accommodation, and integration. The partners struggle to be reasonable and maintain perspective, to affirm complexity and to handle difficult situations with competence and maturity. In contrast to the intense, narrow focus on one another that characterized the first two stages, the couple now opens up more to family and community. Having a child, for example, may serve as a bridge of common concern to repair long-strained relationships with parents; it may become a rite of passage into a more durable adulthood.

The early desire for fusion in the Expansive Stage gives way to close, bitter struggles in the Stage of Contraction. Paradoxically, the blaming and rejection may eventually lead to a sense of perspective. For example, a statement uttered in close, angry combat, like "I'm not at all like you," may usher in a realization of genuine difference: "We really are different." With this realization comes alienation, then at least tolerance and possibly acceptance, followed by a flood of relief.

For a moment the struggle seems over. What had seemed mean in one's partner now seems tolerable. Relief follows, and renewed optimism often comes in its wake. At this point the couple frequently moves forward into another Expansive Stage; but

> **Conflict is not an aberration that can be ignored or cured in couple's lives.**

just as quickly, they can be thrown back into contraction, with each partner feeling disappointed, as if the whole experience had been an illusion.

This moment of increased perspective represents a foray into resolution. The accumulation of these moments of realization, from contraction into resolution, put the couple past a threshold that consolidates their growth. The forays overwhelm the experience of contraction—which comes to seem like a crabby, limited view. The couple moves forward.

Couples try to hold onto their new perspective and the optimism that follows, but they invariably fail. The progress of expansion, contraction, and resolution is a spiral through time: stages cascade one after the other. The character of the couple, as distinguished from the character of the individual partners, is shaped more by the overall cycles than by any single stage. Cycles can be precipitated by a wide number of crises and events.

At first, the promise of the Expansive Stage and the fears of the Stage of Contraction remain relatively separate; but with each turn of the cycle, they become more integrated. Each revolution brings new information into the couple's domain. One partner's terrible and characteristic rages, for example, which show up in other domains, may suddenly emerge in the relationship after years of life together, and eventually become acknowledged and worked into their ways of being together. So, too, with many positive traits, such as capacities that emerge only in response to dangerous situations, such as courage in the face of danger.

For those couples who survive many turnings of the cycle, the Stage of Resolution tends to broaden in content and lengthen in time. Couples spend more and more time in it, and its qualities of tolerance and accommodation increasingly come to define their character.

The character of couples is shaped as much by the rhythm of the cycles as by the content of their stages. In this, couples vary greatly. Some couples, for example, move through wild swings: everything's great, then everything's awful; then there is a brief moment of reconciliation, after which everything's better (or worse) than ever. For others, the stages pass more subtly and their cycles are relatively smooth. Some couples move slowly out of one stage into another; others seem to cycle all the time.

Every couple has a Home Base, a stage in which they generally reside. This habitual stage represents both its public persona and its evolved self-image, but not its full character. Those who reside in contraction, for instance, think of themselves as conflicted and troubled, even though they have moments in expansion and resolution. Once a couple has settled into a stage as its Home Base, its cycles will tend to begin and end there. The couple in contraction might climb out through one compromise or another, relax momentarily in resolution, which feels good enough to revive some old romantic feelings reminiscent of expansion. But with its first minor disappointment, fall back to their familiar Home Base in contraction.

19. Forecast for Couples

After the first few cycles the stages in each couple's repertoire become more like different states of being. The couple can enter them, know them as familiar, and then move on. In this sense the stages become a relatively constant, autonomous reality in the relationship.

But it is a couple's first turn through the cycle that imparts a distinctive style that will tend to endure. We develop our characteristic ways of loving and being loved, of being warm and affectionate, in our first time through expansion. Subsequent expansive moments will usually bring back the memory and flavor of these patterns. Similarly, the fights we had in our first cycle usually recur over and over again through our relationship. No new fight seems all that new, but looks like a variation on the old one. Later, in our first passage through resolution, we develop our characteristic ways of solving problems—our distinctive ways of talking, negotiating, tolerating, and accepting.

Conflict and Resolution

The character of couples is forged through regular cycles of conflict and resolution. Conflict is not an aberration that can be ignored or cured; it is inherent in couples' lives. It stems from real dilemmas that couples must acknowledge and resolve. In relationships, conflict often appears as a choice: individual versus collective good; women's rights versus male entitlement; one partner's style of upbringing versus the other's.

As they continue to cycle, couples struggle for a perspective that can embrace both the good and the bad and help them move ahead. But the perspectives they reach, and the solutions they attain, are always partial: they resolve enough so they can move on, but they rarely resolve disputes completely. Core conflicts hang around, serving as sources of new antagonisms.

Just as we feel we have resolved a conflict about sex, money, or children, our solution unravels or another problem appears. Partners need to negotiate everything, from how to structure child care to how and when to make love—and who should initiate it. Couples will be frustrated if they expect to solve their conflicts once and for all. But if they learn to recognize their cycles of conflict and resolution and adapt to them, they may survive the hard times, grow together, and thrive.

Turning Points and Transformations

At some point, almost all couples find themselves in a profoundly disturbing and immovable impasse. No matter what they do, they cannot escape; there are no more areas of conversation to open up, no more strategies to try, no more activities to limit. They feel totally stuck. Many couples separate at this point. Many others, perhaps only through inertia or devotion to children or to the idea of marriage, stay together. Most couples simply endure, emerging diminished but essentially unchanged after their ordeal.

But some couples are transformed by these terrifying crises. Instead of simply enduring, the partners manage to give up their blaming and bitterness but remain in the relationship. They realize they cannot get what they want by demanding, by manipulating, or even by negotiating. In despair and exhaustion, they finally stop trying to change their partner, and stop trying to make themselves over as well. Giving up this fight has a paradoxical effect: for a moment, the partners may experience one another in a new, fresh, and undefined way.

This experience is so dramatic it often takes on a spiritual dimension. The partners feel enhanced—better known and accepted for who they are, joined anew. They feel as if they have awakened. Beyond the conflict—and their own selfish version of what's right—they can sense a deeper meaning of their relationship.

This awakening becomes a great divide in the history of their relationship, separating a time of truth from one of ignorance. The partners can then return emotionally to one another and share the wisdom and inner strength they've now gained.

Not every couple goes through this trying time of transformation. Nor can the experience be taken on willfully. It has to emerge through the difficulties of life. Still, there is something heroic about people who have the capacity to sustain crushing disappointment, undergo repeated tests of their relationship, yet feel enhanced by their commitment to each other.

We are strongly moved, deeply impressed by the energy and courage of couples who refuse their own dissolution and who seek instead to explore the potential for fulfillment in their relationship.

Reproduction

- Birth Control (Articles 20–25)
- Pregnancy and Childbirth (Articles 26–29)

While human reproduction is as old as humanity, many aspects of it are changing in today's society. Not only have new technologies of conception and childbirth affected the *how* of reproduction, but personal, social, and cultural forces have also affected the *who*, the *when*, and the *when not*. Abortion remains a fiercely debated topic, and legislative efforts for and against it abound. The teenage pregnancy rate in the United States is one of the highest in the Western world. The costs of teenage childbearing can be high for the parents, the child, and for society. Yet efforts to curb U.S. teenage pregnancies have not been effective.

This unit also addresses the issue of birth control. In light of the change of attitude toward sex for pleasure only, birth control has become a matter of prime importance. Even in our age of sexual enlightenment, some individuals, possibly in the height of passion, fail to correlate "having sex" with pregnancy. In addition, even in our age of astounding medical technology, there is no 100 percent effective, safe, or aesthetically acceptable method of birth control. Before sex can become safe as well as enjoyable, people must receive thorough and accurate information regarding conception and contraception, birth and birth control, and make a mental and emotional commitment to

Unit 4

the use of an effective method of their choice. Only this can make every child a planned and wanted one.

Despite the relative simplicity of the above assertion, abortion and birth control remain emotionally charged issues in American society. While opinion surveys indicate that most of the public supports family planning and abortion, at least in some circumstances, there are individuals and groups strongly opposed to some forms of birth control and to abortion. Within the past few years, voices for and against birth control and abortion have gotten louder, and on a growing number of occasions have led to overt behaviors, including disobedience. Some Supreme Court and legislative efforts have added restrictions to the right to abortion, while others are striving to increase the access to abortion and reproductive choice. Voices on both sides are raised in emotional and political debate between "We must never go back to the old days" (of illegal and unsafe back-alley abortions) and "The baby has no choice."

Many of the questions raised in this unit about the new technologies of reproduction and its control are likely to remain among the most hotly debated issues for the remainder of this century. It is likely that various religious and political groups will continue to posit and challenge basic definitions of human life, as well as the rights and responsibilities of women and men associated with sex, procreation, and abortion. The very foundations of our pluralistic society may be challenged. We will have to await the outcome.

The opening article in the *Birth Control* subsection, "Choosing a Contraceptive," is an excellent integration of up-to-date information about the range of contraceptive methods with many of the other factors—individual and couple preferences, lifestyle, aesthetic factors, and so on—that go into decisions about this intimate, important issue. The next three articles provide information on three of the newest forms of birth control: the female condom, the Norplant implant, and the very controversial (and as of the printing of this *Annual Edition* not yet approved in the United States) RU 486. The last two articles address American and international issues and actions involving family planning and the availability of legal and safe abortions.

The first article in the *Pregnancy and Childbirth* subsection, "Making Babies," addresses the current state of fertility treatment. This highly technologically assisted industry has grown so rapidly that laws, regulations, and, some say, ethics and values have been unable to keep up. Gina Kolata's article, "Reproductive Revolution Is Jolting Old Views," follows and provides some of the latest technological possibilities for human reproduction. The next article explores sexuality and intimacy after miscarriage. Finally, "Unnecessary Cesarean Sections" deplores the upward trend in cesarean deliveries and suggests that expectant couples should take an active role in reversing this trend.

Looking Ahead: Challenge Questions

In your opinion, what are the most important characteristics of a contraceptive? Why?

What personal feelings or expectations make you more likely to regularly use contraception?

Under what circumstances have (or would) you not use contraception and risk an unintentional pregnancy?

How do you feel that contraceptive responsibilities should be assigned or shared between men and women?

How have recent events in the abortion rights/access arena affected you? Have they changed your beliefs and/or attitudes?

Have you found a fairly comfortable way to talk about contraception and/or pregnancy risk and prevention with a partner? If so, what is it? If not, what do you do?

Do you know anyone who has experienced infertility? If so, what was it like for them? If you were unable to have a child, what treatments would you consider? Are there any you would not? Why? As a current and future taxpayer, do you feel infertility treatments should be insurance-covered expenses? Why or why not?

Choosing a Contraceptive

Merle S. Goldberg

Choosing a method of birth control is a highly personal decision, based on individual preferences, medical history, lifestyle, and other factors. Each method carries with it a number of risks and benefits of which the user should be aware.

Each method of birth control has a failure rate—an inability to prevent pregnancy over a one-year period. Sometimes the failure rate is due to the method and sometimes it is due to human error, such as incorrect use or not using it at all. Each method has possible side effects, some minor and some serious. Some methods require lifestyle modifications, such as remembering to use the method with each and every sexual intercourse. Some cannot be used by individuals with certain medical problems.

Spermicides Used Alone

Spermicides, which come in many forms—foams, jellies, gels, and suppositories—work by forming a physical and chemical barrier to sperm. They should be inserted into the vagina within an hour before intercourse. If intercourse is repeated, more spermicide should be inserted. The active ingredient in most spermicides is the chemical nonoxynol-9. The failure rate for spermicides in preventing pregnancy when used alone is from 20 to 30 percent.

Spermicides are available without a prescription. People who experience burning or irritation with these products should not use them.

Barrier Methods

There are five barrier methods of contraception: male condoms, female condoms, diaphragm, sponge, and cervical cap. In each instance, the method works

---

Disease Prevention

For many people, the prevention of sexually transmitted diseases (STDs), including HIV (human immunodeficiency virus), which leads to AIDS, is a factor in choosing a contraceptive. Only one form of birth control—the latex condom, worn by the man—is considered highly effective in helping protect against HIV and other STDs. Reality Female Condom, made from polyurethane, may give limited protection against STDs but has not been proven as effective as male latex condoms. People who use another form of birth control but who also want a highly effective way to reduce their STD risks, should also use a latex condom for every sex act, from start to finish.

In April 1993, FDA announced that birth control pills, Norplant, Depo-Provera, IUDs, and natural membrane condoms must carry labeling stating that these products are intended to prevent pregnancy but do not protect against HIV infection and other sexually transmitted diseases. In addition, natural membrane condom labeling must state that consumers should use a latex condom to help reduce the transmission of STDs. The labeling of latex condoms states that, if used properly, they will help reduce transmission of HIV and other diseases. ■

20. Choosing a Contraceptive

Male Condom

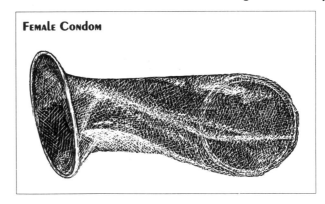

Female Condom

Only latex condoms have been shown to be highly effective in helping to prevent sexually transmitted diseases.

by keeping the sperm and egg apart. Usually, these methods have only minor side effects. The main possible side effect is an allergic reaction either to the material of the barrier or the spermicides that should be used with them. Using the methods correctly for each and every sexual intercourse gives the best protection.

Male Condom

A male condom is a sheath that covers the penis during sex. Condoms are made of either latex rubber or natural skin (also called "lambskin" but actually made from sheep intestines). Only latex condoms have been shown to be highly effective in helping to prevent STDs. Latex provides a good barrier to even small viruses such as human immunodeficiency virus and hepatitis B. Each condom can only be used once. Condoms have a birth control failure rate of about 15 percent. Most of the failures can be traced to improper use.

Some condoms have spermicide added. This may give some additional contraceptive protection. Vaginal spermicides may also be added before sexual intercourse.

Some condoms have lubricants added. These do not improve birth control or STD protection. Non-oil-based lubricants can also be used with condoms. However, oil-based lubricants such as petroleum jelly (Vaseline) should not be used because they weaken the latex. Condoms are available without a prescription.

Female Condom

The Reality Female Condom was approved by FDA in April 1993. It consists of a lubricated polyurethane sheath with a flexible polyurethane ring on each end. One ring is inserted into the vagina much like a diaphragm, while the other remains outside, partially covering the labia. The female condom may offer some protection against STDs, but for highly effective protection, male latex condoms must be used.

FDA Commissioner David A. Kessler, M.D., in announcing the approval, said, "I have to stress that the male latex condom remains the best shield against AIDS and other sexually transmitted diseases. Couples should go on using the male latex condom."

In a six-month trial, the pregnancy rate for the Reality Female Condom was about 13 percent. The estimated yearly failure rate ranges from 21 to 26 percent. This means that about 1 in 4 women who use Reality may become pregnant during a year.

Sponge

The contraceptive sponge, approved by FDA in 1983, is made of white polyurethane foam. The sponge, shaped like a small doughnut, contains the spermicide nonoxynol-9. Like the diaphragm, it is inserted into the vagina to cover the cervix during and after intercourse. It does not require fitting by a health professional and is available without prescription. It is to be used only once and then discarded. The failure rate is between 18 and 28 percent. An extremely rare side effect is toxic shock syndrome (TSS), a potentially fatal infection caused by a strain of the bacterium *Staphylococcus aureus* and more commonly associated with tampon use.

Diaphragm

The diaphragm is a flexible rubber disk with a rigid rim. Diaphragms range in size from 2 to 4 inches in diameter and are designed to cover the cervix during and after intercourse so that sperm cannot reach the uterus. Spermicidal jelly or cream must be placed inside the diaphragm for it to be effective.

The diaphragm must be fitted by a health professional and the correct size prescribed to ensure a snug seal with the vaginal wall. If intercourse is repeated, additional spermicide should be added with

Barrier methods, which work by keeping the sperm and egg apart, usually have only minor side effects.

4. REPRODUCTION: Birth Control

Birth Control Guide

Efficacy rates given in this chart are estimates based on a number of different studies. They should be understood as yearly estimates, with those dependent on conscientious use subject to a greater chance of human error and reduced effectiveness. For comparison, 60 to 85 percent of sexually active women using no contraception would be expected to become pregnant in a year. This chart should not be used alone, but only as a summary of information in the accompanying article.

Type	Male Condom	Female Condom	Spermicides Used Alone	Sponge	Diaphragm with Spermicide	Cervical Cap with Spermicide
Estimated Effectiveness	About 85%	An estimated 74–79%	70–80%	72–82%	82–94%	At least 82%
Risks	Rarely, irritation and allergic reactions	Rarely, irritation and allergic reactions	Rarely, irritation and allergic reactions	Rarely, irritation and allergic reactions; difficulty in removal; very rarely, toxic shock syndrome	Rarely, irritation and allergic reactions; bladder infection; very rarely, toxic shock syndrome	Abnormal Pap te vaginal or cervic infections; very rarely, toxic shoc syndrome
STD Protection	Latex condoms help protect against sexually transmitted diseases, including herpes and AIDS	May give some protection against sexually transmitted diseases, including herpes and AIDS; not as effective as male latex condom	Unknown	None	None	None
Convenience	Applied immediately before intercourse	Applied immediately before intercourse; used only once and discarded	Applied no more than one hour before intercourse	Can be inserted hours before intercourse and left in place up to 24 hours; used only once and discarded	Inserted before intercourse; can be left in place 24 hours, but additional spermicide must be inserted if intercourse is repeated	Can remain in place for 48 hour not necessary to reapply spermicic upon repeated intercourse; may be difficult to insert
Availability	Nonprescription	Nonprescription	Nonprescription	Nonprescription	Rx	Rx

20. Choosing a Contraceptive

	Implant (Norplant)	Injection (Depo-Provera)	IUD	Periodic Abstinence (NFP)	Surgical Sterilization
97%–99%	99%	99%	95–96%	Very variable, perhaps 53–86%	Over 99%
Blood clots, heart attacks and strokes, gallbladder disease, liver tumors, water retention, hypertension, mood changes, dizziness and nausea; not for smokers	Menstrual cycle irregularity; headaches, nervousness, depression, nausea, dizziness, change of appetite, breast tenderness, weight gain, enlargement of ovaries and/or fallopian tubes, excessive growth of body and facial hair; may subside after first year	Amenorrhea, weight gain, and other side effects similar to those with Norplant	Cramps, bleeding, pelvic inflammatory disease, infertility; rarely, perforation of the uterus	None	Pain, infection, and, for female tubal ligation, possible surgical complications
None	None	None	None	None	None
Pill must be taken on daily schedule, regardless of the frequency of intercourse	Effective 24 hours after implantation for approximately 5 years; can be removed by physician at any time	One injection every three months	After insertion, stays in place until physician removes it	Requires frequent monitoring of body functions and periods of abstinence	Vasectomy is a one-time procedure usually performed in a doctor's office; tubal ligation is a one-time procedure performed in an operating room
Rx	Rx; minor outpatient surgical procedure	Rx	Rx	Instructions from physician or clinic	Surgery

4. REPRODUCTION: Birth Control

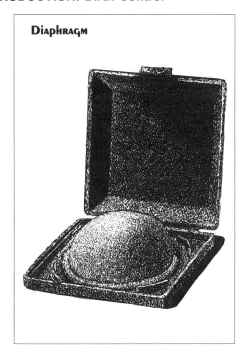

the diaphragm still in place. The diaphragm should be left in place for at least six hours after intercourse. The diaphragm used with spermicide has a failure rate of from 6 to 18 percent.

In addition to the possible allergic reactions or irritation common to all barrier methods, there have been some reports of bladder infections with this method. As with the contraceptive sponge, TSS is an extremely rare side effect.

Cervical Cap

The cervical cap, approved for contraceptive use in the United States in 1988, is a dome-shaped rubber cap in various sizes that fits snugly over the cervix. Like the diaphragm, it is used with a spermicide and must be fitted by a health professional. It is more difficult to insert than the diaphragm, but may be left in place for up to 48 hours. In addition to the allergic reactions that can occur with any barrier method, 5.2 to 27 percent of users in various studies have reported an unpleasant odor and/or discharge. There also appears to be an increased incidence of irregular Pap tests in the first six months of using the cap, and TSS is an extremely rare side effect. The cap has a failure rate of about 18 percent.

Hormonal Contraception

Hormonal contraception involves ways of delivering forms of two female reproductive hormones—estrogen and progestogen—that help regulate ovulation (release of an egg), the condition of the uterine lining, and other parts of the menstrual cycle. Unlike barrier methods, hormones are not inert, do interact with the body, and have the potential for serious side effects, though this is rare. When properly used, hormonal methods are also extremely effective. Hormonal methods are available only by prescription.

Birth Control Pills

There are two types of birth control pills: combination pills, which contain both estrogen and a progestin (a natural or synthetic progesterone), and "mini-pills," which contain only progestin. The combination pill prevents ovulation, while the mini-pill reduces cervical mucus and causes it to thicken. This prevents the sperm from reaching the egg. Also, progestins keep the endometrium (uterine lining) from thickening. This prevents the fertilized egg from implanting in the uterus. The failure rate for the mini-pill is

Methods of hormonal contraception, when used properly, are extremely effective.

1 to 3 percent; for the combination pill it is 1 to 2 percent.

Combination oral contraceptives offer significant protection against ovarian cancer, endometrial cancer, iron-deficiency anemia, pelvic inflammatory disease (PID), and fibrocystic breast disease. Women who take combination pills have a lower risk of functional ovarian cysts.

The decision about whether to take an oral contraceptive should be made only after consultation with a health professional. Smokers and women with certain medical conditions should not take the pill. These conditions include: a history of blood clots in the legs, eyes, or deep veins of the legs; heart attacks, strokes, or angina; cancer of the breast, vagina, cervix, or uterus; any undiagnosed, abnormal vaginal bleeding; liver tumors; or jaundice due to pregnancy or use of birth control pills.

Women with the following conditions should discuss with a health professional whether the benefits of the pill outweigh its risks for them:
- high blood pressure
- heart, kidney or gallbladder disease
- a family history of heart attack or stroke
- severe headaches or depression
- elevated cholesterol or triglycerides
- epilepsy
- diabetes.

Serious side effects of the pill include blood clots that can lead to stroke, heart attack, pulmonary embolism, or death. A

20. Choosing a Contraceptive

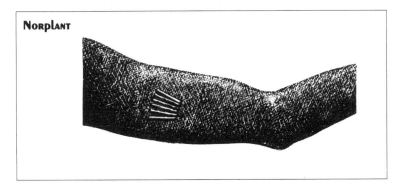

Norplant

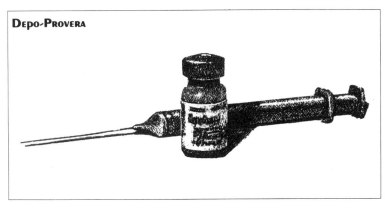

Depo-Provera

Intrauterine Devices

clot may, on rare occasions, occur in the blood vessel of the eye, causing impaired vision or even blindness. The pills may also cause high blood pressure that returns to normal after oral contraceptives are stopped. Minor side effects, which usually subside after a few months' use, include: nausea, headaches, breast swelling, fluid retention, weight gain, irregular bleeding, and depression. Sometimes taking a pill with a lower dose of hormones can reduce these effects.

The effectiveness of birth control pills may be reduced by a few other medications, including some antibiotics, barbiturates, and antifungal medications. On the other hand, birth control pills may prolong the effects of theophylline and caffeine. They also may prolong the effects of benzodiazepines such as Librium (chlordiazepoxide), Valium (diazepam), and Xanax (alprazolam). Because of the variety of these drug interactions, women should always tell their health professionals when they are taking birth control pills.

Norplant

Norplant—the first contraceptive implant—was approved by FDA in 1990. In a minor surgical procedure, six matchstick-sized rubber capsules containing progestin are placed just underneath the skin of the upper arm. The implant is effective within 24 hours and provides progestin for up to five years or until it is removed. Both the insertion and the removal must be performed by a qualified professional.

Because contraception is automatic and does not depend on the user, the failure rate for Norplant is less than 1 percent for women who weigh less than 150 pounds. Women who weigh more have a higher pregnancy rate after the first two years.

Women who cannot take birth control pills for medical reasons should not consider Norplant a contraceptive option. The potential side effects of the implant include: irregular menstrual bleeding, headaches, nervousness, depression, nausea, dizziness, skin rash, acne, change of appetite, breast tenderness, weight gain, enlargement of the ovaries or fallopian tubes, and excessive growth of body and facial hair. These side effects may subside after the first year.

Depo-Provera

Depo-Provera is an injectable form of a progestin. It was approved by FDA in

Only two IUDs are presently marketed in the United States; both have a 4 to 5 percent failure rate.

1992 for contraceptive use. Previously, it was approved for treating endometrial and renal cancers. Depo-Provera has a failure rate of only 1 percent. Each injection provides contraceptive protection for 14 weeks. It is injected every three months into a muscle in the buttocks or arm by a trained professional. The side effects are the same as those for Norplant and progestin-only pills. In addition, there may be irregular bleeding and spotting during the first months followed by periods of amenorrhea (no menstrual period). About 50 percent of the women who use Depo-Provera for one year or longer report amenorrhea. Other side effects, such as weight gain and others described for Norplant, may occur.

Intrauterine Devices

IUDs are small, plastic, flexible devices that are inserted into the uterus through the cervix by a trained clinician. Only two IUDs are presently marketed in the United States: ParaGard T380A, a T-shaped device partially covered by copper and effective for eight years; and Progestasert, which is also T-shaped but contains a progestin released over a one-year period. After that time, the IUD should be replaced. Both IUDs have a 4 to 5 percent failure rate.

It is not known exactly how IUDs work. At one time it was thought that the IUD affected the uterus so that it would be inhospitable to implantation. New evidence, however, suggests that uterine and tubal fluids are altered, particularly in the case of copper-bearing IUDs, inhibiting the transport of sperm through the cervical mucus and uterus.

The risk of PID with IUD use is highest in those with multiple sex partners or with a history of previous PID. Therefore, the IUD is recommended primarily for women in mutually monogamous relationships.

In addition to PID, other complications include perforation of the uterus (usually at the time of insertion), septic abortion, or ectopic (tubal) pregnancy. Women may also experience some short-term side effects—cramping and dizziness at the time of insertion; bleeding, cramps and backache that may continue for a few days after the insertion; spotting between periods; and longer and heavier menstruation during the first few periods after insertion.

Periodic Abstinence

Periodic abstinence entails not having sexual intercourse during the woman's fertile period. Sometimes this method is called natural family planning (NFP) or "rhythm." Using periodic abstinence is dependent on the ability to identify the approximately 10 days in each menstrual cycle that a woman is fertile. Methods to help determine this include:

• **The basal body temperature method** is based on the knowledge that just before ovulation a woman's basal body temperature drops several tenths of a degree and after ovulation it returns to normal. The method requires that the woman take her temperature each morning before she gets out of bed.

• **The cervical mucus method,** also called the Billings method, depends on a woman recognizing the changes in cervical mucus that indicate ovulation is occurring or has occurred. There are now electronic thermometers with memories and electrical resistance meters that can more accurately pinpoint a woman's fertile period. The method has a failure rate of 14 to 47 percent.

Periodic abstinence has none of the side effects of artificial methods of contraception.

Surgical Sterilization

Surgical sterilization must be considered permanent. Tubal ligation seals a woman's fallopian tubes so that an egg cannot travel to the uterus. Vasectomy involves closing off a man's vas deferens so that sperm will not be carried to the penis.

Vasectomy is considered safer than female sterilization. It is a minor surgical procedure, most often performed in a doctor's office under local anesthesia. The procedure usually takes less than 30 minutes. Minor post-surgical complications may occur.

Tubal ligation is an operating-room procedure performed under general anesthesia. The fallopian tubes can be reached by a number of surgical techniques, and, depending on the technique, the operation is sometimes an outpatient procedure or requires only an overnight stay. In a minilaparotomy, a 2-inch incision is made in the abdomen. The surgeon, using special instruments, lifts the fallopian tubes and, using clips, a plastic ring, or an electric current, seals the tubes. Another method, laparoscopy, involves making a small incision above the navel, and distending the abdominal cavity so that the intestine separates from the uterus and fallopian tubes. Then a laparoscope—a miniaturized, flexible telescope—is used to visualize the fallopian tubes while closing them off.

Both of these methods are replacing the traditional laparotomy.

Major complications, which are rare in female sterilization, include: infection, hemorrhage, and problems associated with the use of general anesthesia. It is estimated that major complications occur in 1.7 percent of the cases, while the overall complication rate has been reported to be between 0.1 and 15.3 percent.

The failure rate of laparoscopy and minilaparotomy procedures, as well as vasectomy, is less than 1 percent. Although there has been some success in reopening the fallopian tubes or the vas deferens, the success rate is low, and sterilization should be considered irreversible.

Merle S. Goldberg, a writer in Washington, D.C., has also been involved in contraceptive services for women, both in the United States and developing countries, for the last 25 years.

The Female Condom

Reality is all about women protecting themselves

BETH BAKER

"It was really weird," says Deborah Keaton, of Phoenix, Arizona, recalling her initial reaction to the new female condom, Reality. "It was like . . . BIG. I showed it to my friends, and they couldn't believe it."

"Hilarious!" says Felicia Bembower, of Virginia Beach, Virginia, who also participated in the trial study for the new device. "It made for a lot of laughs."

But Dr. Mary Ann Leeper, senior vice president at Wisconsin Pharmacal, isn't laughing. She's responsible for developing the female condom for the United States market, with women's safety in mind. Reality has two flexible rings, one on either end of a six-inch polyurethane sheath. The sheath is wider than the male condom, but not longer. To use the condom, a woman must squeeze the ring at the closed end of the sheath and insert it into her vagina (similar to fitting a diaphragm). The other ring, at the open mouth of the sheath, extends about an inch beyond the vaginal opening. During intercourse, the prelubricated sheath fills the vagina while the outer ring lies flat against the labia, thus shielding the partners from skin-to-skin contact. Reality requires no prescription and may be used only once. One size fits all.

According to Dr. Leeper, the condom can be used in any coital position (except standing up) and, ideally, won't be felt by either partner during intercourse. The key is the proper amount of lubricant. Eighty percent of the women in the study said they were unaware of the condom during intercourse and some even said it actually increased their pleasure. Overall, 71 percent of the women liked Reality.

Then there's the other 29 percent: "It made me feel like an alien," says Michelle Smith, of Chesapeake, Virginia. "I tried to put it in in advance and the plastic swished when I walked. It's like having a plastic Baggie stuck in you." Others complained that the condom occasionally can become twisted or slip, and that the lubricant makes it messy. Some men said that they were more aware of the sheath than they are of the male condom.

If the Food and Drug Administration (FDA) gives the new device the go-ahead as expected, Reality, to be sold over the counter, will be in general distribution this summer. Each condom costs $2.50 and will be available in packets of three, with a tube of extra lubricant and a detailed instructional leaflet.

The price—nearly three times that of the male condom—reflects the higher cost of polyurethane compared to the latex used in most male condoms. (Perhaps the high price is also a result of the $7 million spent by Pharmacal over the past four years for research and testing to meet FDA requirements.)

Despite this higher cost, Reality does have advantages over the male condom. Polyurethane is a stronger yet thinner material that is a better conductor of heat than latex. Because it is not dependent on a male erection, a woman can insert the device ahead of time. But the most significant factor in its favor is that in covering the labia and the base of the penis, the female condom has the potential to offer the greatest protection against sexually transmitted diseases (STDs), including HIV/AIDS and herpes. As a result, women's health advocates and AIDS organizations have joined forces to put

4. REPRODUCTION: Birth Control

Reality on a fast track for FDA approval.

Last December, despite concerns about insufficient data verifying Reality's effectiveness against STDs, the FDA advisory panel recommended approval of the device, citing the "moral imperative" of HIV prevention. To ensure maximum protection against HIV, some health professionals suggest using a spermicide with the condom; Reality has a 15 percent failure rate. Although some may be tempted to use male and female condoms simultaneously, Dr. Leeper says it can't be done and not to try it.

Reality's biggest selling point is that it will be the first protection against disease—short of abstinence—that women can control themselves. The issue of control extends to its proper use. "With a male condom, I might not know if my partner has put it on properly," says a nurse who participated in the trial study in Virginia. "But with the female condom, I'd know right away if it were misplaced. I'd feel the ring inside me."

Assuming that the female condom is a reliable way to prevent pregnancy and the spread of disease, the big unknown remains public acceptance. Unfortunately, the product's targeted market was not included in the trials. Wisconsin Pharmacal wanted to test the product among women with multiple partners or among couples in which one partner is HIV-positive, but the FDA said no. "We were told, 'It's just not done,'" says Dr. Leeper. Instead, the FDA insisted that the trial study be conducted with monogamous, disease-free couples, whom the agency felt would be more likely to follow the test protocol.

As a result, a valuable opportunity to see how the new device would be accepted by the women most at risk was lost. The research was conducted among married couples, who presumably feel at ease with each other and can more readily deal with the woman emerging from the bathroom with her new appendage. But will young single women have enough self-confidence to use the condom? If her partner is convulsed with laughter, or if his ardor is cooled by her appearance, or if he just plain refuses, will she have the gumption to insist on using it? And will she be able to afford it?

Dr. Denese Shervington, a psychiatrist with Louisiana State University Medical Center, is optimistic. She conducted focus groups among low-income African American women in New Orleans and reports that the women knew the cost, and were still enthusiastic. Wisconsin Pharmacal, the distributor of Reality in the U.S., Mexico, and Canada, has agreed to sell the condom at a discount to public health providers in the U.S. A similar stance was taken by the primary international distributor as well.

Beth Baker is a writer living in Takoma Park, Maryland.

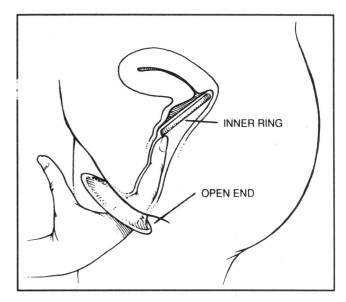

The Femidom In Europe

The female condom is marketed under the trade name Femidom in Europe. It is currently available in Austria, Holland, Switzerland, and the United Kingdom, but by the end of 1993, should be available in most of Western Europe. Femidom sells for about 30 to 45 cents less than Reality. Manufactured by the British firm Chartex International, Femidom is distributed through grocers, drugstores, pharmacies, and supermarkets.

Still new to women, Femidom has sparked an array of responses; initial reactions reflect a mixture of anxiety and naïveté. Women have described Femidom as "gross," "disgusting," "ridiculous," "huge," and "big enough for an elephant or rhinoceros"—responses not unlike their initial responses to other barrier methods.

On first use, women report that Femidom is slippery and sometimes hard to insert. Some find the lubricant too much and others say not enough. Some women complain that the outer ring rubs on the labia and clitoris, while others don't notice the ring at all. Then too there are reports that Femidom does not stay in place during intercourse.

Aesthetics are another consideration: many women do not like the way the device hangs out of the vagina and cite that as inhibiting foreplay. Some point out that it also can be noisy.

Reported advantages are that it virtually never breaks, women can insert it before lovemaking, and some men like this method better. Everyone agrees it takes some getting used to—at least three or four tries before couples feel comfortable and competent. Those who like Femidom best have had problems with other methods of birth control.

—The Boston Women's Health Book Collective

Our Bodies, Ourselves

The Norplant Debate

In Baltimore and a dozen states, a birth-control device raises a hard issue: should poor women be urged—or forced—not to have more kids?

BARBARA KANTROWITZ
AND PAT WINGERT

The Paquin School is a simple brick box of a building in a working-class Baltimore neighborhood. It doesn't look like the setting for a social experiment, but it is. Paquin's 300 students are all pregnant teens or new mothers; last month they became the first students in America to be offered the implantable contraceptive Norplant at school. For these girls, many of whom have babies at 13 or 14, Norplant's promise—no pregnancies for five years—could mean a second chance. "Our dream is to prevent them from getting pregnant again until they're at least 21," says Gracie Dawkins, a Paquin counselor.

But Melvin Tuggle, a black Baltimore minister, thinks making Norplant available at Paquin is genocide. "One third of us are in jail and another third is killing us and now they're taking away the babies," he says. "If the community, the churches and our white brothers don't stand up for us, there won't be any of us left." This week Tuggle and other black leaders say they'll use a Baltimore city council hearing to protest plans to expand Norplant to schools throughout the city. "We won't let a 12-year-old have a drink," he says, "and in the Baltimore school system, a 12-year-old needs a letter from her parents to go to the zoo. She needs permission to get aspirin but she needs nothing to get Norplant."

Or get pregnant. Consuelo Laws's mother didn't like her boyfriend, so she insisted that Consuelo's little brother go with her everywhere. He wasn't much of a contraceptive device. Consuelo was 13 when the doctor told her she was pregnant, "and with that," she says, "my teenage years were over." Now 19 and the mother of two, Consuelo thinks girls should get Norplant as soon as they start menstruating. "My mother thought she could protect me," Consuelo says. When her mother asked how it happened, Consuelo told her: "My little brother would ... play with my boyfriend's brothers and sisters, while we went in another room." A Paquin senior, Consuelo has had Norplant for a year. "Without it I'd probably have more children," she says. "I want to complete my education."

When it was approved by the Food and Drug Administration in 1990, Norplant was heralded as the first innovation in birth control since the pill and the IUD in the 1960s. It's turned out to be as controversial as it is revolutionary. Norplant is being touted as a cure not only for teen pregnancy, but also for welfare dependency, child abuse and drug-addicted mothers.

It's a heavy burden for a half dozen little sticks to carry. The Norplant system, as it is called, consists of six matchstick-size capsules surgically implanted in the arm that slowly release a low dosage of levonorgestrel, the same synthetic hormone used in several versions of the birth-control pill. Norplant is as good as the pill in preventing pregnancy, and its long-lasting effectiveness—up to five years—makes it especially attractive to younger women who want to delay childbearing. It's teenager-proof; girls don't have to worry about remembering to take the pill or use a diaphragm. Its only serious medical drawback: couples still need to use latex condoms to prevent the spread of sexually transmitted diseases, including AIDS. A recent survey of 21,276 women who received Norplant through Planned Parenthood found that the vast majority, 89 percent, were under 30, and 22 percent were 19 or younger. "We need long-term methods like this because failure rates are high" for younger women using other forms of contraception, says Laurie Schwab Zabin of the Johns Hopkins School of Hygiene and Public Health.

NORPLANT — Its capsules are as effective as the pill but last for up to five years. The only serious drawback: couples still need latex condoms to avoid sexually transmitted diseases.

But even Norplant's most ardent supporters are troubled by the way the contraceptive has become the focus of an emotional debate about the fertility of poor women and teenagers. The issues here stretch well beyond poverty to the power of the state to regulate—or coerce—the reproductive choices of women. Put simply, if the state is expected to pay to support the children of the poor, do taxpayers have a say in whether the children will be conceived?

"There are all sorts of reasons why policies that might achieve a good goal—like the reduction of welfare costs and fewer poor babies—give too much authority to the government," says Arthur Caplan, a bioethicist at the University of Minnesota. "I'm not saying the goal is bad, but the means to get there will come at a terrible price, a scary price." Caplan thinks Norplant could be just the beginning of a whole range of efforts to cut public costs through control of reproduction. "I can see us mandating the genetic testing of embryos and fetuses," he says. "If we're willing to put Norplant into a 16-year-old today to contain costs, then why couldn't there be a government official saying you can't be a parent because you're likely to create a kid whose needs will cost society too much?"

To some lawmakers, Norplant is more than a contraceptive; it's a panacea. Last month, in his state-of-the-state speech, Maryland Gov. Donald Schaefer suggested requiring mothers on welfare to get Norplant or get off the dole. "The simple truth is, we've run out of money," he says. "We may be forced to make mothers take care of themselves." He's not alone. In the past two years, according to The Alan Guttmacher Institute, legislators in 13 states have proposed nearly two dozen bills that aim to use Norplant as an instrument of social policy. In Tennessee, officials wanted to pay women on welfare $500 to get Norplant and $50 a year for each year they kept it. The bill was approved by the state House with amendments offering a $500 incentive to men on Medicaid who got vasectomies. The measure foundered in the Tennessee Senate, but the sponsor, state Rep. Steve

4. REPRODUCTION: Birth Control

McDaniel, plans to reintroduce it in the next two weeks. Legislators in other states have proposed requiring Norplant for mothers convicted of felony drug abuse and mothers who have given birth to drug-addicted infants.

Politicians say they've turned to Norplant out of desperation. Last year Walter Graham, a state senator in Mississippi, proposed that his state require the contraceptive for women with at least four children who wanted any kind of government support. His legislation didn't pass, he says, because it got combined with another bill. But a new session has just begun and Graham thinks it will eventually be approved. "The taxpayer is willing to support one child, maybe even two children," he says. "But there's a point where if people want to continue to receive assistance, they will have to have an implant ... Everyone supports the idea of helping the person who *cannot*. They just don't support the concept of helping the person who *will* not."

Norplant isn't just a social issue; it's also an economic one. Because the contraceptive is so expensive—$365 plus $200 or more to have it implanted and then an additional $100 or so to have it taken out—many women can't afford to pay for it out of their own pockets. Medicaid, the health insurance for people on welfare, covers Norplant in all 50 states. That means that only two groups of American women can generally afford Norplant: rich women and very poor women. In the Planned Parenthood survey, 69 percent of women who received Norplant at the organization's clinics had the cost covered by Medicaid. There are no studies of overall Norplant use, but Dr. Michael Policar, Planned Parenthood's vice president of medical affairs, estimates that roughly half of the 500,000 women in this country who use Norplant are covered by Medicaid.

The U.S. distributor of Norplant, Wyeth-Ayerst Laboratories, set the price in this country when it obtained FDA approval at the end of 1990. In 13 other countries where Norplant is used, the cost is much less, as low as $23. All contraceptives are more expensive in the United States than in other countries where medicine is often supported by government and international aid. But Wyeth-Ayerst says its price is fair because it has had to pay the cost of training 26,000 medical practitioners to implant Norplant, at a cost of $1,000 each. There's also the potential cost of litigation—a concern for anyone in the contraceptive business after lawsuits by IUD users.

Given these numbers, Norplant is increasingly viewed as a contraceptive for women on public assistance. That has helped to revive a long-dormant debate over who should control the fertility of poor women. If taxpayers are paying for

T. L. LITT—IMPACT VISUALS

Monica Irving, a mother at 15: 'If someone had talked to me about Norplant before, I would have used it. I think it's the best birth control out.'

their health care and paying for the support of their children through welfare, isn't the government entitled to set limits? In the first half of this century, compulsory sterilization was legal in the majority of states. Dr. Allan Rosenfield, dean of Columbia University's School of Public Health, estimates that between 1907 and 1945, as many as 45,000 Americans—most poor or mentally incompetent—were compelled to be sterilized. Since then, more progressive social policies—combined with the introduction of the pill and the IUD—presented a host of voluntary contraceptive choices. In 1974, the federal government issued guidelines for federally funded sterilizations that mandated counseling, informed consent and a ban on the procedure for minors. Those guidelines became law four years later. Now, sterilization by choice—normally tubal ligation—is the second most popular form of birth control for American couples, after the pill.

Before the introduction of Norplant, most available forms of reversible birth control were by nature voluntary. There's no way to make a woman swallow a pill every day or to force a man to use a condom during intercourse. But Norplant is different. Once inserted, it remains in a woman's arm until a medical practitioner takes it out. "Because it's the closest thing to sterilization, folks have seized on this and tried to impose it on the women who have the least power in our society," says Julia Scott of the National Black Women's Health Project. "They see it as social control for those women who they believe are responsible for all of our social issues."

That certainly wasn't the intent of Norplant's developers, the Population Council, a nonprofit group based in New York that sponsors contraceptive research. Doctors supported by the council began work on an implantable contraceptive in 1966. After years of clinical trials involving 40,000 women in 43 countries around the world, Finland became the first country to approve Norplant in 1983. A Finnish pharmaceutical company, Leiras Oy, is the only manufacturer of Norplant. Over the next decade, Norplant was approved in 23 countries, including Indonesia, Thailand and Colombia.

The United States is the only country where coercion has emerged as a serious issue. Just after Norplant was approved by the FDA, an editorial writer for The Philadelphia Inquirer suggested linking Norplant use to welfare payments. Within a week, after protests inside and outside the newsroom, the paper published an apology. A few months later, a California judge tried to make Norplant a condition of probation for a mother of five convicted of child abuse. The woman appealed, but the case was dismissed last year when she violated another condition of her probation and was sent to jail. In Texas, a judge also ordered a convicted child-abuser to use Norplant. She didn't appeal, but developed medical problems and eventually had a tubal ligation instead.

While policymakers debate coercive use of Norplant, the contraceptive has gained popularity with women. Rosetta Stitt, the principal of the Paquin School in Baltimore, says many students chose Norplant even before it became available at her school. Some Paquin students see Norplant as a way to break a cycle of early motherhood. "The girls we have in our school today have seen two decades of what their aunts and mothers went through because they were teenage mothers, all the things they couldn't do because they had a baby," says Stitt. "They do not want to go through that, too."

Monica Irving, a 15-year-old ninth grader with a 3-month-old son, asked for Norplant right after her baby was born. She lives at home with her mother, Michele, 33. Monica's grandmother looks after the baby while Monica is at school. "I always thought I would get married and then have a child," Monica says. "I never thought of it happening when I was this young. If somebody had talked to me about Norplant before, I would have used it." Michele Irving supports her daughter's choice. "It broke my heart when Monica got pregnant," she says, "but I tried not to let her know that. Some mothers throw their daughters out when this happens. But I didn't. I have tried to back her up. But I am very happy she's on Norplant now."

Most studies have shown that women who use Norplant like it better than their previous contraceptives—mostly because it's much more reliable and effortless once it's implanted. The Planned Parenthood survey of women who received Norplant at the organization's clinics in the last two years showed a very high degree of satisfaction. Only 3.5 percent of women asked to have it removed, a much lower rate than the 10 to 25 percent in earlier Norplant studies. The women who complained generally cited the most common side effects: irregular bleeding, headaches, mood changes and weight gain.

But another study, of Norplant users in Texas, indicated that Norplant's growing popularity could have some negative consequences. Nearly half of former condom users said they wouldn't use condoms now that they have Norplant, putting them at risk for sexually transmitted diseases, especially HIV, the virus that causes AIDS. "If I had my druthers with kids, I would put them in a head-to-toe condom until they were 21," says Columbia's Rosenfield. "The best of both worlds would be if they would use Norplant and a condom, but I know it's hard to get them to use one."

Norplant's supporters say both the coercion issue and the risk of STDs without condoms point to the need for increasing sex education and adequate counseling for women—especially teenagers—who are considering Norplant. "This is one of the most critical issues in the adolescent population," says Dr. David Kessler, the commissioner of the FDA. "I think the risk of people using Norplant and not a condom is very real."

Some of Norplant's critics worry that for teenagers, the implant is a license for promiscuity. "If a girl has Norplant in her arm, the boy is going to say he can screw her because she's protected, she's not going to have a baby," says Tuggle. But Norplant researchers disagree. Dr. Philip Darney, a professor of obstetrics and gynecology at the University of California, San Francisco, participated in the clinical trials of Norplant. He's now studying teens and Norplant. "Some of the girls we have on Norplant are no longer sexually active," he says. "Twenty-five percent of them say they have no current partner—and they would have quit using birth-control pills. They keep the Norplant because they want to be protected if they do find a partner ... In that sense, Norplant makes it easier for them to be sexually responsible." In truth, these are all just tentative conclusions; hard facts remain elusive. One area ripe for study: a comparison of the sexual activity of girls who start on Norplant while still virgins with those who come to it later.

"We're still going to be saying: abstain, abstain, abstain," says Gracie Dawkins, the Paquin guidance counselor. "But if you're not going to abstain, Norplant is an option." For their part, the students say they need better sex education in the earlier grades, too. "My mom had talked to me about my period, but it was all Greek to me," says 21-year-old Kimberly Lucas, the mother of an 8-year-old and a 2-year-old. "I remember her giving me a book with body parts in it, but I didn't realize what the big deal was." Kimberly learned about sex from her boyfriend. "When you're in middle school, and you run into a boy who's 19 and cute, he can teach you about sex in a few minutes. You don't want him to be the one who teaches your kid about sex, but if you don't, he will."

After listening to these sad truths all day, Rosetta Stitt, Paquin's principal, gets angry when she hears people criticize Norplant—or any other method of birth control that can help get young mothers' lives back on track. "Morality is one thing, reality is another," she says. "We have to deal with reality here every day." The reality is that there's no silver bullet to stop teen pregnancy. But Norplant can be an effective weapon in an arsenal that includes sex education, better health care and the possibility of a future worth waiting for.

NORPLANT: A WEARER'S GUIDE

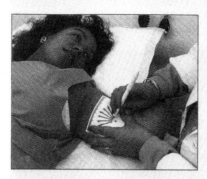

INCISION: After giving the patient a local anesthetic, the nurse makes a 2-millimeter incision on the underside of the upper arm. The nurse then puts the tip of a metal tube through the incision, beneath the skin.

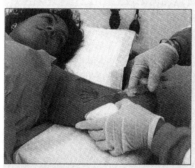

INSERTION: She then loads the first Norplant cylinder into the tube, pushing it with the tube's plunger until it is in place. She inserts the next five Norplants through the tube, arraying them in an arc.

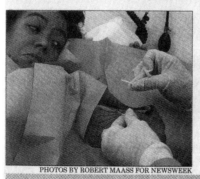

BANDAGING: The nurse removes the tube and presses together the edges of the incision, closing it with a bandage and wrapping the patient's arm in gauze to prevent infection. The procedure has taken about 10 minutes.

PHOTOS BY ROBERT MAASS FOR NEWSWEEK

NEW, IMPROVED AND READY FOR BATTLE

THE ABORTION PILL is finally coming to the U.S., and a breakthrough that eliminates the follow-up shots will make it simpler to use

Jill Smolowe

Abortion is never easy. There is the anguish of the decision, the invasive nature of the procedure, and sometimes an ugly confrontation with right-to-life forces lying in wait outside the clinic door. But imagine if abortion could be a truly private matter. Say, something as easy as visiting a doctor, getting a few pills, returning home to swallow them, then checking back a few days later to make sure that all went as planned.

Science and politics are now conspiring to make that scenario—scary to some, a godsend to others—a reality, one that could allow abortion to be a truly private decision, albeit still not an easy one. Doctors have reported on a pivotal breakthrough in the use of the controversial French abortion drug known as RU 486: a woman who takes the drug will no longer have to go to a clinic for a follow-up injection to induce contractions. Instead, the entire procedure will involve simply taking two sets of pills. Concurrently, President Clinton has firmly signaled a willingness to reconsider the policies of the Reagan and Bush Administrations, which barred RU 486 from the U.S.

The resulting social upheaval could transform one of the nation's most divisive political debates by making abortion far more difficult to regulate. And eventually it could mean abortions will become simpler, safer and more accessible not only throughout the U.S. but also around the world.

Dr. Etienne-Emile Baulieu, the inventor of RU 486, and his French colleagues describe the successful tests of the no-injection method in the *New England Journal of Medicine*. "This new regimen," they conclude, "is simpler and potentially allows greater privacy than any other abortion method." In a tough accompanying editorial, the *Journal* brands efforts to block use of the drug in the U.S. a "disgrace."

Those political barriers, however, are quickly crumbling. Two days after his Inauguration, President Clinton ordered his Administration to "promote the testing, licensing and manufacturing" of RU 486. Until then, the French manufacturer of the drug, Roussel Uclaf, and its German parent company, Hoechst AG, had steadfastly shied away from becoming involved in the American market for fear of infuriating antiabortion activists. But in April, at the instigation of the U.S. Food and Drug Administration, Roussel announced a compromise: it agreed to license RU 486 to the U.S. Population Council, a nonprofit organization based in New York City, which in turn would run clinical tests.

As a result, the abortion pill could become available through a testing program later this year. The Oregon and New Hampshire legislatures have already volunteered their states as test sites, and the FDA is enthusiastic. Says commissioner David Kessler: "If there is a safe and effective medical alternative to a surgical procedure, then we believe it should be available in this country." Although testing a new drug generally takes seven to 10 years, RU 486 has been so widely used in France that U.S. approval could come in as little as two to three years. In the meantime, the testing will enable at least 2,000 women to use the pill.

These developments could change the nature of abortion and even of birth control by eventually permitting the widespread distribution of pills. Though the Supreme Court's *Roe v. Wade* decision of 1973 made abortion legal in the U.S., the ruling was rendered moot in some places by the dearth of doctors willing to perform the procedure and by the fervor of demonstrators who frightened women away from clinics. Now the battleground may shift to the FDA, drug manufacturers and state legislatures.

"We will not allow anti-choice zealots to deny RU 486 to American women," vows Pamela Maraldo, president of the Planned Parenthood Federation of America. The pro-life forces are no less determined. "When they invent new ways to kill children, we will invent new ways to save them," warns the Rev. Keith Tucci of Operation Rescue National. A coalition of antiabortion forces has scheduled a demonstration in front of the French embassy in Washington on June 18, just three days before Roussel Uclaf holds its annual meeting in Paris.

> "This new regimen is simpler and potentially allows greater privacy than any other abortion method."
> —Dr. Etienne-Emile Baulieu, Inventor of RU 486

> "When they invent new ways to kill children, we will invent new ways to save them."
> —The Rev. Keith Tucci, Operation Rescue National

23. New, Improved and Ready for Battle

THE ABORTION DRUG HAS BEEN A source of controversy ever since its invention was announced in 1982 by Baulieu, a French physician who worked as a researcher at Roussel Uclaf. The concept was rather simple: RU 486, an antiprogestin, could break a fertilized egg's bond to the uterine wall and thus induce a miscarriage. An injection two days later of prostaglandin, a hormone-like substance, would force uterine contractions and speed the ejection of the embryo. It took six more years and tests on more than 17,000 women before the French government announced that RU 486 would be made available for public use.

The news spawned furious reaction in the press, an outpouring of outraged letters from Roman Catholic doctors, and a church-sponsored protest through the streets of Paris. A month later, a shaken Roussel Uclaf yanked the drug from the market, saying the company did not want to engage in a "moral debate."

Doctors around the world certainly did. Thousands of physicians had convened that month at a medical congress in Rio de Janeiro, and most of them signed a petition demanding that the French government reverse Roussel's decision. Within 48 hours, Health Minister Claude Evin declared that once government approval had been granted, "RU 486 became the moral property of women," and he ordered Roussel to resume distribution. In 1989 RU 486 was made available to all licensed abortion clinics and hospitals in France. The results proved encouraging, save for a freak incident in 1991 when a woman who was an avid smoker suffered a heart attack while trying to use RU 486 to abort her 13th pregnancy. After that mishap, the government banned use of the pill by heavy smokers and women age 35 and older, who have a greater than usual risk of complications.

Using RU 486 was less painful, carried less risk of infection and gave women greater control over the process than a surgical procedure. Over the next 3½ years, 100,000 Frenchwomen used it successfully. Of those who made the decision early enough, about 85% chose RU 486 over surgery. (The pill is currently used in France only within seven weeks of the first day of a woman's last menstrual period; there is now talk of extending usage to a 10-week interval.) Almost all judged the method satisfactory.

Such promising results persuaded both Sweden and Britain to license RU 486; India is testing the drug. China is manufacturing clones that as yet are not widely available. Other countries, most notably Canada, are waiting for the U.S. to take the lead. "The U.S. is the leader in advanced research, the main source of development funds and the heart of worldwide networks that can allow RU 486 to help women everywhere," explains Baulieu.

HOW RU 486 WORKS

Progesterone, a hormone produced by the ovaries, is necessary for the implantation and development of a fertilized egg.

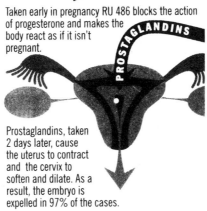

Taken early in pregnancy RU 486 blocks the action of progesterone and makes the body react as if it isn't pregnant.

Prostaglandins, taken 2 days later, cause the uterus to contract and the cervix to soften and dilate. As a result, the embryo is expelled in 97% of the cases.

In 1991 the French began testing the new method of using RU 486 that does not require going to a clinic for a follow-up shot. An oral prostaglandin, commercially marketed as Cytotec by the American manufacturer G.D. Searle, enabled women to abort simply by swallowing a combination of pills. The efficiency rate rose from 95.5% to 96.9%, and the speed of the procedure improved. In 61% of the cases, the uterine contents were expelled within four hours after taking Cytotec, in contrast to 47% in the case of prostaglandin injections. Although there were instances of nausea and diarrhea, which are also common side effects with injections, those who took the pills reported considerably less pain. "Women tolerate it much better," says Dr. Elisabeth Aubeny of the Broussais Hospital in Paris, a testing ground for RU 486 in 1984. For French taxpayers, who foot 80% of the bill for each abortion through their national healthcare system, there is also an advantage: a dose of Cytotec costs only 72¢, vs. $22 for the prostaglandin shot.

Once again, controversy erupted. When Baulieu first began experimenting with RU 486 in combination with an oral prostaglandin, Roussel balked. As a result, Baulieu had to persuade French public health officials to defray insurance costs. After preliminary trials, the government compelled Roussel to participate, arguing that the proposed testing of an oral prostaglandin was important for women. Although Searle raised no objections, its executives remain uncomfortable about being linked to the abortion business. "Searle has never willingly made [Cytotec] available for use in abortion," a company official wrote in a letter to the *Wall Street Journal* in February. "It is not Searle's intention or desire to become embroiled in the abortion issue." Searle's reservations echo that of Hoechst president Wolfgang Hilger, who has been open about his ethical objections to RU 486.

The uses of RU 486 could extend well beyond dealing with some of the 37 million abortions carried out around the globe each year. European studies have shown that it is an effective morning-after pill, inducing less nausea or vomiting than other drugs used for the same purpose. There are also indications that RU 486 can combat endometriosis, a leading cause of female infertility, and fibroid tumors, a condition that often necessitates hysterectomy. Thus the same drug that can help some women end unwanted pregnancies may enable others to bear children. Assorted studies have found that RU 486 may also combat breast cancer and Cushing's syndrome, a life-threatening metabolic disorder.

Despite the many potential uses for RU 486 and its effectiveness as an abortion method, efforts to legalize it in the U.S. have met with repeated failure. Last year a pro-choice group called Abortion Rights Mobilization decided to force a court challenge of the import ban imposed on RU 486 by the Bush Administration in 1989. The organization helped Leona Benten, a pregnant 29-year-old California social worker, fly to England, obtain a dose of RU 486, then try to bring it into the U.S. through New York City's Kennedy Airport. Customs officials seized the pills. The ensuing legal battle went up to the Supreme Court, which refused to order the government to return the pills. Benten subsequently had a surgical abortion.

The Clinton Administration has not yet revoked the ban, but its significance is minor. Because distribution of the pills is tightly controlled in Europe and they cannot easily be purchased and imported, the real issue is how quickly the Administration will encourage the manufacture and marketing of the drug in the U.S.

When the pill does become available in America, abortion will not be as easy as going to the doctor and taking some of the tablets home—at least not right away. In

4. REPRODUCTION: Birth Control

France, for instance, a woman is required to pay four visits over a three-week period to one of the country's 800 licensed clinics or hospitals. The first step is a gynecological exam. Doctors make sure the pregnancy is in its early stages, and a social worker or psychologist discusses with her the decision to abort. Then the woman is sent home for a weeklong "reflection" period.

When she returns, she is required to sign a government form requesting the abortion. She must also sign a Roussel form that confirms her understanding that a malformed fetus might result if she does not see the abortion through to completion. (As yet no defects have been found in the small number of babies born to women known to have taken RU 486.) At that point, the woman is given three aspirin-like RU 486 tablets, each containing 200 mg of the drug. After swallowing the pills, she again goes home.

Except in the rare instance where the RU 486 is enough to induce a quick abortion, the woman must take two 200-mg Cytotec pills within the next 48 hours. Because the timing is critical and doctors want to monitor the effects of this contraction-inducing drug, women are required to return to the clinic. They are encouraged to remain for four hours, even if the expulsion happens earlier. Eight to 10 days later, they must pay a final visit for an exam to make sure no part of the egg remains.

Even with all these steps, the procedure seems blessedly simple to most women. "Taking a pill seems far less murderous and violent to the child than using a vacuum cleaner," says a 31-year-old woman who has had both types of abortion. "You feel so helpless when they put you to sleep and you know they're going to be using their tubes and knives on you." Some women, however, become traumatized by the thought of performing an abortion with their own hand. After her experience with RU 486, Joelle Mevel, 34, vows that if there is a next time, she will choose surgery. "I spent the whole time worrying that I would see the child in the basin, that I would be able to discern something human in the blood," she says. "I would rather have gone to sleep and awakened later knowing it was all over."

American abortion-rights advocates talk of boiling France's time-consuming RU 486 procedure down to just two visits to the doctor. It would be possible, though controversial, for the government to let RU 486 be administered in any doctor's office or possibly even by trained nurse practitioners. If that happened, many women could avoid running a gauntlet of protesters outside an abortion clinic. Still, it won't take all the anguish out of the procedure. "It's insulting to women to say that abortion now will be as easy as taking aspirins," says Baulieu. "It is always difficult, psychologically and physically, sometimes tragic."

—Reported by
***J. Madeleine Nash/Chicago, Frederick Painton/
Paris, Janice C. Simpson/New York and
Tala Skari/Paris***

Before Roe v. Wade
DESPERATION
After Roe Is Reversed

LINDA ROCAWICH

Linda Rocawich is the Managing Editor of The Progressive.

Rusty coathangers are not what I'm talking about. Yes, women have used coathangers. Knitting needles. Weight-lifting. Jumping up and down. When they are pregnant and don't want to be—it's the wrong man, the wrong time, the wrong place—they will do what they need to do to abort. *Desperation* is what I'm talking about. The Supreme Court can overturn *Roe v. Wade*, and probably will, but women will have abortions. They had illegal ones before *Roe* and they will again, if that's what they must do.

Women and men who are too young to remember what we women did before January 22, 1973, are the ones who will be most affected if the Court outlaws abortion once again next June or July, because those young people are the ones in their prime years of fertility.

When young women need abortions, will they have them safely, legally, in a sterile setting, performed by a trained professional whose reputation they can check? Whose name they know? Who will follow up afterward if something goes wrong? Or will they be forced to resort to the harrowing means sought by their mothers and older sisters? Will they be made to feel like criminals or—as some in the anti-abortion movement would have it—actually become criminals?

In her new book *Moving the Mountain: The Women's Movement in America since 1960*, Flora Davis includes a chapter on the movement leading up to *Roe*. "The Relegalization of Abortion," she calls it, because abortions *were* legal in the United States until around the turn of the Twentieth Century. And she cites estimated statistics on the 1960s: More than a million illegal abortions were done every year. And, as she writes, "They were performed by moonlighting clerks, salesmen, and barbers, and—less often—by doctors willing to risk imprisonment. Every year, more than 350,000 women who had an illegal abortion suffered complications serious enough to be hospitalized; 500 to 1,000 of them died."

Many of us who are over forty remember an illegal abortion in vivid detail, whether our own or one we helped a friend through, but many younger people haven't thought much about that experience. This is what we went through.

I started this story with calls to friends. What struck me was that every woman I called had a story. Then, over the past few months, their friends started calling me. Some of these women have been active in the movement to save abortion rights, but some were talking about those abortions for the first time in twenty or more years.

4. REPRODUCTION: Birth Control

Each individual included here stands in for the thousands of others like her. I've used only their first names; some names are real and some are not.

Carol was sixteen and pregnant, living in a rural area of upstate New York in 1966. Her mother tried but failed to find a doctor nearby who would provide an abortion. But Carol's grandparents, who lived in south Florida, helped. No one who lived in the Miami area and read the papers could fail to know about the illegal-abortion rings there.

I know, because I lived there, too. I was a senior in a high school for Catholic girls; at least a third of my classmates were Cuban refugees, and their new situation was a source of constant discussion. Many of those early refugees—my classmates' fathers—were professionals. Cuban physicians—many if not most of whom had trained in U.S. medical schools—were cut off from practicing legally in Florida because the state required U.S. citizenship for licensing. Most took whatever legal work they could find, but many did illegal abortions. When the police busted a "clinic," it hit the papers.

Carol's grandfather arranged an appointment; she doesn't know how.

Carol and her mother met a Cuban man in a Miami shopping center's parking lot. He was, she says, "pleasant but businesslike and aloof." She got in the front seat of his car. When her mother tried to get in also, he said, "No, you can't come," and drove off. "He made me hide on the floor," she says.

"We drove around for what *seemed* like hours," she recalls. "But it was at least an hour, lots of turns. He was making sure we weren't tailed and probably making sure I'd never be able to find the place again. I was alone. All I had was a little plastic purse with $1,500 cash in it, to pay for the abortion.

"He asked me what I was going to do for birth control in the future. I was scared. I said something like, 'Oh, I guess I just won't do it anymore.' He just looked at me and rolled his eyes.

"We were in a residential neighborhood, upper-middle-class houses. He turned into a driveway and the garage closed behind him. We went into the house through a side door. He took me into this room. It was a child's room, just like every little girl always wanted. Pretty white bedspread, frilly curtain, stuffed animals. And it was clean. Really clean.

"Another man came in, and they took the spread off the bed, and did something that turned it into a hospital bed, an operating table. He took out his instruments. It happened so fast. One minute I'm in this little girl's room, the next it's like a hospital.

"He gave me sodium pentathol. I wasn't out very long, but it was over. Then we drove around again for a long time, taking me to meet my mother. I was only gone from her for three or four hours.

"There were no complications, a little cramping is all. I'd had periods that were worse.

"It turned out fine. But I was never allowed to discuss it. It was hush-hush secrecy at home. I was not allowed to tell *anybody*. It was years before I could talk about it and not break out in a cold sweat."

Carol's safe, illegal abortion was still a horrifying experience. Carol has a daughter of her own now, who will be sixteen in a few years. I spoke with Carol a second time, a month or two later, and asked if she thought she could put her daughter in a car with a strange man and let her drive off.

"Never."

In the interim, she had spoken with her mother. She told me how agonizing it had been for her mother, that she was backed up against the wall, that "her facts then differed from mine now. She didn't have any choices, any more than I did."

Desperation is what she was talking about.

Deborah's story is a lot like Carol's. She was seventeen and pregnant in 1967, and responsible since her father's death a few years earlier for the care of some younger children. And, she says, "There was no way I was gonna marry this guy." An older brother, a soldier, arranged the abortion.

She met a man in a restaurant parking lot, late at night in a city in the Middle West. He blindfolded her and drove around—for about half an hour, she thinks—before arriving at a three-story house on a street where all the houses looked alike.

Her abortionist, too, was a trained physician, but he had a legitimate practice in another part of town. "It was not painful," she says. "I don't remember pain." He gave her a number to call if there were complications.

She did need treatment for infections—which she attributes not to a faulty procedure but to a lifelong propensity to infection she blames on both parents' being victims of medical experiments in Nazi death camps. The physician took care of her infections at his legitimate office, as a regular patient.

She spoke of being terrified, but she had friends whose experiences were much worse.

She is terrified again. She does not want her friends and business associates to know about her abortion, but she has a sixteen-year-old daughter. I found Deborah through an outreach worker at a legal clinic in the city where she lives, where she does what she can, anonymously and privately, to make sure her daughter and other young women never have to relive her story.

When she asked me to call her Deborah, she said, "That's as in 'Awake, Deborah, awake.' You might not know, but that's from the Old Testament."

Gertrude is younger. Legal abortion has been an option for her, but she knows what her mother did in the 1950s. Her mother had one-year-old Gertrude and a set of two-year-old twins to take care of, and she was pregnant again. She couldn't cope with the thought of another child just then.

With money tight and no way to find a safe way out, Gertrude's mother used a case of soda, one of those old wooden crates with twenty-four heavy, full glass bottles. She lifted it up, down, up, down, until she miscarried. Gertrude's sisters remember the blood.

Barbara had an abortion in Philadelphia in the 1960s, legally, but she needed certification from a psychiatrist that turned it into a "therapeutic" procedure. The gynecologist who was to perform the abortion referred her to a colleague. "You know," she said, "I had to break down and cry. Say my life will be ruined. I'm the sole support of my family. Everything will fall apart." She said it didn't take long, probably not even half an hour. The first time we talked about it, I had asked if she had exaggerated. "Oh, I embellished," she said, "but not by much." Other women who went through such experiences—some states required evaluations by two psychiatrists, though Barbara only had to see one—have described that part as worse than the abortion itself.

Ginnie was a college classmate of mine, nineteen and pregnant by another classmate whom she would later marry. Both felt they weren't ready for marriage or a family. Both also *knew* that their parents would cut them off if they didn't finish college first and wouldn't help with an abortion. So their friends pitched in to send Ginnie to a doctor in Canada. We called it passing the hat; it wasn't the first time and wouldn't be the last that friends helped out, knocking on friends' doors in her dorm and his fraternity house saying, "Ginnie needs a doctor. Can you help?" People knew what we meant, and they always helped.

Many rape victims have faced the prospect of bearing the rapist's child. In most places, there were ways around the abortion prohibitions in such a situation, but there was much less openness about rape twenty-some years ago. So I was sure there were stories to be told. I didn't find anyone who would talk

24. Desperation

though a few women said they had a friend or a cousin who had confided in them.

I called Kathryn Marshall, a writer in Livingston, Montana. I remembered that when we both lived in Austin, Texas, in the 1970s, there had been a serial rapist on the loose, a man who was climbing through women's bedroom windows and not limiting himself to any particular part of town. Katy often spent her nights sitting up reading in the rotunda of the Capitol, which is open and lighted twenty-four hours a day and has lots of armed guards wandering around. She wasn't getting much sleep.

She had been raped in Dallas in 1971, and she wasn't taking chances. She survived, and she wasn't pregnant, but abortions were illegal. I asked what she would have done if she had needed one.

"I would have gone through *any* channels," she said quietly. "No matter what the cost, no matter what horror stories I'd heard from friends." And she reminded me that her novel *My Sister Gone*, published a few years later, has a particularly harrowing scene patterned on a friend's bungled abortion.

Of course, we are talking here about a middle-class woman with resources, who could have gotten on a plane for a legal abortion in New York or California, and probably would not have risked an illegal abortion in Dallas. But still, she says, if a back-alley butcher had been her only alternative, she'd have taken her chances rather than carry the fetus.

There were nurses and midwives in rural areas who took care of poor women, especially in the South. Some of their stories involve herbal abortifacients, potions to be drunk, and descriptions of what sounds like hypnosis, most likely to deal with pain.

By the late 1960s and early 1970s, there were networks to help poor women by a sort of underground railroad. The Jane Collective in Chicago is now well known, but the feminists there weren't alone. One friend told me about a Texas group that regularly drove women to a clinic just across the Mexican border, a safe, clean place with a sympathetic physician. Another mentioned a North Carolina woman who arranged trips to New York when its doctors offered the nearest legal abortions.

Those networks involved less trauma for a pregnant woman than a midnight rendezvous with a man she'd never seen before and a confusing drive to a house in the suburbs. Nevertheless, this method meant a drive of hundreds of miles, many hours in a car. Going one way, she's pregnant, often in her morning-sickness days; on the way back, she's just had an abortion and may be in pain.

But women will organize and use such networks again if they have to, if they can't go to a legal clinic in their own town or some nearby city.

By 1970, for less money than Carol spent on her illegal Miami abortion, a woman could try the services of the London Agency, Inc., in Springfield, Massachusetts. The agency didn't provide then-illegal abortions in its home state. It made arrangements for a package deal in England: airfare, passports, health certificates, hotel accommodations, and an abortion in a private London hospital, for $1,250. It required a letter from the woman's American physician providing "satisfactory evidence that an abortion was necessary, and [that the woman was] physically and mentally able to accept the operation."

When the legality of its advertising was challenged and it faced a hearing before the Massachusetts Supreme Court, the agency's attorney explained, "We are acting only as a kind of travel agency which expedites trips and makes the passenger as comfortable as possible."

This service or one like it is the sort of thing we can expect to pop up again if *Roe* is overturned. A woman with money, in a state where abortion is illegal, will be able to check the Yellow Pages to find an entrepreneur willing to profit from her need. Poor women will be stuck.

Another kind of pre-*Roe* abortion service is also likely to reappear. The politicians—and the people who like to legislate morality—should have learned from the Prohibition era that when they outlaw a product or service that people want, organized crime gets involved.

In February 1970, for example, a Detroit cop was appointed to be police chief of Cleveland. A few days later, newspapers in Cleveland and Detroit revealed that in his years on the street he had earned an extra $1,000 a week in protection money. What he was protecting was a Mafia-run abortion ring.

Both the chief and the mayor who appointed him denied the charge. But the chief resigned a week after his appointment.

Restrictions on abortion such as that faced by Barbara, who needed a psychiatric evaluation that, as she put it, makes you persuade them that having the baby will make you crazy, offer profitable opportunities somewhere in between the travel agency and the Mafia. New York was one of the places that required two corroborating evaluations. A nurse who worked in a New York City hospital in the 1950s recalls that young, underpaid, attending psychiatrists "were in heaven" when an ob-gyn physician called them in. Neither bothered doing much of an interview, but they split the $500 fee they charged the pregnant woman.

There are another thousand variations on the theme of what women did before *Roe v. Wade*.

Today, women who confront situations of the sort their older sisters or their mothers faced twenty or thirty years ago can look to safe, legal alternatives. They don't have to feel like criminals. But they aren't stupid and they know that their younger sisters, their daughters, may not have the same options when their time comes.

At Reproductive Health Services in St. Louis, the well-respected clinic that challenged Missouri's restrictions on abortions in the 1989 *Webster* case, patients come back three weeks after their abortions for medical check-ups. They are also then asked to answer some questions about why they chose abortion and what they think they would have done if they couldn't have done so safely and legally.

Amelia McCracken, who handles the clinic's community outreach, shared some of the responses with me. The clinic asks the women to sign these questionnaires only with their first names and zip codes—the zip code because the compiled responses are a part of Reproductive Health's continuing efforts to persuade elected officials not to criminalize abortion. The copies she made for me had that information blanked out. I haven't fixed the grammar or spelling.

Women's reasons for seeking abortions never change:

¶ "I'm not married. Not financially capable at this time."

¶ "Not wanting any *more* children."

¶ "I was raped and couldn't mentally deal with having this man's child."

¶ "We were so hongry and winter was coming I couldn't see how I would take care of a baby. I believe you shouldn't bring into this world what you can't take care of. This is very hard for me to fill out I am crying. This child was wanted."

¶ "I had an abortion because I was taking prescribed drugs, that were harmful to a baby without knowing I was pregnant. I'm not financially enough secure or responsible enough to take care of a child, and at the time I found out I was pregnant, I was kicked out of the house."

Most of these women would not have had access to legal abortions under the restrictive laws sought by the anti-abortion movement.

Another question asks, "What would you have done if abortion were not available in a safe and legal clinic?"

¶ "I had an abortion because I was taking prescribed drugs, that were harmful to a baby without knowing I was pregnant. I'm not financially enough secure or responsible enough to take care of a child."

¶ "I would have gotten a back alley abortion or taken an over dosage of downers to have a miscarriage. Something, because I don't want a baby right now."

¶ "Either commit suicide to lose the baby or fount some kind of drug or way to misscarrage. I would've tried my best 1 way or another to lose it."

121

4. REPRODUCTION: Birth Control

¶ "I would have gone to an unsafe/illegal one and probably be dead or in the hospital right now. Or I would have found some way to abort it myself."

¶ "I would have sought out an illegal abortion or I could have possibly become an abusive parent just because you are forced to have a child by no means makes it wanted or loved."

¶ "I would have looked for an illegal abortion service or choose an easy way to comitt suicide."

¶ "(1) Flown to another country where abortion is legal. (2) Sought out a person to perform it. (3) Attempted to do procedure myself."

All of these responses are taken from one recent two-day period. The most striking thing about them is this: Every woman talks about seeking some unsafe, illegal way to get what she wants. Or she mentions suicide.

The final question is: "What would you say to those who want to make abortion illegal and close down all abortion clinics?"

¶ "It hurts enough without you. We don't need you to hurt us. Our parent, our boyfriend and our friend hurt enough. . . . We don't need pro-lifers to make our decisions."

¶ "They have no right to decide whats best for other people."

¶ "Remind them that abortion is as old as sex. . . . Hand them a bumper sticker saying AGAINST ABORTION? DON'T HAVE ONE."

¶ "I think if they want to do this and succeed, they will have a bigger problem than they realize if women are forced to take other measures. If a person is vulnerable and have their mind set, a woman will do almost anything. Please don' make us go through that."

¶ "Theres a lot Id like to say but there not enough paper. There is one thing Abortions are a very important thing fo females and they should be legal *forever.*"

Women have a message for the men and women who think they have a right to tell us what to do when we're desperate.

A twenty-one-year-old woman who had an abortion at Reproductive Health a few months ago can send that message. When asked what she would have done without a safe, legal way, she wrote, "I don't know perhaps suicide. I do know though if abortion is illegal any doctor is welcome to perform safe abortions at my house. I won' let girls stick hangers up them."

The global politics of abortion

Why do lawmakers, clerics, husbands—not women—have the right to choose?

Jodi Jacobson

Voltaire once described history as a "set of fables agreed upon." Much the same can be said of the current public policy debate on abortion.

The abortion controversy is often portrayed as a conflict between black and white views, with virtually every person firmly planted on one side or the other, unswervingly for or against the right to choose abortion. But on this issue the canvas of social morality is actually painted in every imaginable hue. Social discomfort with abortion often exists alongside the notion that abortion may best be viewed as a "lesser evil" and a necessary adjunct to public health and women's freedom of determination.

In most societies with liberal abortion laws, public opinion polls bear this out. In the United States, for example, many surveys show that while the majority of people are to varying degrees uncomfortable with abortion, they are opposed to governmental interference in a woman's right to choose to end an unwanted pregnancy. Half the adults surveyed by *The New York Times* in 1989 supported the availability of abortion as specified by *Roe v. Wade*. Only 9 percent felt abortions should not be permitted at all. Likewise, four out of five Britons believe the decision should be a private matter between a woman and her physician. In those developing countries where opinion surveys have been carried out, similar sentiments are evident, especially among urban dwellers and people with higher levels of income and education.

Abortion politics, however, has been heavily influenced by those who seek to completely ban abortions, except perhaps in cases where a woman's life is at stake. More and more, the public policy debate on abortion has been shaped by a series of myths, based on a kind of "moral absolutism" that is perpetuated by abortion rights opponents. This absolutist view is blind to the vast public health and social costs of restrictive abortion policies.

The first and most pervasive myth is that there is theological unanimity regarding a woman's right to end an unwanted pregnancy. In fact, religious doctrines have been interpreted differently at different periods in history and by different theologians.

Whether the religion in question is Catholicism, Islam, or Judaism, historical evidence indicates a diversity of opinion and practice regarding induced abortion. Early Christians condemned abortion but did not view the termination of a pregnancy to be an abortion before "ensoulment," which was equated with "quickening" and generally taken to mean the end of the first trimester. While the distinctions between "formed and unformed" fetuses were eliminated by Pope Pius IX in 1869, therapeutic abortion for medical reasons was not explicitly or publicly condemned by any Roman Catholic authority before 1895. Today, Catholic canon law assigns embryonic life equal importance to that of the mother from the moment of conception.

No other major religion has a consistent or unified position on this issue. Islamic law, for example, allows abortion through the fourth month of pregnancy, although few fundamentalist Muslim countries grant women this right. Within Judaism, the Orthodox and Hasidic sects prohibit abortion, while the Reform and Conservative branches do not.

In spite of strict teachings, women of every faith have defied dogma to end unwanted pregnancies. Illegal abortion is widespread throughout heavily Catholic Latin America, for example, and in

4. REPRODUCTION: Birth Control

the United States, 32 percent of all abortions are obtained by Catholic women—who account for only 22 percent of the female population.

The second myth is that criminal laws will eliminate abortion, which provides the underlying justification for the modern-day crusade to ban this procedure. But why focus on banning abortions when history has proven that laws cannot eliminate them but can only make them less safe and more costly?

Try as it might, no government has ever legislated abortion out of existence. In Romania under Nicolae Ceausescu, policies made preventing unwanted pregnancies virtually impossible. Contraceptives were outlawed. A special arm of the secret police force, Securitatae—dubbed the "pregnancy police"—oversaw monthly checkups of female workers. Pregnant women were monitored, married women who did not conceive were kept under surveillance, and a special tax was levied on unmarried people over 25 and on childless couples who could not give a medical reason for infertility. No Romanian woman under 45 with fewer than five children could obtain a legal abortion. Despite the law, both abortion and abortion-related mortality rates in Romania rose precipitously.

Abortion rights cannot be separated from broader struggles to gain equality in all facets of women's lives.

Recent estimates indicate that more than 1.2 million clandestine abortions were performed each year in Romania—a country of 23 million—as compared with some 1.6 million legal procedures carried out annually in the United States, a country with 11 times as many people. One survey found that Bucharest Municipal Hospital alone dealt with 3,000 failed abortions in 1989; other sources indicate that well over 1,000 women died within that city each year due to complications of botched procedures. Legalization of abortion in Western Europe and the United States, by contrast, has produced the world's lowest abortion-related mortality rates. Moreover, in several European countries the widespread availability of family planning information and supplies has precipitated a drop in the number of abortions.

Another myth holds that abortion is not a method of family planning, a notion that even abortion rights advocates have unwittingly helped to further. This myth serves a dual purpose. For one thing, it allows moral absolutists to perpetuate social discomfort about the practice of abortion. Second, it provides a convenient escape for politicians who want to straddle both sides of the fence by supporting "family planning" but not abortion.

Ignoring abortion's critical role in the spectrum of family planning services is counterintuitive and is counterproductive to the goal of reducing the number of induced abortions. (Contraceptives reduce, but do not eliminate, the need for abortion as a backup to their own failure: 7 out of 10 women using a 95 percent effective method of birth control, such as the mini-pill or the IUD, would still require at least one abortion in their lifetimes to achieve a two-child family.)

Unfortunately, the "abortion is not an acceptable form of family planning" myth is so strong it has permeated every level of public policy. In the United States, it has been used as a justification for denying poor women federal funding of abortions, even in cases of rape and incest. The Reagan and Bush administrations have used this dubious reasoning as the basis for severing ties between publicly funded providers of contraceptives and providers of abortion.

Internationally, private voluntary groups providing abortion services or counseling as part of their programs are prohibited from receiving U.S. funds unless they sign a contract promising to end these activities. Although a few, such as the International Planned Parenthood Federation, can afford to refuse U.S. contributions, most smaller organizations are compelled by financial need to sign on.

Apart from curtailing already limited access to safe abortion services throughout the Third World, U.S. policy has other far-reaching and insidious effects on women's health. The interpretation of "abortion services" has been read by both the Reagan and Bush administrations to prohibit giving advice and information about medical reasons for abortion and its legal availability, as well as lawful lobbying for abortion rights.

Abortion myths allow politicians to avoid dealing with the effects of limiting access to safe procedures on public health and on reproductive freedom. Moreover, absolutist arguments against abortion rights provide cover to those groups opposed to or threatened by the empowerment of women in full possession of their reproductive rights. Looking behind the myths reveals the genuine tensions in the abortion debate.

What, then, is the abortion debate really about? As abortion rights activist and medical doctor Warren Hern writes in *Birth and Power: Social Change and the Politics of Reproduction* (Westview Press, Boulder, 1990), it is a struggle over "who runs our society . . . self-determination . . . individual choice, personal freedom and responsibility." This is particularly and painfully true for women. The struggle for abortion rights cannot be separated from the broader struggles of women to gain equality in all facets of life, from family and domestic issues to parity in the workplace.

For most groups, the term "pro-life" translates

plainly into "anti-family planning." In California, groups pressured the state to slash funding from 500 family planning clinics that provide health care to the poor. According to one study, the funding cuts—the $24 million budget was originally slashed to zero and then restored to $20 million—would have led to nearly 86,000 more pregnancies in California, at least half of which would have been aborted. Clearly, abortion itself is not the only target of attack.

Curiously, once a child is born, he or she might not get so much attention from these crusaders. The commitment of the so-called pro-life movement to social services for disadvantaged children is hardly evident. Where, for example, is the international outcry over the thousands of children in Romanian warehouses? In the United States, there is little simultaneous push among opponents of abortion for adequate prenatal care, maternal and infant health care, day-care services, or increased access to contraceptives among groups most at risk of unwanted pregnancy.

Black women fight silence on abortion

ON APRIL 9, 1989, MORE THAN 300,000 PEOPLE CONVERGED ON the White House in what has been called the largest women's rights march in history. On that day, feminists, lesbians, grandmothers, flight attendants, and movie stars stood shoulder to shoulder, demanding full reproductive rights for all women.

It was a landmark day—but it did not take a polster to see that something was missing. The sea of women that washed over the city was overwhelmingly white and middle class. Where were the women of color?

Ultimately, the answer depends on how the question is asked. White women ask, "Why aren't more black women involved in the pro-choice movement?" Black women ask, "Why doesn't the movement involve more black women?"

No one disputes the need to involve more women of color, but many fail to realize that any genuine effort to expand African-American participation will have to take into account the reality of our everyday lives. Most black women have not read the statistics on abortion—they have lived them.

According to recent studies, about 3 percent of American women aged 15 to 44 end an unwanted pregnancy in abortion. Of those, minority women are more than twice as likely as white women to have an abortion. The figures also show that nearly half of all black females become pregnant as teenagers—90 percent of them while they are unmarried.

Contrary to any stereotype, an abortion has never been a decision of convenience for most black women. The death of a fetus never takes place without guilt, shame, and emotional upheaval, especially confronted as we are with the very survival of our race. We have not forgotten the days when abortion was illegal, and those memories have forced us into a conspiracy of silence. Abortion is seldom discussed in our communities, even among families who know a relative or neighbor who has had an abortion.

This painful conspiracy of silence pervades the black community, stifling our voices, denying us the shared wisdom of our own experiences. "How can a silent community be a committed one?" asks Loretta Ross at the National Black Women's Health Project.

Many women active in the reproductive rights movement blame the black church for playing a conspiratorial role in promoting this silence. Black ministers denounced abortion when it was illegal, preaching warnings of hell and brimstone from their pulpits while member after member of their congregations—often members of their own immediate families—sought abortions in the still of the night. Now, in the era of legal abortion, the church continues its warnings of eternal damnation, laying the burden of sin at the feet of its poorest parishioners.

Black activists point out that the church gave birth to the civil rights movement, and they argue that it should be at the forefront once again in the struggle for freedom of choice.

With or without the church, black activists say they will continue to fight for their human rights—including their reproductive rights. But the question remains: How can we break our silence and join ranks with white women in the struggle for freedom of choice?

Above all, black activists stress that whites who genuinely want to work with them must place abortion and reproductive rights in a wider context. "White organizations will have to offer broader issues that directly affect us, and then offer position statements that are relevant," says Brenda Williamson, former director of the North Carolina Religious Coalition for Abortion. "Those issues must demand more than the right to abortion. They must demand the right of a woman to have a healthy baby if she chooses, a decent job, and a good education."

Evelyn Coleman
Southern Exposure

Excerpted with permission from the regional magazine Southern Exposure *(Summer 1990). Subscriptions: $16/yr. (4 issues) from Southern Exposure, Box 531, Durham, NC 27702.*

4. REPRODUCTION: Birth Control

Men & abortion

A few questions to think about

EVERY WOMAN, WHETHER SHE EVER CON-ceives a child or not, has to live with this possibility in a way men do not. Yet many men feel compelled to impose on women their beliefs about reproductive rights, and I cannot escape the feeling that part of the reason is these men's desire to have power over others.

I am intimately aware—through my four children—of the costs of conceiving, bearing, birthing, and raising children. Many articles detail the financial costs of raising a child; the emotional and psychological costs are not as easy to catalog. Love and responsibility are not free, but cost a great deal of energy, anguish, self-doubt, groping for wisdom, and flat-out pain. With each child I made a 20-year leap of faith that I would be able to meet the child's needs, integrate one more complicated person into the family, and enter into a lifelong commitment to that unknown person. The first few years after conception involve a physical dependency that at times makes it seem that the woman will never regain her own physical autonomy. With diminishing physical dependency come increasingly complicated emotional and mental demands.

Several years ago, I had a bad strep infection and came close to dying. I missed a period and panicked. I had all that I could handle in my life at that time. I had reached a point where finally my body was not a physical necessity for any of my children. I was starting to make some long-term plans to better my working conditions and explore ways my talents could be used. The thought of another pregnancy and the long sweep of time between birth and my even partial emancipation, so recently hard won, left me angry, resentful, distrustful of my husband's ability to cope, and so very weary. Were this another baby coming, how long would I have to wait before I could put some energy into my own independent self? How can you, if you have never been in my position, or (in the case of men) if you never can be in my position, judge my reactions? In the name of God/Goddess, do not dismiss us as selfish if we cannot face the burden and responsibility of giving and maintaining life.

In my struggle to bring something creative out of my pain, frustration, and raging response, I wrote these queries for men and any others who might find them useful.

IF YOU FEEL OVERWHELMED BY ALL MY QUE-ries, be gentle with yourself, answer one or two in your heart. Create queries yourself and talk about them with another man. Start or join a men's consciousness-raising group.

To be against abortion is not enough. As a man, you must find creative ways to take away the occasion for abortion. Take these concerns to your brothers. Until men learn to behave responsibly, women will have to continue to make choices that may involve abortion.

In the workplace: Do I give more than lip service to supporting quality day care for parents of both sexes who need that service? Do I support a living wage for day-care workers? Do I expect the men with whom I work to be concerned and involved fathers? Am I willing to sacrifice some of my ambitions at work to provide the nurturance my children need? Do I insist on furthering my own career at the expense of my partner's pursuit of her own career, because I expect her to do a better job of nurturing our children?

At home: Am I "helping her out" or assuming half the responsibility for my child(ren)? If I am not the primary care giver, do I find ways to acknowledge the intelligence and patience, etc., of the person or people meeting my child(ren)'s needs and wants?

In the world: Am I careful to keep myself from making or tolerating sexist jokes and comments? When I have sex, am I assuming my share of the responsibility for contraception? Do I ask myself if I would honestly be willing to co-raise a child to maturity, should one result from our union? Am I willing to take comparable risks with my health, sanity, and personal career that I expect women to take when I insist that the women carry all babies conceived to term? *And raise them over a 20- to 30- year period?*

—Laurie Eastman
Friends Journal

Excerpted with permission from the Quaker publication Friends Journal *(July 1989).*

In developing countries, the issue of "choice" is often not so much about a "better" life as it is about the fundamental right to life itself. In societies where cultural constraints on women are strong and they remain economically and politically subservient to men, reproduction is simply one among many aspects of their lives in which women lack self-determination.

From childhood well into their reproductive years individual women have little—if any—power to determine at what stage they become sexually active, whom they are bonded to, when sex will take place, or when and how to bear children. Women may be forced into unwanted sexual contact and unwanted pregnancies through violent attack, sexual coercion, or the more socially acceptable arranged and often forced marriage. Once a woman is wed, decisions on the timing and number of births are more often the prerogative of her husband and family members than her own. Even where the means to prevent pregnancy are available to women in the Third World (and this is relatively rare), lack of spousal support often leads to high rates of contraceptive failure.

Equally reprehensible is the subjugation of women's lives to those of state-enforced pro-natalism. Nicolae Ceausescu is only one among many heads of state who have relegated women to reproduction in the interest of "national security." As demographer Judith Bruce states quite plainly, for women who by no choice of their own face unwanted pregnancy, "abortion is the final exit from a series of enforced conditions." For a large share of the millions of women whose only option is illegal abortion, that final exit is death.

The tremendous social gains to be reaped from eliminating illegal abortions cannot be ignored. First among them is a reduction in abortion-related maternal mortality of at least 25 percent and in related illnesses of far more. Reductions in illegal abortions and unwanted pregnancies would save billions in social and health care costs, freeing these resources for other uses.

Only by increasing access to family planning information and supplies, offering couples a wider and safer array of contraceptives, and placing abortion within an improved and comprehensive public health care system can the number of abortions be reduced. Some countries have already chosen this common-sense approach. Italy, for example, now requires local and regional health authorities to promote contraceptive services and other measures to reduce the demand for abortion, while Czech law aims to prevent abortion through sex education in schools and health facilities and through the provision of free contraceptives and associated care. Some countries now require post-abortion contraceptive counseling and education; some mandate programs for men as well.

Many of these efforts registered success quickly. On the Swedish island of Gotland, for example, abortions were nearly halved in an intensive three-year program to provide information and improved family planning services. Similar results have been seen in France and elsewhere.

The steps needed to make these gains universal are plain. Decriminalization and clarification of laws governing abortion would secure the rights of couples around the world to plan the size and spacing of their families safely.

While the way is evident, the will is lacking. The missing ingredient is political commitment. Natural allies—representatives of groups concerned with women's rights, environmental degradation, family planning, health, and population growth—have failed to mount a concerted effort to dispel abortion myths. And despite the overwhelming evidence of the high human and social costs incurred by restrictive laws, abortion politics remains dominated by narrowly drawn priorities that reflect only one set of beliefs and attitudes. Respect for both ethical diversity and factual accuracy is a precondition for a truly "public" policy on the question of abortion. Reforming restrictive laws may stir opposition. Failing to do so exacts an emotional and economic toll on society—and sentences countless women around the world to an early grave.

MAKING BABIES: MIRACLE OR MARKETING HYPE?
Risks, Caveats and Costs

The hype is that women are frivolous, if not downright evil, frittering their time away in law school and then demanding fertility treatment

Elayne Clift

News stories call them "a worldwide epidemic," "a public health problem," and "an economic issue." Successful parents call them "a dream come true." They are the babies—often several per birth—resulting from reproductive technology, a booming business in this country and elsewhere. In the United States alone, approximately 270 fertility clinics exist and infertility care is estimated to be a $1 to 2 billion-a-year business, with each attempt at assisted reproduction ranging from $2,500 to $10,000, depending on the method tried.

Based on surveys over the last decade, the U.S. Public Health Service says there are at least 2.3 million "infertile" couples in America who have not managed to conceive after one year of unprotected intercourse. (The World Health Organization (WHO) uses two years as its standard definition.) Since 1978, when the famous "test-tube baby" Louise Brown was born in England, approximately 20,000 babies have been born in the U.S. through treatments ranging from drug therapy (Clomid and Perganol are the most commonly prescribed) to in-vitro fertilization (IVF), and other techniques in which eggs are harvested, fertilized and implanted.

More than 3,100 babies have been born through IVF in each of the last two years. But that number can lead to false optimism. Estimates of successful outcome (which many call "a take-home baby") vary widely, with 9 to 14 percent being the accepted range.

The rapid proliferation of clinics established to attract and serve infertile couples does raise serious health, economic and ethical questions. For example, what are the health risks to multiple-birth babies? Are clinics profit-driven, promoting more expensive techniques, offering IVF to women who are not suitable candidates, or encouraging those interventions which insurance companies will cover? What are the ethical issues raised by storing thousands of frozen human embryos, the "leftovers" from high-tech fertility treatment? Is the high rate of multiple births that result from reproductive interventions placing an unfair burden on neonatal intensive care units in hospitals? (Overall, according to Dr. Louis Keith, professor of obstetrics at Northwestern University Medical School in Chicago, the number of twins in the U.S. rose 33 percent from 1975 to 1988; the number of triplets increased by a staggering 101 percent during the same time period.) In an interview with the *New York Times* last May, Keith said, "This is a public health problem because we are producing an incredible number of children who are at grave risk for prolonged stays in the neonatal intensive care unit and all of the complications of prematurity." In the same article, Dr. Emile Papiernik, who until recently had worked at the hospital that produced the first French test-tube baby in 1982, called the rise in multiple births due to fertility-enhancing drugs a "worldwide epidemic." Papiernik said that at the hospital where he practiced, half the babies transferred from maternity to neonatal intensive care are from fertility-induced pregnancies. The cost of such care has led countries with national healthcare such as England and France to restrict the number of embryos implanted during IVF to three, to prevent multiple births.

The questions being asked are important, but are they the right ones? Barbara Katz Rothman, sociologist, women's health advocate, and author of *In Labor: Women and Power in the Birthplace,* is uncomfortable with conventional queries arising from the boom of hi-tech fertility. "Insofar as access to any medical service is an issue, then yes, it's an economic issue," Rothman says. "But whether or not IVF and its consequences are expensive is not really relevant to the question." The discussion, she says, should be driven by women's health concerns, and not the marketing interests of clinics, whose advertising techniques Rothman sees as the real problem. "The hype is that women are frivolous, if not downright evil, frittering their time away in law school and then demanding fertility treatment long before they need it to produce exceptional babies. It is a victim-blaming discussion." In fact, says Rothman, "the economic system doesn't permit women to have a child when they want one; it creates a world in which it's very hard for women to do what they want, and then blames the women."

In addition to positioning women as cranky and demanding, marketing techniques often contribute to a couple's perception that they are infertile, and media coverage frequently suggests that fertility techniques have more to offer than they really do. Says Rothman, "You always see the two smiling women with their six babies, never the one woman who made it and the six who didn't. Or the mother with breast cancer and we don't know why. Or the woman whose triplets have died."

20,000 babies have been born in the U.S. through treatments ranging from drug therapy to in-vitro fertilization

---

Health risks do exist, and are underplayed, for both mothers and infants. Ann Pappert, adoptive mother, health journalist, and author of *Cruel Promises: Inside the Reproductive Technology Industry* (Simon & Schuster, 1993), sees clever marketing as leading to the notion of "miracle babies." But, she says, "the health problems are buried in the warm glow of promotion which positions the doctor as brave and humanistic and the client as lucky and adoring." Both Pappert and Rothman think IVF and other techniques are used inappropriately in many cases and that they are "fraught with problems for mothers and babies." For example, what are the long-term risks for women taking Clomid, Perganol and other drugs, including Lupron, which is approved by the Food and Drug Administration only for treatment of endometriosis but which is often prescribed for infertility? IVF also raises questions. As Pappert points out, "No clinical trials anywhere in the world have measured the safety and efficacy of IVF, but since there are over 200 clinics providing it, we say, 'Oh, it must be safe.'" WHO agrees that IVF and related technologies have not been adequately evaluated. A recent report states that "serious risks are associated with IVF. The ovarian hyperstimulation syndrome occurs in 1 to 2 percent of women treated with ovulation-inducing drugs. Multiple gestation occurs in approximately 25 percent of IVF pregnancies. The perinatal mortality rate for IVF babies is four times and the neonatal mortality rate twice that of the general population. The rate of very low birthweight among IVF babies is over 11 times higher than in the general population."

In Australia, where every IVF attempt and every IVF birth are followed, data reveal that less than 5 percent of babies resulting from IVF are considered to be healthy, primarily because of their low birthweight and related problems. Considered an international standard, the Australian registry results are dismissed by Dr. Duane Alexander, Director of the National Institute of Child Health and Human Development (NICHHD) at the National Institutes of Health. According to Dr. Alexander, NICHHD has a voluntary IVF registry and in a one-year follow-up study of 100 children, no physical or developmental problems were identified. The Australian data, he says, have been "dismissed," and problems with low-birthweight babies are a "blip on the screen." IVF studies in the U.S. have been curtailed since a NICHHD Ethics Advisory Board was disbanded in 1980 by then-President Ronald Reagan, leaving bioethical research tied to fetal tissue in limbo.

Still, as one recent article in *Newsweek*'s "Business" section put it, "The in-vitro fertilization business is taking off," adding credence to the concerns of Rothman and Pappert about aggressive marketing techniques. It is chilling to hear one executive of a holding company which supports a chain of IVF clinics say that "the market has barely been scratched," or to think of clinics becoming "the Burger King of baby making," both statements reported in *Newsweek*. Says one physician, "There's a certain amount of merchandising in IVF." A *New York Times* article last summer cited IVF America Inc. for its "ambitious growth plans" to become "the McDonald's of the baby-making business."

Claims like these prompted Rep. Ron Wyden (D-OR) to introduce legislation last year to regulate IVF clinics. In a statement to a Congressional Subcommittee on Health and the Environment, Wyden said, "Couples seeking help for an infertility problem are bombarded with advertising claims which have touted success rates of 30, 40, 50 percent or more. They don't know that a minority of clinics are responsible for the most successful IVF births, let alone *which* clinics have the best track record in treating patients with their specific infertility problem. And they don't even know that there's no one watching to make sure that these facilities meet even *minimal* quality controls." Wyden's bill, which was passed by Congress in October, calls for fertility clinics to report their pregnancy success rates (definitions of success vary but are usually defined as the percentage of IVF treatments that result in a live birth) and for the federal government to publish these rates annually along with the names of reliable embryo laboratories being used by clinics. In addition, a model program for the inspection and certification of embryo labs would be promulgated for states to adopt. Any state failing to comply with this code would be cited in the annual consumer guide book to be produced by the Department of Health and Human Services.

Wyden's legislation is supported by the American Fertility Society, an organization of health professionals concerned with infertility, the Society for Assisted Reproductive Technology, which conducts its own yearly review of fertility clinics, and Resolve, Inc., a national advocacy group for infertile couples. Amy Hill, Resolve's Twin Cities Chapter Board chair and a consumer member of the Ethics Committee of Abbott Northwest Hospital in Minneapolis, thinks the legislation has helped give needed exposure to the issues of reproductive technology. "The risks are real," she says, "and truly informed consent is critical."

Hill agrees with Rothman and Pappert that potential risks and consequences must be put on the table for potential clients. "It is a very expensive, very stressful process with so many unknowns," she says. But she shies away from an analysis which frames problems primarily from a feminist perspective, and believes that most clinics behave responsibly. "We need to address the needs of infertile couples and their health issues," Hill says. "I see my mission with Resolve as giving access to others."

Hill is active in trying to increase insurance coverage for infertility treatment. Currently only 10 states require insurers to provide limited coverage. Because private health insurance companies have been reluctant to pay for IVF, consumers must bear substantial costs. With each treatment costing anywhere from $2,500 to $10,000, to obtain a baby a couple may have to go through several treatments with a final total of over $20,000. According to government estimates, Americans spend $1 billion a year to combat infertility.

Whatever the risks, the caveats, the expense, women like Amy Hill continue to employ assisted reproductive technologies, despite the limited success rate. Hill is also one of the lucky ones. She will give birth this year to her second child. "The joy," she says, "is inexplicable."

Reproductive Revolution Is Jolting Old Views

Gina Kolata

Suppose that Leonardo da Vinci were suddenly transported to the United States in 1994, says Dr. Arthur Caplan, an ethicist at the University of Minnesota. What would you show him that might surprise him?

Dr. Caplan has an answer. "I'd show him a reproductive clinic," he said. "I'd tell him, 'We make babies in this dish and give them to other women to give birth,'" And that, Dr. Caplan predicted, "would be more surprising than seeing an airplane or even the space shuttle."

It was only 15 years ago that the world's first baby was born through the use of in vitro fertilization, the method of combining sperm and egg outside the body and implanting the embryo in the uterus. That step ushered in a new era of reproductive technology that has moved so far, so fast that ethicists and many members of the public say they are shaken and often shocked by the changes being wrought.

Although its aims are laudable—helping infertile couples to have children—the new reproductive science is raising piercing challenges to longstanding concepts of parenthood, family and personal identity.

It is now possible for a woman to give birth to her own grandchild—and some women have. It is possible for women to have babies after menopause—and some have. There may soon be a way for a couple to have identical twins born years apart. It may also become feasible for a woman to have an ovary transplanted from an aborted fetus, making the fetus the biological mother of a child.

These events are frightening, Dr. Caplan said. "I never thought that technology would throw the American public into a kind of philosophical angst, but that's what's going on here," he said.

Dr. Mark Siegler, director of the clinical ethics program at the University of Chicago, said that it does not matter so much whether large numbers of people use the new technologies. Their very existence throws long-held values into deep confusion. They introduce, he said, "the idea that making babies can be seen as a technological rather than a biological process and that you can manipulate the most fundamental, ordinary human process—having children."

In the most recent stunning development, a scientist at Edinburgh University said last week that he was working on transplanting the ovaries of a fetus into infertile women, adding that he can do this already in mice. The scientist, Dr. Roger Gosden, said that a 10-week-old human female fetus has six million to eight million eggs. He estimated that it might take a year for a newly transplanted fetal ovary to grow and start producing mature eggs. Several ethicists pointed out that an ovary transplant could lead to a bizarre situation: women who donated their fetus's ovaries might become grandmothers without ever being mothers.

Dr. Gosden said he also expected to be able to help women undergoing chemotherapy, which can destroy the ovaries, by freezing strips of a patient's own ovary and then putting them back after the treatment was over. He has perfected this method in sheep, he said. In principle, both of these ovary transplant methods could also allow women to avoid menopause altogether, remaining potentially fertile until the end of their days.

A few months ago, researchers at George Washington University said they had cloned human embryos, splitting embryos into identical twins or triplets. The method could enable couples to have identical twins born years apart.

But ovary transplants and clones are in the future. Already here are other methods that were unimaginable just a short while ago. For example, doctors routinely freeze extra human embryos that are produced by in vitro fertilization, storing them indefinitely or until the couple asks for them again. Thousands of embryos rest in this frozen limbo in freezers around the nation.

THE MEANING OF 'MOTHER'

Using donated eggs, women can now have babies after menopause. And egg donors are enabling women to give birth to babies to whom they bear no genetic relationship.

Each new development gives rise to new questions of personal identity, experts say. For example, several ethicists said they were repelled by the idea of using eggs from a fetus to enable an infertile woman to become pregnant. They asked, for example, how a child would feel upon discovering its genetic mother was a dead fetus.

The cloning experiment elicited questions of what it says about the uniqueness of individuals if embryos can be split into identical twins or triplets. Just as copying a work of art devalues it, might copying humans devalue them?

Egg donors raise the question of what it means to be a biological mother. Is the mother the woman who donated her egg or is it the woman who carried the baby to term? At least one ethicist and lawyer, George Annas of Boston University, argues that the only logical answer is that both are the mother. This means, he added, that for the first time in history, children can have two mothers.

Pregnancy after menopause threatens the ancient concept of a human life cycle. Until now, there was a time when a woman's reproductive life came to an end. Now, she can be fertile indefinitely. "It makes the life course incoherent," Mr. Annas said.

4. REPRODUCTION: Pregnancy and Childbirth

The growing hordes of frozen embryos call into question the status of these microscopic cell clusters. Do they deserve some consideration as potential humans or is it acceptable to simply discard them if they go unclaimed? Are they potential brothers and sisters of children already born, or are these just spare cells, not much different from blood in blood banks?

"It really is a brave new world," said Jay Katz, a law professor at Yale University.

TROUBLED BY TECHNOLOGY

Yet, said Dr. Susan Sherwin, a professor of philosophy and women's studies at Dalhousie University in Nova Scotia, "the profound question is whether we treat these developments as profound or take them for granted." Although, she said, "our technologically oriented society is fairly good at adapting quickly," the new reproductive technologies give her pause.

For example, Dr. Sherwin said, the possibility of pregnancy after menopause means that there no longer is a point when an infertile couple must give up their quest for a child. If women can have ovarian transplants, will it be even less acceptable for them to grow old without fighting it through medical intervention?

"An enormous industry has grown up in recent years to postpone or prevent menopause through hormone replacement therapy; now reproductive life can also be prolonged," Dr. Sherwin said. She added, "There are questions of what we value in women."

Making babies can now be seen as a technological, not a biological, process.

Mr. Annas said he hopes that people in the next century will look back on these reproductive technologies and find them "as strange as can be." He added: "We have a couple of billion more people on this planet than we need. What are we doing trying to figure out new ways to let couples have more babies? On a societal level, we have to ask, 'What's driving this?'"

But others said they thought the new technologies are heralding a future so astonishing as to be almost unbelievable.

"We're starting to see hints of 21st century reproductive technology," Dr. Caplan said. "We're getting early sightings and we can start to talk about them."

DEATH AND BIRTH

Dr. Siegler predicted that the technologies would change views of the start of life just as profoundly as the developments in the past quarter century changed views of life's end.

A generation ago, Dr. Siegler said, "death was not optional," adding, "When people reached the natural end of their lives thorough a failure of their heart or liver or kidneys or lungs, they died." But then researchers discovered kidney dialysis, machines that could take over kidneys and allow people to live even when their kidneys had failed. And they developed respirators that allowed people to keep breathing when they no longer could draw a breath on their own. Those machines, Dr. Siegler said, "opened a new rang of options about life and death."

"By what process do people decide to use them?" he asked. "Who is in control? When do you start? When do you stop?"

As a consequence of these technologies, Dr. Siegler added, "the first 25 years of medical ethics has been devoted heavily to death and dying and informed consent."

COPING WITH MISCARRIAGE

Grieving for an unborn child is natural,
and emotional recovery takes time.

Dena K. Salmon

Dena K. Salmon *is a free-lance writer based in Queens, New York, where she lives with her husband, daughter, and new baby.*

The day before Mardi Gras in New Orleans, the city is like a person on a bender: boozy, familiar, and disorderly. Most people are off from work, and those in a party mood add to the general madness. It is not a good time to have contracts drawn up or mail delivered. Nor is it a good time to have a medical emergency.

My obstetrician was out of town when I noticed the first sign of trouble. My husband and I instead consulted his medical group's senior partner, Michael Prince,* an elderly, delicate man with two hearing aids. Despite his infirmity, the exam was quick and thorough.

"When did the bleeding start?" he asked.

"Last night, around three o'clock." He cupped a hand around his ear.

"Three o'clock!" I shouted.

"Cramping?"

"Yes!"

"How many weeks pregnant did you say you were?"

"Fourteen!"

He shook his head and looked grim. "You feel about five weeks along. Are you sure of your dates?"

I nodded miserably. Although my husband had some hope that the bleeding would not lead to miscarriage, I was pessimistic. Recently, I had begun to feel somehow less pregnant: My breasts seemed to have shrunk, I could suddenly bear the smell of fish again, and my energy level seemed higher. I was uneasy and didn't know why. Now I knew. The pregnancy was over.

An ultrasound exam revealed tiny, identical twins, far too small for their gestational age. There were no heartbeats, but Prince suggested we wait a few days until my regular obstetrician returned. The bleeding and cramping continued for the rest of the week and were ended by the inevitable D&C (a procedure in which tissue is scraped from the lining of the uterus). Although this is not always the case, at my hospital the tests, surgery, and anesthesia related to my miscarriage proved more costly than when I had given birth to my daughter there two years earlier. Never had an irony seemed so cruel; never had a loss seemed so devastating.

Rationally, I knew that there must have been something very wrong with my twins and that I was mourning two babies who were never meant to be. But logic gave me no comfort. I still felt a rage and a sorrow that seemed inappropriate—for wasn't it a blessing that the pregnancy had ended then and not after I had given birth to two nonviable babies? Sensible thoughts only spurred my guilt for not being able to conquer the depression. "I should be over this already," I'd repeat to myself as gray days turned into weeks. But the full sense of my loss kept washing over me, whether or not I accepted my feelings. Months later, when I finally began talking about my experience with other women who had had miscarriages, I realized that my feelings were natural and not at all unusual.

A lonely experience.

Many women report that almost from the moment a pregnancy is confirmed, a baby exists in their imagination. I, too, had the sense of carrying a real baby, even when it was merely an embryo, its cells dividing and multiplying. My husband and I had very much wanted another baby. Also, I was aware of the pregnancy from very early on. Each moment of lethargy, each wave of morning sickness, reminded me that I had invited an unknown and unpredictable element into my future. It was frightening and exhilarating all at once. But since it was *my* future that was altered and *my* body that experienced the changes of pregnancy, my grief and sense of failure were not understood by those for whom the baby was an abstract concept.

Kindly meant, tactless comments delivered a subtle message that I was overreacting, somehow blowing the event out of proportion. I began to feel defensive about my sadness. This reaction is commonplace, but no one understands it so well as one who has been through a similar experience. (See "When Someone You Know Miscarries.")

Amy, a friend who miscarried in her third month, recalls, "People would try to say comforting things, that I was young, that I could still have another baby. I found it totally beside the point that I could have another one. All the hopes and plans I had had for this baby were gone. No one would try to discourage a person from grieving for a husband or a parent; no one would dream of saying at a funeral, 'You'll have another husband one day.' But that's

*Some names and identifying comments in this story have been changed in the interest of privacy.

4. REPRODUCTION: Pregnancy and Childbirth

the way some people dealt with my miscarriage."

"Why me?"

Many of the women I spoke to found themselves asking, "Why me?" They worried that maybe it was something they had done that had caused the miscarriage. Could it have been that glass of wine? Too much exercise? Not stopping the birth control pills early enough? (See "Miscarriage Facts and Fallacies.")

Meredith Vieira, a *60 Minutes* correspondent, feels that she wasted a tremendous amount of time blaming herself for her four miscarriages. But once she pushed aside guilt and focused on her grief, she was able to look back on her miscarried pregnancies more positively. "It's been four years since the first miscarriage, and there is still an aching," says Vieira. "You can't lose life and be cavalier about it. The baby is part of you; it's your family, and then it's gone. You grieve for it. That's human—that's healthy. But for me, underneath the sadness are a lot of warm memories attached to being pregnant, and attached to the life that was inside of me. Pregnancy made me suddenly feel optimistic, young, vibrant—like something wonderful was happening, because it was. When I think in those terms, those babies live for me in a very real way. They affected my life in ways that will never go away. They have made me stronger, more reflective, more loving toward the child that I have now."

Seeking support.

Being able to remember the lovely, hopeful moments of a pregnancy that ended in miscarriage is a stage not easily reached. Fresh from the loss, you find it unlikely that something positive could ever come from it. Yet women report that they have become stronger and more sensitive after the experience and more understanding of others' sorrows. This is why so many women find it helpful to seek out those who have gone through a similar experience. Judy, who lives in upstate New York, found comfort and support by joining a group in Schenectady called Haven. "I walked around for weeks feeling as if someone had stabbed me in the heart, it ached so much," Judy says. "But once my husband and I joined the group, we met others going through the same thing. Just talking about your experience is so important after a prenatal loss, and so is listening to how others are coping. It helped me live through it."

Some women who mourn deeply for a miscarried child find that they have trouble letting go of their grief, as if coming to terms with the miscarriage and going on with their life means turning away from the child. Letting go of sadness can be difficult for some, since it is all that is left to remember the baby by. Betsy, a teacher, suffered four miscarriages before she finally carried a baby to term. By the fourth miscarriage, her sense of loss had become unbearable. "I had cried so much that I was amazed I had any tears left or the energy to shed them. One night, as I sat on the beach, I looked up at the sky and had a kind of mystical experience: I talked with the baby. I explained that I didn't know what had happened to prevent him from being here but that I had to let him go—I couldn't hold on to his image anymore. And if he ever wanted to come back, the family would love to have him. That's all I could say, but it helped me stop crying. It helped me deal with the reality that there wasn't going to be this little baby. It helped me accept the fact that there had been a death."

Not everyone feels a profound sense of loss after a miscarriage. Anne, a physician who has a ten-year-old daughter, became pregnant accidentally and miscarried after ten weeks. "I was very ambivalent about having a second child at that point and had many concerns about how I would manage a new baby and my career. My miscarriage was like the fulfillment of a secret wish. The

---

When Someone You Know Miscarries

● Always acknowledge what has happened. Avoiding the subject will alienate your friend.
● It is better to address the issue directly by saying something like "I'm sorry to hear about your miscarriage" than it is to say something trite, such as "It's really all for the best." Some other things *not* to say:
"A miscarriage is nature's way of sparing you an imperfect child."
"Don't be so sad. At least you didn't lose a child."
"You are still young—you can always have another."
"At least it happened now, before you felt life."
"Miscarriages are so common. Try not to obsess about it."
"Try to forget about the miscarriage. Your husband and children need you."
● Be a sounding board. Allow your friend to retell the events of her pregnancy and miscarriage. Letting her air her feelings is often the most helpful thing you can do.
● Do not encourage her to dull her emotions by using alcohol or drugs.
● Encourage her to be good to herself—to take vitamins and iron if her doctor has recommended them, eat a good diet, exercise when she feels strong enough, and indulge in little luxuries, such as massages, facials, or sauna baths.
● Buy her a book about miscarriage. Here are some recommendations:

Empty Cradle, Broken Heart: Surviving the Death of Your Baby, by Deborah L. Davis, Ph.D. (Fulcrum; $12.95)

Preventing Miscarriage: The Good News, by Jonathan Scher, M.D., and Carol Dix (Harper & Row; $18.95)

When Pregnancy Fails: Families Coping With Miscarriage, Stillbirth, and Infant Death, by Susan Borg and Judith Lasker (Bantam; $8.95)

● Suggest that she find a support group for bereaved parents. Often, obstetricians and hospitals can recommend one. Also, the following national bereavement groups have local chapters throughout the country. The central agency can provide information on the group nearest you.

The Compassionate Friends
P.O. Box 3696
Oak Brook, IL 60522-3696
708-990-0010

Resolve, Inc.
5 Water Street
Arlington, MA 02174-4818
1-800-662-1016

SHARE
Saint Elizabeth's Hospital
211 S. Third Street
Belleville, IL 62222
618-234-2415

—D.K.S.

worst part was feeling guilty over my lack of maternal feeling. Even though I am a doctor and know that there is no single *correct* way to handle the emotions of a miscarriage, it took me a while to stop being so self-critical."

Some women become reclusive after a miscarriage, preferring to share their feelings with only their husband and family. Emily, who always kept her personal life and professional life separate, says, "None of my colleagues knew I was pregnant, and none of them knew that my 'minor surgery' was really a D&C. I missed only two days of work because physically, I felt okay. But if everyone had known what had happened, I couldn't have coped. I needed privacy. This miscarriage was a personal tragedy, and my husband and I went through it alone, with the support of our families. We hadn't been married that long and had never had to deal with a crisis like this; we were worried for each other and cared for each other. As horrible as the miscarriage was, it was good to know that we could pull together the way that we did."

A husband's feelings.

Since husbands are expected to "be strong" for their wives, men's feelings are often lost in the shuffle. Also, they may not experience the miscarriage with equal intensity, since they don't have the physical experience of pregnancy. But some men who keep their feelings private do grieve over the loss. "I didn't want to add to my wife's troubles," says Kevin. "And I didn't think that she could have handled what I was going through, on top of what she felt. Unfortunately, this caused problems down the line because she thought I didn't care." Men who discussed their feelings of loss with their wives frequently observed that their marriage became closer and more open after the grieving period.

Talking and spending time with my own husband helped me through the most difficult times. But I also needed to express my feelings to others and to have my loss acknowledged. When I returned, a week later, to my job as a high school English teacher in a boys' preparatory school, I gave my students a real-life lesson on how to deal with someone who has experienced a loss. The boys had many questions about miscarriage, and quite a few volunteered stories about sad experiences they had had. The conversation that ensued with each class was so affecting that two weeks later, during parent-teacher conferences, almost all the parents who came to see me mentioned how moved their sons had been by the lesson. Some mothers became so involved in reliving miscarriages of ten to fifteen years ago that they never got around to asking about their son's progress. One woman even took my hand and, with tears in her eyes, urged me not to try to get pregnant before the worst of the emotional ordeal was behind me.

By the end of the evening, I was astonished and moved by the tremendous outpouring of emotion that I had heard from women still affected by their miscarriages; after all this time, they still wanted to talk about them. I realized then that time does heal wounds, but not completely, and that even having other children does not make one forget about the child who was never born.

Miscarriage Facts and Fallacies

The facts:
- A study of women trying to conceive, published in the *New England Journal of Medicine*, found that 31 percent of pregnancies never make it to term and that 22 percent of these pregnancies end even before the women know they are pregnant.
- About 80 percent of all miscarriages occur in the first trimester of pregnancy. Many of these are due to a chromosomal abnormality in the fetus, a random event that is unlikely to be repeated in subsequent pregnancies.
- Women who have had only one miscarriage are likely to have other pregnancies that do go to term.
- Women who have had two consecutive miscarriages or a total of three miscarriages are more likely to have another pregnancy loss and should be evaluated by a high-risk–pregnancy specialist.
- Miscarriages that occur during the thirteenth to twentieth weeks of pregnancy are generally caused by abnormalities in the uterus or the cervix rather than in the fetus. After twenty weeks, a fetal death is called a stillbirth, not a miscarriage.
- Women are more likely to miscarry if they are over 35.
- Smoking cigarettes or drinking alcohol has been found to increase the risk of miscarriage.

The fallacies: None of the following activities contribute to a miscarriage.
- Reaching your arms up high over your head, for example, to hang a picture.
- Taking a warm bath.
- Riding a bike.
- Driving over a bumpy road.
- Having sexual intercourse. —**D.K.S.**

Unnecessary Cesarean Sections:

Halting a National Epidemic

Public Citizen's Health Research Group released its third report on unnecessary cesarean sections, entitled *Unnecessary Cesarean Sections: Halting a National Epidemic* on May 12, 1992. The report disclosed that, for the first time, the national rate of cesarean births actually decreased. Based on 2.8 million deliveries in 1990 and 2.6 million in 1989, there was a statistically significant decrease in the proportion of deliveries done by cesarean section, from 23.0 percent in 1989 to 22.7 percent in 1990. The national c-section rate had almost quintupled in 18 years — from 5.5 percent in 1970 to 24.7 percent in 1988 (according to data based on a much smaller number of deliveries from the National Center for Health Statistics - NCHS). Now, based on the Health Research Group's much larger sample (more than two-thirds of all deliveries in the U.S.), the cesarean epidemic seems to have turned the corner. If hopes are realized the decline will continue until the rate reaches 12 percent, or about one-half the 23.5 percent reported by NCHS for 1990 (the most recent year for which finalized data is available).

HRG's latest report represents the most comprehensive source of information on hospital and state cesarean section rates available. It includes cesarean rates for 2,657 hospitals in 34 states and statewide data for 47 states and the District of Columbia (every state except Colorado, Oklahoma and South Dakota). Here are some highlights:

❏ *The five highest statewide cesarean rates* were for Arkansas (27.1 percent in 1989 and 27.8 percent in 1990), Louisiana (27.3 percent in 1990), New Jersey (27.0 percent in 1989), and the District of Columbia (26.6 percent in 1990).

❏ *The five lowest statewide cesarean rates* were for Minnesota (17.6 percent in 1990), Wisconsin (17.5 percent in both 1989 and 1990), and Alaska (15.3 percent in 1990 and 15.2 percent in 1989).

❏ *The 10 hospitals with the highest cesarean rates* were — Abrom Kaplan Memorial Hospital, Louisiana (57.5 percent); Williamson Appalachian Regional Hospital, Kentucky (55.4 percent); Southern Baptist Hospital, Louisiana (50.3 percent); Bunkie General Hospital, Louisiana (50.0 percent); Bogalusa Community Medical Center, Louisiana (49.8 percent); Tyrone Hospital, Pennsylvania (49.4 percent); Highland Hospital, Louisiana (48.4 percent); Hialeah Hospital, Florida (48.3 percent); University of Connecticut Health Center — John Dempsey, Connecticut (46.9 percent); and Mt. Sinai Medical Center, Florida (46.6 percent).

(It is noteworthy that five of the 10 hospitals with the highest c-section rates were in Louisiana which ranked third among states with abnormally high rates.)

Despite a significant decrease in the national c-section rate from 1989 to 1990, cesarean section continues to be the most frequently performed major surgical operation in the U.S., and the most frequently performed unnecessary surgery. It is

> *Cesarean section continues to be the most frequently performed major surgical operation in the U.S., and the most frequently performed unnecessary surgery*

unconscionable that every day thousands of needless cesareans are performed, squandering millions of scarce health care dollars (to say nothing of putting women unnecessarily at risk) while nearly 40 million Americans lack access to care they really do need. Moreover, the HRG report clearly shows that most

29. Unnecessary Cesarean Sections

U.S. physicians, hospitals, and insurers continue to ignore more than 10 years of evidence undeniably showing that American women are undergoing an onslaught of unnecessary and dangerous surgery.

Not only does recent research clearly demonstrate that c-section rates are too high, but there is little doubt that an optimal, much lower rate could be achieved while preserving or even improving maternal and infant health. Public Citizen Health Research Group agrees with rates proposed by Dr. Edward Quilligan, Dean of the School of Medicine at the University of California at Irvine and editor of the *American Journal of Obstetrics and Gynecology*. Dr. Quilligan has set targets of 7.8 to 17.5 percent for hospitals, the lower number for those serving low-risk patients and the higher number for those serving high-risk patients, with a range of 12 to 14 percent for states and the country as a whole.

These numbers represent conservative goals well above rates already achieved in hospitals with comprehensive programs designed to prevent unnecessary cesareans, as discussed later in our story. Using the conservative target rate of 12 percent, we estimate that 480,520 of the 982,000 cesareans performed in 1990 (48.9 percent) were unnecessary — more than 1,300 *a day* — and cost the economy more than $1.3 billion.

Four diagnoses have been associated with the recent explosive growth of cesarean section use and accounted for about seven-eighths of all cesareans done in 1989: *repeat cesareans,* 35.6 percent of all cesareans; *dystocia* (abnormal progress of labor), 28.9 percent; *breech position,* 12.3 percent; and *fetal distress,* 9.9 percent.

Growth in the rate of automatic repeat cesarean sections has started to slow recently as "VBAC" (vaginal birth after cesarean) has gained popularity. VBAC, which accounted for 12.6 percent of all births in 1988, rose to 18.5 (an increase of 46 percent) in 1989 and to 20.4 percent (a further 10 percent increase) in 1990. Though the 1988-89 increase was substantial, the VBAC rates for both 1989 and 1990 were still far too low. Approximately 80 to 90 percent of all women with a cesarean history should be candidates for VBAC.

The 20 percent VBAC rate in 1990 suggests that most hospitals and doctors are still not heeding important new recommendations by the American College of Obstetricians and Gynecologists (ACOG). In October 1988 ACOG revised its guidelines on vaginal (normal) delivery, scrapping the outdated philosophy of "once a cesarean, always a cesarean." ACOG's revised guidelines recommend that, in the absence of medical complications of pregnancy, all women with previous cesarean sections be encouraged to attempt VBAC. These recommendations were based on studies demonstrating VBAC's safety and success.

Many other studies have demonstrated that indiscriminate use of fetal monitoring increases the cesarean rate without added benefit for infants, while the availability of non-invasive tests to identify *real*

Using the conservative target rate of 12 percent, we estimate that 480,520 of the 982,000 cesareans performed in 1990 (48.9 percent) were unnecessary

fetal distress has improved. Improvements in management of dystocia and in the appropriate management of breech babies have also been made, although ideal strategies in these areas remain unclear. Unfortunately, it seems that most of these advances have not been integrated into most physicians' practice patterns.

The good news is that a growing number of initiatives in various places around the country are doing their part to stop this epidemic.

❑ Cesarean reduction programs at several U.S. hospitals have successfully reduced c-section rates while maintaining or even improving infant and maternal health. St. Luke's Hospital in Denver, Mt. Sinai Medical Center in Chicago, University Medical Center in Jacksonville, West Paces Ferry Hospital in Atlanta, North Central Bronx Hospital in New York City (where all primary care for birthing women is provided by certified nurse midwives), and the Kaiser Permanente health maintenance organization's hospitals in southern California have developed strategies aimed at preventing unnecessary cesareans and other interventions, as well as improving women's overall childbirth experiences.

❑ Consumer groups are working to make information on cesarean rates routinely available to women and their partners. In 1985, through the work of C/SEC (Cesarean Support, Education, and Concern) and other organizations, the Massachusetts legislature passed a bill (Chapter 714) mandating that hospitals provide consumers with hospital cesarean rates and other important maternity information. This bill served as a model for the maternity information bill eventually passed by New York state in 1989 (S.2803-j), thanks to the work of ICAN (International Cesarean Awareness Network), the New York Public Interest Research Group (NYPIRG), and the National Women's Health Network of New York, as well as other groups. Several state data agencies, including those in Illinois, Iowa, Maryland, Massachusetts, Nevada, and Vermont, also publish brochures or reports containing this vital information.

❑ Cesarean support/education groups around the country, such as ICAN, formerly the Cesarean Prevention Movement, have grown tremendously in membership and activism. Many of these groups publish newsletters, hold conferences, produce their own

4. REPRODUCTION: Pregnancy and Childbirth

educational materials, and work toward preventing unnecessary cesarean surgery in their area.

❏ A handful of Blue Cross/Blue Shield programs around the country, including those in Illinois, Kansas, Minnesota, North Carolina, Pennsylvania, and Rhode Island, now reimburse physicians at the same rate for vaginal delivery as for cesarean.

❏ As reported in *Health Letter* (vol. 7, no. 6), an exciting new study conducted at Jefferson Davis Hospital in Houston found that the reassurance and support during labor and delivery from trained companions (called *doulas*, a Greek word) who know about childbirth from personal experience

---

Reassurance and support during labor and delivery from trained companions dramatically reduced the use of obstetrical interventions

---

dramatically reduced the use of obstetrical interventions such as epidural anesthesia, forceps, and cesarean delivery (confirming ancient wisdom that childbirth educators and midwives have known all along). This study verified the importance of a supportive atmosphere that responds to a woman's many emotional and social needs during childbirth.

❏ A new-format United States Standard Birth Certificate introduced in 1989 will permit virtually every state to monitor indications for cesarean section and cesarean rates, as well as other obstetrical procedures. Disseminating such information to the public, however, will continue to be a major challenge.

Questions for the Doctor

Below are questions pregnant women should ask their obstetricians about obstetrical interventions, including cesarean section, and other issues that may affect their childbirth experience. Also included is information on a number of organizations around the country that work on childbirth issues, including cesarean section (see page 141). We encourage women and their partners to contact such organizations if they have pregnancy or childbirth-related questions.

1. *What is your c-section rate?* (Ideally, doctors with high-risk practices should have rates of no more than 17 percent, and doctors with low-risk practices should have rates under 10 percent. If possible, try to avoid doctors with cesarean rates above the national rate of approximately 24 percent.)

2. *Do you offer "trial of labor" (attempted vaginal delivery) to women who have had a previous c-section? If so, what percentage of them deliver vaginally?* (Ideally, approximately 80-90 percent of women with a prior cesarean should be encouraged to undergo a trial of labor, and approximately 60 percent to 90 percent of these women should be able to deliver vaginally, meaning that 50-80 percent of women with prior cesareans could subsequently have a normal delivery.)

3. *Do you consider an independent second opinion for elective c-sections good medical practice?* (If your doctor gets angry or defensive when you ask questions, including this one, consider changing physicians — if you have that option.)

4. *How do you monitor labor of low-risk patients? Of high-risk patients? Do you routinely use electronic fetal monitoring (EFM) to monitor certain groups of patients? Do you use fetal blood sampling or fetal stimulation tests to confirm fetal distress indicated by EFM?* (Routine use of EFM may only be useful for high-risk patients. At least one other test should be used to confirm fetal distress unless a serious emergency is noted.)

5. *If a patient presents with a fetus in breech position, do you attempt to turn the fetus manually to a head-first position* (called "external cephalic version") *after 37 weeks?*

6. *Are you concerned about the high c-section rate in this country (or at this hospital)? Do you follow policies designed to reduce this rate?*

Not all of these questions may pertain to every woman's situation, but women have a right to ask their obstetricians these and any other questions they wish. Effective communication is an important part of the doctor-patient relationship, especially if women intend to negotiate for limitation of obstetrical interventions, including cesarean section.

Cesarean Section Rates for 47 States and the District of Columbia

State	Year	Total Cesareans	Total Births or Deliveries	Cesarean Rate
Alabama	1989	15,901	61,913	25.7%
	1990	16,443	63,420	25.9%
Alaska	1989	1,773	11,659	15.2%
	1990	1,813	11,883	15.3%

29. Unnecessary Cesarean Sections

State	Year	Total Cesareans	Total Births or Deliveries	Cesarean Rate
Arizona	1990	11,802	59,280	19.9%
Arkansas	1989	9,378	34,616	27.1%
	1990	9,819	35,300	27.8%
California	1989	126,504	551,625	22.9%
	1990	127,105	592.817	21.4%
Connecticut	1990	10,828	49,431	21.9%
Delaware	1989	2,791	10,492	26.6%
	1990	2,800	11,073	25.3%
District of Columbia	1989	2,895	11,567	25.0%
	1990	3,139	11,802	26.6%
Florida	1990	44,254	167,603	26.4%
Georgia	1989	25,566	110,272	23.2%
	1990	25,808	115,833	22.3%
Hawaii	1990	4,162	20,218	20.6%
Idaho	1990	3,113	16,491	18.9%
Illinois	1989	39,626	181,018	21.9%
Indiana	1989	17,532	82,834	21.2%
Iowa	1989	4,440	22,007	20.2%
Kansas	1989	8,963	37,817	23.7%
	1990	8,865	37,965	23.4%
Kentucky	1989	12,393	52,409	23.6%
	1990	12,623	53,108	23.8%
Louisiana	1990	19,635	72,046	27.3%
Maine	1990	3,576	16,080	22.2%
Maryland	1990	16,990	69,671	24.4%
Massachusetts	1989	21,285	92,836	22.9%
Michigan	1989	33,860	148,164	22.9%
	1990	33,744	153,080	22.0%
Minnesota	1989	12,005	66,509	18.1%
	1990	11,962	67,798	17.6%
Mississippi	1990	11,266	42,915	26.3%
Missouri	1990	18,793	80,608	23.3%
Montana	1989	2,308	11,148	20.7%
	1990	2,323	11,118	20.9%
Nebraska	1990	4.738	24,184	19.6%
Nevada	1989	3,828	16,761	22.8%
New Hampshire	1989	3,665	16,681	22.0%
	1990*	1,874	8,448	22.2%
New Jersey	1989	31,617	117,075	27.0%
New Mexico	1989	5,033	26,934	18.7%
	1990	4,982	26,980	18.5%
New York	1989	68,397	290,771	23.5%
	1990	70,494	298,702	23.6%
North Carolina	1990	24,044	104,347	23.0%
North Dakota	1990	1,927	9,980	19.3%

4. REPRODUCTION: Pregnancy and Childbirth

State	Year	Total Cesareans	Total Births or Deliveries	Cesarean Rate
Ohio	1989	40,160	164,894	24.4%
Oregon	1989	7,920	37,756	21.0%
Pennsylvania	1989	38,424	167,234	23.0%
	1990	37,437	171,053	21.9%
Rhode Island	1989	3,166	15,237	20.8%
	1990	3,144	15,737	20.0%
South Carolina**	1989/90**	11,547	50,583	22.8%
Tennessee	1989	18,840	77,119	24.4%
	1990	18,372	74,870	24.5%
Texas	1989	52,919	204,318	25.9%
Utah	1989	6,516	35,377	18.4%
	1990	6,686	37,295	17.9%
Vermont	1989	1,517	7,948	19.1%
	1990	1,518	7,817	19.4%
Virginia	1989*	12,040	50,021	24.1%
	1990	23,313	96,626	24.1%
Washington	1990*	6,996	33,280	21.0%
West Virginia	1989	5,014	19,716	25.4%
	1990	5,251	19,964	26.3%
Wisconsin	1989	12,212	69,828	17.5%
	1990	12,791	73,080	17.5%
Wyoming	1989	1,200	6,484	18.5%
	1990	1,283	6,544	19.6%

* indicates data for a 6-month period
** indicates data for a fiscal year

The Cutting Edge

The 16 Hospitals with the Highest Cesarean Section Rate
45 Percent and Over

State	Name of Facility	Total Births or Deliveries	Cesarean Rate
Louisiana	Abrom Kaplan Mem. Hosp.	120	57.5%
Kentucky	Williamson App. Regional Hosp.	101	55.4%
Louisiana	* Southern Baptist Hosp.	1,913	50.3%
Louisiana	Bunkie General Hosp.	62	50.0%
Louisiana	Bogalusa Community Med. Ctr.	251	49.8%
Pennsylvania	Tyrone Hosp.	77	49.4%
Louisiana	Highland Hosp.	395	48.4%
Florida	Hialeah Hosp.	1,890	48.3%
Connecticut	U. of CT Health Ctr. — John Dempsey	407	46.9%
Florida	Mt. Sinai Med. Ctr.	1,244	46.6%

29. Unnecessary Cesarean Sections

The 16 Hospitals with the Highest Cesarean Section Rate
45 Percent and Over

State	Name of Facility	Total Births or Deliveries	Cesarean Rate
Louisiana	* Lakeside Hosp.	2,899	46.5%
Louisiana	St. Anne Hosp.	353	46.5%
Louisiana	St. Tammany Parish Hosp.	285	46.3%
New Jersey	St. James Hosp.	576	46.2%
New York	Carthage Area Hosp.	266	45.9%
New York	Victory Mem. Hosp.	924	45.6%

* indicates this facility has a neonatal intensive care unit (nicu).
Data are for either 1989 or 1990.

What You Can Do

Contact the following helpful organizations.

Cesarean Section Support Groups:
International Cesarean Awareness Network (ICAN, formerly the Cesarean Prevention Movement, CPM)
PO Box 152
Syracuse, NY 13210
Telephone (315) 424-1942
This group publishes a newsletter and has approximately 80 chapters across the country.

Cesarean Support, Education, and Concern (C/SEC)
22 Forest Road
Framingham, MA 01701
Telephone: (508) 877-8266
This group no longer publishes a newsletter, but back copies of its letter are still available, and the group will continue responding to mail and phone calls, directing women to resources and groups in their area.

Childbirth Educators:
American Society of Psychoprophylaxis in Obstetrics (ASPO/Lamaze)
1840 Wilson Blvd., Suite 204
Arlington, VA 22201
Telephone: (703) 524-7802

International Childbirth Education Association (ICEA)
PO Box 20048
Minneapolis, MN 55420-0048
Telephone: (612) 854-8660

Certified Nurse-Midwives:
The American College of Nurse-Midwives (ACNM)
1522 K Street NW, Suite 1120
Washington, DC 20005
Telephone: (202) 347-5445

Midwives Alliance of North America (MANA)
30 South Main
Concord, NH 03301
Telephone: (603) 225-9586

Consortium for Nurse-Midwifery, Inc. (CNMI)
1911 West 233rd St.
Torrance, CA 90501
Telephone: (213) 539-9801
Information about nurse-midwifery in California

Sexuality Through the Life Cycle

- Youth and Their Sexuality (Articles 30–32)
- Sexuality in the Adult Years (Articles 33–37)

Individual sexual development is a lifelong process that begins at birth and terminates at death. Contrary to popular notions of this process, there are no latent periods during which the individual is nonsexual or noncognizant of sexuality. The growing process of a sexual being does, however, reveal qualitative differences through various life stages. This section devotes attention to these stages of the life cycle and their relation to sexuality.

As children gain self-awareness, they naturally explore their own bodies, masturbate, display curiosity for the bodies of the opposite sex, and show interest in the bodies of mature individuals such as their parents. Such exploration and curiosity are important and healthy aspects of human development. Yet it is often difficult for adults (who live in a society that is not comfortable with sexuality in general) to avoid making their children feel ashamed of being sexual or showing interest in sexuality. When adults impose their ambivalence upon a child's innocuous explorations into sexuality, or behave toward children in sexually inappropriate ways, distortion of an indispensable and formative stage of development occurs. This often leaves profound emotional scars that hinder full acceptance of self and sexuality later in the child's life.

Adolescence, the social status accompanying puberty and the transition to adulthood, proves to be a very stressful period of life for many individuals as they attempt to develop an adult identity and forge relationships with others. Because of the physiological capacity of adolescents for reproduction, sexuality tends to be heavily censured by parents and society at this stage of life. Yet individuals and societal attitudes place tremendous emphasis on sexual attractiveness—especially for females, and sexual competency—especially for males. These physical, emotional, and cultural pressures combine to create confusion and anxiety in adolescents and young adults about whether they are okay or normal. Information and assurances from adults can alleviate these stresses

Unit 5

and facilitate positive and responsible sexual maturity if there is mutual trust and willingness in both generations.

Sexuality finally becomes socially acceptable in adulthood, at least within marriage. Yet routine, boredom, stress, pressures, the pace of life, extramarital sexual activity, and/or lack of communication can exact heavy tolls on the quantity and quality of sexual interaction. Sexual misinformation, myths, and unanswered questions, especially about emotional and or physiological changes in sexual arousal/response or functioning, can also undermine or hinder intimacy and sexual interaction in the middle years.

Sexuality in the later years of life is also socially and culturally stigmatized because of the prevailing misconception that sex is for young, attractive, and married adults. Such an attitude is primarily responsible for the apparent decline in sexual interest and activity as one grows older. Physiological changes in the aging process are not, in and of themselves, detrimental to sexual expression. A life history of experiences, health, and growth can make sexual expression in the later years a most rewarding and fulfilling experience.

Youth and Their Sexuality begins with an article about young children and their sexuality. It is written for parents about understanding, accepting, and utilizing the natural, healthy curiosity of children to begin positive and healthy communication about sex. The next articles focus on sexuality and sexual behavior in the adolescent years. "Truth and Consequences: Teen Sex" focuses on the changes in adolescents and adolescent sexual activity, and the American culture's response to it over the last 30 years. It discusses some current and controversial approaches to reducing teenage pregnancy, including media campaigns, peer education, and condom distribution and family planning clinic services in schools. "Single Parents and Damaged Children" focuses on single parents (mostly young and handicapped by poverty and/or incomplete education) and their children. Findings cited about the damage to these children underscore the need for America to do something about our rate of teenage pregnancy and childbearing, which exceeds the rates of most other developed countries.

The articles in the *Sexuality in the Adult Years* subsection deal with a variety of issues for individuals and couples who are twentysomething, thirtysomething, and all the way to ninetysomething and older. "Let the Games Begin" looks at the early 30s age cohort who have been caught somewhat in the middle between their upbringing that was not characterized by the sexual explicitness of MTV and the ticking of their sexual and biological clocks that reminds them to get it while you can, you are not getting any younger, and try it, you will like it. The next article offers "25 Ways to Make Your Marriage Sexier" and includes advice from a variety of experts and from some of "America's sexiest wives" identified from a research study. The next article focuses on sexual intimacy, fidelity, and infidelity. "Beyond Betrayal" addresses common myths about extramarital sexual behavior while examining the common scenarios, experiences, and effects.

The final two articles focus on sexuality in what can be viewed as the sunset years. The first presents what the doctor-author, who lists his age as the late 70s, asserts doctors and others need to know about human sexuality and aging. In the process, he does a lot of myth-busting and normalizing of sexuality as a lifelong human dimension. The final article, "Sexuality and Aging," is a special edition of the Mayo Clinic newsletter written for senior citizens and anyone who cares about them. It provides a wealth of information with a very human side through the inclusion of some personal experiences by people in their 60s, 70s, and beyond.

Looking Ahead: Challenge Questions

Do you remember trying to get answers about your body, sex, or similar topics as a young child? What was your parent's response? How did you feel?

As an adolescent, where and why did you get answers to your questions about sex? Are you still embarrassed at any lack of information you have? Why or why not?

How do you view sex and sexuality at your age? In what ways is it different than 5 to 10 years ago? Are there things you feel you have missed? Do not want to miss? What are they?

Close your eyes and image a couple having a pleasurable sexual interlude. When you are done, open your eyes. How old were they? If they were younger than middle age, can you replay your vision with middle-aged or older people? Why or why not? How does this relate to your expectations regarding your own romantic and/or sexual life a few decades from now?

What do you know about the aging process and how it affects sexuality? Do you worry about growing older and "losing it"? About sexual attractiveness?

List some things you have heard about infidelity and extramarital sex. How would you deal with finding yourself in bed with a pounding hangover and someone other than your spouse?

Raising Sexually Healthy Kids

Parents who very early on send their kids positive messages about sexuality will help them develop a healthy attitude toward sex.

Elizabeth Fishel

Elizabeth Fishel is the author of *Family Mirrors: What Our Children's Lives Reveal About Ourselves* (Houghton Mifflin).

Your two-year-old son wakes up and announces that his penis is "standing up." You discover your four-year-old daughter and her best friend engaged in a game of "doctor"; the patient has her tights around her ankles, and the doctor is coaching her through a very noisy labor. As you pass the open door of your seven-year-old son's bedroom, you hear him enthusiastically telling friends a joke (something about a farmer's daughter); he lowers his voice as you pass.

Although such incidents are as much a part of childhood as loose teeth and skinned elbows, it can be a bit startling when children, so innocent and naive, openly show their sexuality. But if you have a good understanding of your child's stage of sexual development, you can use such situations to help her develop a healthy attitude toward sex.

All parents, whether they grew up in an era of secrecy and sexual taboos or in the permissive sixties, want to raise sexually healthy children. We want to help our kids become adults who feel good about their bodies, can treat people of both genders with respect, and are comfortable expressing love. To do so, it is important for us to understand that sexuality is a continuum, a part of everyone's life from birth. Although each child will develop at her own pace, there are common milestones that all children eventually reach—from an infant's gentle sensuality to a preschooler's curiosity, from a school-age child's need for privacy to a young adult's first awkward attempts at experimenting with sex. How and when you discuss sex with your children will depend on your family's beliefs and values. To help you prepare for those discussions, here is a stage-by-stage guide—from infancy to the teen years—to your child's developing sexuality.

Age of exploration.
The foundation of healthy sexuality begins in infancy, according to Debra Haffner, executive director of SIECUS, the Sex Information and Education Council of the United States, in New York City. Indeed, children are sexual from birth—infant boys have erections, and infant girls lubricate vaginally. Both traits are healthy signs that the sexual-response system has begun to develop normally.

Babies also come into the world with a "skin hunger" as basic as their need for food. From the first moments after birth, a baby needs to be touched, gazed at, and held. Cuddling your baby helps him thrive and helps build a bond that provides the base for later attachments.

"From the beginning, parents can convey a positive message about the human body," says Pamela Wilson, author of *When Sex Is the Subject: Attitudes and Answers for Young Children* (Network Publications). "When you teach your child the names of various body parts, what message does your facial expression give? Do you smile when you name the eyes and nose but look nervous when you label the penis or vagina? Even if you are conveying a positive message verbally, your ambivalent or negative feelings about sexual parts of the body can come through.

"Your positive verbal and nonverbal messages let your child know that his body—including the genitals—is a source of pleasure throughout life," Wilson continues. Allowing children to explore their genitals, rather than pushing their hands away, is part of recognizing them as sexual beings.

Many families use nicknames to describe body parts or rely on euphemisms such as "privates" and "down there." Although this may be the most comfortable way to handle such discussions, Haffner stresses that parents should be sure not to omit teaching their children the correct names for their body parts: penis, testicles, and scrotum; or vagina, vulva, and clitoris.

The naked truth.

As babies turn into toddlers, they become more active, more aware of the world around them, and more curious about their own bodies. Around fifteen to eighteen months, most toddlers become especially interested in exploring and stimulating their genitals. They will often do so anytime and anywhere for comfort or pleasure—or both. By the time kids are two or three—and certainly when they're ready for school—they can learn to understand the difference between appropriate public and private behavior. You might say to your child, "I know that touching yourself feels good, but I would be more comfortable if you did it in the privacy of your room." One friend of mine has repeated this remark so often that now she has only to say "This is not the bedroom" for her daughter to get the message and comply immediately.

Along with this curiosity about their own bodies, toddlers are also beginning to develop gender identity, a sense of themselves as boys or girls. If you are comfortable having your child see you naked, she will have ample opportunity to discover the differences between males and females. In families that are not comfortable with casual nudity, children will still have plenty of chances to establish gender identification by seeing the bodies of their siblings, schoolmates, and even their dolls. "There is no right or wrong on the issue of incidental parental nudity around kids this age—it's up to the family to decide," says Haffner.

As children move from toddlerhood to preschool age, they begin to confront messages about what boys and girls should and should not do: Play with dolls or trucks? Become a doctor or a dancer? Be gentle or rough; quiet or outspoken? Many children this age accept such messages as dogma, and parents who have worked hard to raise nonsexist children are often dismayed to find their boys acting like mini macho men and their girls like little princesses.

This behavior is completely age-appropriate. Try to keep in mind that children *do* outgrow this phase. Meanwhile, says Haffner, "remember to model nonsexist behavior yourself. Ultimately the most powerful messages are the ones received at home—from who does the dishes to who cuts the grass to who comforts a crying child at night."

Playing doctor.

"As toddlers become preschoolers, they ought to know the correct terms for their body parts, have a positive sense of their own bodies, and a sense of privacy—both for themselves and for others," says Haffner. "They can express and label feelings and are comfortable asking their parents questions." Much of their sexual awareness will come from peers. "Giggling over bathroom humor or being curious about each other's body is as much a part of preschool life as riding tricycles and finger painting—and usually just as innocent," says Linda Perlin Alperstein, a social worker and assistant clinical professor in the psychiatry department at the University of California Medical School, San Francisco. Learning about bodies, exploring the differences between boys and girls, and rehearsing adult roles such as doctor and patient and mommy and daddy are all examples of healthy sex play in preschoolers.

Preschoolers' awareness of the physical differences between boys' and girls' bodies continues to grow as they mature. "Which one is better and which one is worser—boys or girls?" my four-year-old son asked me. As I struggled to explain how boys and girls are different—and both special—he eyed me coolly and made a remark that no enlightened mother is happy to hear: "Girls are worser because they have no penis!"

Like my son, children this age want concrete information in response to their questions. Because preschoolers tend to take statements literally, you will want to choose your words with care when you are answering questions about the differences between the sexes. You may need to reassure both boys and girls that boys will never lose their penises; nor will girls develop them. But, you may add, girls have the special ability to have babies when they are women.

When you broach the topic, try to take the child's point of view, and

30. Raising Sexually Healthy Kids

keep in mind that his ability to handle sophisticated information is limited. Bernie Zilbergeld, Ph.D., the author of *The New Male Sexuality* (Bantam), ruefully recalls when his stepson, Ian, then four, developed the age-old male fascination with the relative size of penises. But before gearing up for a lecture on how sexual pleasure was not related to penis size, Zilbergeld asked the boy why he wanted a bigger penis. "Silly!" came the retort. "So I can pee more!"

Besides their blossoming sexual curiosity and hunger for information, preschoolers, Freud noted, can be unabashedly romantic—especially about the parent of the opposite sex—and occasionally competitive with the parent of the same sex. "I like Daddy," my son confided to me at three and a half, "but you're my best sweetheart." As children get older, and certainly by puberty, Freud believed, the romantic focus switches from inside to outside the family. Most boys will identify with their father and transfer their desire for their mother to other females; most girls will identify increasingly with their mother and replace their desire for their father with an interest in other males.

Privacy, please.

Freud called the years between six and nine the "latency period" because he thought that it was a time when sexual interests went underground. He saw it as a quiet time, a kind of calm between the stage of active infantile sexuality and the turbulent storm of adolescence. Most experts now believe that children this age are still curious about sex but hide their interest to avoid adult disapproval. By this age, however, "children ought to have some knowledge about where babies come from and a basic understanding of sexually transmitted diseases, including AIDS," says Debra Haffner. Many children are uncomfortable broaching these topics. And many parents believe that AIDS education can wait until children become sexually active. But you need to initiate these discussions now to ensure your child's future health and well-being.

This is also a time when children begin to bond with friends and to value their peer relationships as much as they value their family connections. Their same-sex friendships may be especially strong during these

5. SEXUALITY THROUGH THE LIFE CYCLE: Youth and Their Sexuality

years. Parents of school-age children may walk in on their kids telling "dirty" jokes or making silly comments about the opposite sex to same-sex friends, and may notice that voices are suddenly lowered and gazes averted.

Kids this age may also be learning to consider another person's point of view for the first time. Your child may think that she's doing you a favor by lowering her voice when she tells certain kinds of jokes around you. As one astute mother put it, "My son needs to act as if I don't know what's going on—and right now I need to pretend that I don't as well."

Since school-age children begin to visit each other's house more often than when they were younger, now might be a good time to remind them that other families' values may differ from those of their own family—without being better or worse. "Parents need to prepare a child for the fact that things may be different in other people's homes," suggests Linda Alperstein.

The members of one family may smother each other with hugs and kisses; in another one, no less loving, they may be more reserved. One family may practice casual nudity, which to them means familiarity; another may cover up because to them, nudity is sexual and should be kept private. Says Alperstein, "You might say to your children, for example, 'In our household we don't walk around without clothes, but in other homes, sometimes people do.' That way you're socializing but not squelching them."

Caught in the act.

The school-age child may begin to show a need for privacy and can be taught to respect yours. But what if, despite your best efforts, your child walks in on you and your partner making love? Don't make a big deal about the interruption, and try to remain calm, suggests Pamela Wilson. It is natural for you to feel uncomfortable. After you catch your breath, you might ask your child what he needed, and then take care of his request. If your child asks what you were doing, you might say something like, "Honey, you walked in on us when we weren't expecting you." If he asks about the noises you were making, reassure him by saying something like, "We weren't fighting; we were hugging and kissing and making each other happy."

If your child still wants to know more, you can offer him information such as, "We were making love. It's something good that we like to do, but we like to do it in private. That's why we had the door closed. So next time, would you please knock before you enter?" But make sure you send your child the message that it's not his fault that he walked in on a private moment between his mom and dad; feeling guilty about what he saw may be more upsetting to your child than catching you in the act.

The onset of puberty.

Between ages ten and twelve, the sexual and reproductive systems mature, hormones start to rage, and the child you once thought you knew so well may seem like another person entirely. Boys begin to produce sperm and ejaculate or have wet dreams; their penises get thicker and longer, and their voices deepen. Girls begin to develop breasts and pubic hair; their hips broaden, and they begin to ovulate and menstruate. The average age for the onset of menstruation is twelve. Although many girls start as young as nine, some girls will not reach full puberty until their middle or late teens. Getting her period either much earlier or much later than her friends can be stressful for a girl and may call for extra support from you and from the family doctor.

"By this age, children ought to un-

When Is Sex Play Reason to Worry?

Sex play should be a concern when there is a significant age or size difference between the children involved, according to Linda Perlin Alperstein, assistant clinical professor of psychiatry at the University of California Medical School, San Francisco. "The warning signs of possible abuse are when sex play is exploitive and secretive or when it's the only game the kids are playing," she says.

So if your three-year-old and his best friend take off their clothes to play doctor, they are probably just curious about what is underneath the clothes—behavior that is usually harmless. But if an older boy in the neighborhood wants to undress your three-year-old to play doctor, you should be suspicious. In this situation, you may want to increase your supervision and offer some alternative games. Later on, talk the matter over with your child. Start by asking him a few open-ended questions, such as "Do you like it when Johnnie comes to play?" "Is there anything he does that you don't like?" "Is there anything he does that you think is naughty?"

If your child describes sex play that sounds inappropriate to you but doesn't seem to bother him, you may want to casually say something such as, "I don't think that's something children should do." If, however, your child seems scared or upset by the experience, you may want to say the same thing but add, "and maybe it would be better if you and he didn't play together for a while." This kind of statement reassures your child that you will step in and protect him from a threatening situation. In either case, be sure that your response doesn't make him feel guilty or ashamed. Then, depending on your relationship with the other child's family, discuss the matter with his parents. Banning the child from visiting should be viewed only as a last resort.

Once you have had a chance to step back from the situation, explain to your child that he has control over who touches his body and where. Let him know that there may be good reasons for some people to touch him—a parent or caregiver bathing him, a doctor examining him, or a grandparent getting him ready for bed. But teach him to trust his own feelings and to understand that if certain touching makes him uncomfortable, he has every right to say no.

One more caution: Children need to know the difference between private and *secret* activities. Urge them not to be enticed into keeping secrets with other kids or adults. You may want to tell them, "If someone wants you to keep a secret from your parents, that's not okay. We want you to tell us about it."
—E.F.

derstand the changes that their bodies are undergoing, and they may begin to deepen their awareness of feelings and relationships," Debra Haffner explains. For parents helping a son or daughter on the roller coaster of puberty, the focus should be on "acknowledging that your child has achieved another landmark in development rather than emphasizing the new burdens that accompany this change," says Pamela Wilson. Comments such as "You're a woman [or man] now and can get pregnant [or get someone pregnant]" can make a young person feel overwhelmed by her newfound maturity. Despite all of the outward changes that your child is experiencing, she probably feels the same on the inside and is struggling with a new body and body image; she needs all the understanding you can provide.

Children this age begin playing kissing games and the like. Your most helpful response is to understand how your child views the experience—before you panic, judge, or give any advice. Open-ended, nonjudgmental questions such as "How do you feel about it?" can help elicit comments from even the most reticent teenager.

Becoming sexually active.

A 1990 survey by the Centers for Disease Control, in Atlanta, shows that by ninth grade, 40 percent of all students have had intercourse; by senior year, 72 percent have done so. By the time our children are teenagers, they will need more detailed and specific information about sex. The sobering prevalence of AIDS makes sexual awareness a life-and-death matter.

Still, despite our best intentions to be approachable, our teenagers may be afraid to ask us vital questions about sex. They may not want to disappoint or alarm us, tarnish their reputation as "perfect" kids, or get into trouble. Luckily, many teens get a good sex education at school, but some still make do with half-truths or errors gleaned from friends. And some are adept at hiding their ignorance behind a know-it-all facade.

If you suspect that your son or daughter is sexually active—or is considering becoming sexually active—revealing your own feelings about the sexual choices ahead will open up conversation and may prompt your child to respond in kind. "I'm concerned about how you may be experimenting," you might say, "and I'd like to give you some information that you may not have. Can we talk about it?" Then be prepared to share your values about sexual responsibility, and sex before marriage.

If you and your partner believe that it is best to wait until marriage or a long-term relationship to have intercourse, you may wish to explain that your teen's version of "skin hunger" may be satisfied in other ways—by touching, kissing, and holding. Be aware, however, that although your teenager may say—and genuinely believe—one day that he agrees with you, he may be tempted to change his mind the next time he is making out and wants to go further. "Teenagers should be educated about birth control and safer sex, regardless of when or whether they'll actually put it into practice," says Haffner.

But teenagers are hungry for the whole complex web of information that makes up mature sexuality—arousal, pleasure, making decisions about relationships, saying no. As you talk, be as personal as you feel comfortable with. Was your first sexual experience particularly joyous or painful? Did you rush into sex too early? Let your child see that each generation confronts some of the same sexual decisions.

You may need to address the issue of whether you are comfortable having your child behave sexually in your home. "Know also that no matter what your limits are, your teenager will push them," says Anne Bernstein, Ph.D., a Berkeley, California, psychologist specializing in family

---

"WILL MY CHILD BE GAY?"
By Maureen Arvai

If the sight of your four-year-old son proudly modeling high heels and a feather boa—or your three-year-old daughter ferociously charging "bad guys" with a sword—makes you uncomfortable, you are not alone.

Many parents worry that stereotypical "boy" behavior in girls (and vice versa) may lead to homosexuality. It seems likely that homosexuality has multiple causes and that the origins can be different for different people. Although much of the research indicates that sexual preference is largely determined by prenatal biological conditions, there is no conclusive answer. (Studies on homosexuality have focused almost exclusively on males, but researchers state that their findings apply to both sexes.) What does seem clear, though, is that popular myths that lead parents to believe that somehow *their* influence will cause a child to be gay are just that—myths.

One common myth about homosexuality suggests that a son becomes gay because he grows up wanting to be like his dominant mother instead of his weak father. According to a study by the Kinsey Institute for Research in Sex, Gender, and Reproduction, however, parents' personalities have little if any effect on their child's sexual orientation. Also ruled out as "causes" of homosexuality were quality of the parents' marriage, birth order, being an only child, the quality of relationships with siblings, and, in the case of sons, the parents' wish for a daughter.

Another myth—that young boys are seduced into becoming homosexual by older males—was also dispelled by the Kinsey Institute study. Adult homosexuals are no more sexually interested in children than heterosexual adults are. Another mistaken belief is that early sex play among boys causes homosexuality. The truth is, among boys and girls, same-sex play in childhood is fairly common and not indicative of ultimate sexual preference.

Experts urge us not to worry that something we do—or fail to do—will cause our sons or daughters to become homosexual. Instead, a father should be accepting of his son's interests and behavior, even if they are not typical. If Dad participates in the non-traditional activities that his son enjoys—such as playing with dolls and other "feminine" toys—and Mom can engage in a little pretend dragon slaying with her daughter, they will boost their children's self-image.

Children who grow up with high self-esteem and strong family relationships have the greatest chance of becoming well-adjusted, productive members of society, regardless of their sexual orientation.

Maureen Arvai *is a free-lance writer in Manassas, Virginia.*

5. SEXUALITY THROUGH THE LIFE CYCLE: Youth and Their Sexuality

therapy, and the author of *Yours, Mine, and Ours* (W. W. Norton). When discussing hot issues, she suggests, use "I" statements that reflect your feelings, and make clear which behaviors are and are not acceptable—for example, "I feel uncomfortable when you snuggle with your girlfriend in front of the family, so it's not something you can do here." This approach is ultimately more persuasive than accusations such as, "You're corrupting your younger brothers."

When negotiating limits—or initiating discussion about most sexual matters—try to be positive and maintain your sense of humor. Says Pamela Wilson, "Teens respond to humor better than they do a serious, heavy talk." Remember to let your children know how joyous sex can be. Tell them that when sex and love come together, it can be among life's best experiences—one that created *them*, and one that you want them to find satisfying too.

Truth and Consequences
TEEN SEX

Douglas J. Besharov with Karen N. Gardiner

Douglas J. Besharov is a resident scholar at the American Enterprise Institute. Karen Gardiner is a research assistant at the American Enterprise Institute.

Ten million teenagers will engage in about 126 million acts of sexual intercourse this year. As a result, there will be about one million pregnancies, resulting in 406,000 abortions, 134,000 miscarriages, and 490,000 live births. Of the births, about 313,000, or 64 percent, will be out of wedlock. And about three million teenagers will suffer from a sexually transmitted disease such as chlamydia, syphilis, gonorrhea, pelvic inflammatory disease, and even AIDS.

This epidemic of teen pregnancy and infection has set off firestorms of debate in school systems from Boston to San Francisco. Last May, Washington, D.C. Mayor Sharon Pratt Kelly announced that health officials would distribute condoms to high school and junior high school students. Parents immediately protested, taking to the streets with placards and angry shouts. And the New York City Board of Education was virtually paralyzed for weeks by the controversy surrounding its plans for condom distribution.

Both sides have rallied around the issue of condom distribution as if it were a referendum on teen sexuality. Proponents argue that teenagers will have sex whether contraceptives are available or not, so public policy should aim to reduce the risk of pregnancy and the spread of sexually transmitted diseases by making condoms easily available. Opponents claim that such policies implicitly endorse teen sex and will only worsen the problem.

The causes of teen pregnancy and sexually transmitted diseases, however, run much deeper than the public rhetoric that either side suggests. Achieving real change in the sexual behavior of teenagers will require action on a broader front.

Thirty Years into the Sexual Revolution

Some things are not debatable: every year, more teenagers are having more sex, they are having it with increasing frequency, and they are starting at younger ages.

There are four principal sources of information about the sexual practices of teenagers: the National Survey of Family Growth (NSFG), a national in-person survey of women ages 15–44 conducted in 1982 and again in 1988; the National Survey of Adolescent Males (NSAM), a longitudinal survey of males ages 15–19 conducted in 1988 and 1991; the National Survey of Young Men (NSYM), a 1979 survey of 17- to 19-year-olds; and the Youth Risk Behavior Survey (YRBS), a 1990 questionnaire-based survey of 11,631 males and females in grades 9–12 conducted by the Centers for Disease Control (CDC). In addition, the Abortion Provider Survey, performed by the Alan Guttmacher Institute (AGI), collects information about abortions and those who provide them.

With minor variations caused by differences in methodology, each survey documents a sharp increase in the sexual activity of American teenagers. All these surveys, however, are based on the self-reports of young people and must be interpreted with care. For example, one should always take young males' reports about their sexual exploits with a grain of salt. In addition, the social acceptability of being a virgin may have decreased so much that this, more than any change in behavior, has led to the higher reported rates of sexual experience. The following statistics should therefore be viewed as indicative of trends rather than as precise and accurate measures of current behavior.

A cursory glance at Figure 1 shows that there was indeed a sexual revolution. The 1982 NSFG asked women ages 15–44 to recall their first premarital sexual experience. As the figure shows, teenagers in the early 1970s (that is, those born between 1953 and 1955) were twice as likely to have had sex as were teenagers in the early 1960s (that is those born 1944 to 1946).

5. SEXUALITY THROUGH THE LIFE CYCLE: Youth and Their Sexuality

The trend of increased sexual activity that started in the 1960s continued well into the late 1980s. According to the 1988 NSFG, rates of sexual experience increased about 45 percent between 1970 and 1980 and increased another 20 percent in just three years, from 1985–1988, but rates have now apparently plateaued. Today, over half of all unmarried teenage girls report that they have engaged in sexual intercourse at least once.

These aggregate statistics for all teenagers obscure the second remarkable aspect of this 30-year trend: sexual activity is starting at ever-younger ages. The 1988 NSFG found that the percentage of 18-year-olds who reported being sexually active increased about 75 percent between 1970 and 1988, from about 40 percent to about 70 percent. Even more startling is that the percentage of sexually experienced 15-year-old females multiplied more than fivefold in the same period, from less than 5 percent to almost 27 percent.

Moreover, the increase in sexual activity among young teens continued beyond 1988. In 1990, 32 percent of ninth-grade females (girls ages 14 and 15) reported ever having had sex, as did 49 percent of the males in the same grade. At the same time, the proportion of twelfth-grade females (ages 17 and 18) who reported ever engaging in sex remained at 1988 levels.

Teenagers are not only having sex earlier, they are also having sex with more partners. According to the NSAM, the average number of partners reported by males in the 12 months preceding the survey increased from 2.0 in 1988 to 2.6 in 1991. Almost 7 percent of ninth-grade females told the YRBS that they had had intercourse with four or more different partners, while 19 percent of males the same age reported having done so. By the twelfth grade, 17 percent of girls and 38 percent of boys reported having four or more sexual partners.

A major component of these increases has been the rise in sexual activity among middle-class teenagers. Between 1982 and 1988, the proportion of sexually active females in families with incomes equal to or greater than 200 percent of the poverty line increased from 39 percent to 50 percent. At the same time, the proportion of females from poorer families who had ever had sex remained stable at 56 percent.

Until recently, black teenagers had substantially higher rates of sexual activity than whites. Now, the differences between older teens of both races have narrowed. But once more, these aggregate figures obscure underlying age differentials. According to the 1988 NSAM, while 26 percent of white 15-year-old males reported engaging in sex compared to 67 percent of blacks, by age 18 the gap narrowed to 71 percent of whites and 83 percent of blacks. A similar trend appears among females. Twenty-four percent of white 15-year-old females have engaged in sex, compared to 33 percent of their black counterparts, reports the 1988 NSFG. By age 16, the proportions increase to 39 percent and 54 percent, respectively. Even by age 17, fewer white females have started having sex (56 percent) than have blacks (67 percent). On the other hand, white teen males reported having had almost twice as many acts of intercourse in the 12 months preceding the 1988 NSAM than did black teen males (27 versus 15). The white males, however, had fewer partners in the same period (2 versus 2.5).

The Social Costs

Among the consequences of this steady rise in teen sexuality are mounting rates of abortion, out-of-wedlock births, welfare, and sexually transmitted diseases.

Abortion. About 40 percent of all teenage pregnancies now end in abortion. (Unmarried teens account for about 97 percent.) This means that of the 1.6 million abortions in 1988, over 400,000—or a

> **The trend of increased sexual activity that started in the 1960s continued well into the late 1980s. . . . Today, over half of all unmarried teen-age girls report that they have engaged in sexual intercourse at least once.**

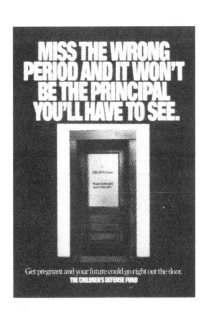

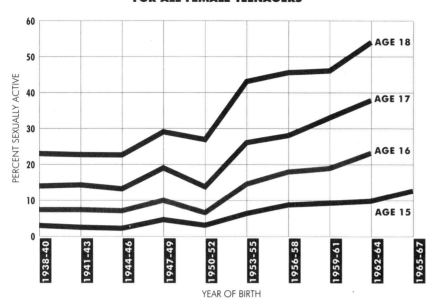

FIGURE ONE: TRENDS IN PREMARITAL SEXUAL ACTIVITY FOR ALL FEMALE TEENAGERS

SOURCE: S. Hofferth, J. Kahn, and W. Baldwin, "Premarital Sexual Activity Among U.S. Women Over the Past Three Decades," *Family Planning Perspectives*, Vol. 19, No. 2, March/April 1987.

31. Truth and Consequences

their children up for adoption as was often done in the past. And yet most are not able to support themselves, let alone their children. Consequently, about 50 percent of all teen mothers are on welfare within one year of the birth of their first child; 77 percent are on within five years, according to the Congressional Budget Office. Nick Zill of Child Trends, Inc., calculates that 43 percent of long-term welfare recipients (on the rolls for ten years or more) started their families as unwed teens.

As Table 1 shows, welfare dependency is more a function of a mother's age and marital status than of her race. White and black unmarried adolescent mothers have about the same welfare rate one year after the birth of their first child. After five years, black unmarried mothers have a somewhat higher rate of welfare dependency than whites (84 percent versus 72 percent), but various demographic factors such as family income, educational attainment, and family structure account for this relatively small difference.

quarter of the total—were performed on teenagers. In the 11 years between 1973 and 1984, the teenage abortion rate almost doubled, from about 24 to about 44 per 1,000 females ages 15–19. (Between 1984 and 1988, the rate stabilized.)

A study by AGI's Stanley Henshaw found that between 1973 and 1988, the abortion rate for girls ages 14 and under increased 56 percent (from 5.6 to 8.6 per 1,000), 62 percent for those ages 15–17 (from 18.7 to 30.3), and among older teens, almost 120 percent (from 29 to 63.5). In absolute numbers, the youngest group had about 13,000 abortions, the middle group had 158,000, and the oldest group had 234,000.

Out-of-Wedlock Births. Over 300,000 babies were born to unwed teenagers in 1988. That's three-fifths of all births to teenagers. Although the total number of births to teenagers declined between 1970 and 1988, the percentage born out of wedlock more than doubled (from 29 percent to 65 percent), and the teenage out-of-wedlock birth rate increased from about 22 per 1,000 to 37 per 1,000. Over 11,000 babies were born to children under 15 years old in 1988.

Welfare. Few teen mothers place

Disease. Over three million teenagers, or one out of six sexually experienced teens, become infected with sexually transmitted diseases each year, reports the Centers for Disease Control (CDC). One Philadelphia clinic administrator laments that she used to spend $3 on contraceptives for every $1 on disease screening and related health issues. Today, the ratio is reversed. Susan Davis, a contraception counselor at a Washington, D.C. area Planned Parenthood clinic, explains, "The risk of infection is greater than the risk of pregnancy for teens." These diseases can cause serious

5. SEXUALITY THROUGH THE LIFE CYCLE: Youth and Their Sexuality

TABLE ONE: PERCENT OF ADOLESCENT MOTHERS ON AFDC

	BY FIRST BIRTH	WITHIN ONE YEAR OF BIRTH	WITHIN FIVE YEARS OF BIRTH
All	7%	28%	49%
Married	2	7	24
Unmarried	13	50	77
White	7	22	39
Black	9	44	76
White, Unmarried	17	53	72
Black, Unmarried	10	49	84

SOURCE: Congressional Budget Office, Sources of Support for Adolescent Mothers, Government Printing Office: Washington, D.C., 1990

problems if left untreated. The CDC estimates that between 100,000 and 150,000 women become infertile every year because of sexually transmitted disease-related pelvic infections.

The recent explosion of these diseases is in large measure caused by the sexual activity of teenagers; sexually transmitted disease rates decline sharply with age. Take gonorrhea, for example. According to AGI, there were 24 cases per 1,000 sexually experienced females ages 15–19 in 1988. Among women ages 20–24, the rate declined to 15 and fell rapidly with age. For women ages 25–29, 30–34, and 35–39, the rates are 5, 2, and 1 per 1,000, respectively. Except for AIDS, most sexually transmitted diseases follow a similar pattern.

AIDS has not reached epidemic proportions in the teen population—yet. According to the Centers for Disease Control, fewer than 1,000 cases of AIDS are among teenagers. However, there are 9,200 cases among 20–24 year-olds and 37,200 cases among 25–29 year-olds. Given the long incubation period for the AIDS virus (8–12 years), many of these infections were probably contracted during adolescence.

According to Lawrence D'Angelo and his colleagues at the Children's National Medical Center in Washington, D.C., the rate of HIV (the virus that causes AIDS) infection among teenagers using the hospital increased rapidly between 1987 and 1991. For males, the rate increased almost sevenfold, from 2.47 per 1,000 in 1987 to 18.35 per 1,000 in 1991. The female rate more than doubled in the same period, from 4.9 to 11.05. These statistics only reflect the experience of one hospital serving a largely inner-city population, but they illuminate what is happening in many communities.

Use, Not Availability

Many people believe that there would be less teen pregnancy and sexually transmitted diseases if contraceptives were simply more available to teenagers, hence the call for sex education at younger ages, condoms in the schools, and expanded family planning programs in general. But an objective look at the data reveals that availability is not the prime factor determining contraceptive use.

Almost all young people have access to at least one form of contraception. In a national survey conducted in 1979 by Melvin Zelnik and Young Kim of the Johns Hopkins School of Hygiene and Public Health, over three-quarters of 15- to 19-year-olds reported having had a sex education course, and 75 percent of those who did remembered being told how to obtain contraception.

Condoms are freely distributed by family planning clinics and other public health services. They are often sitting in a basket in the waiting room. Edwin Delattre, acting dean of Boston University's School of Education and an opponent of condom distribution in public schools, found that free condoms were available at eight different locations within a 14-block radius of one urban high school.

And, of course, any boy or girl can walk into a drug store and purchase a condom, sponge, or spermicide. Price is not an inhibiting factor: condoms cost as little as 50¢. Although it might be a little embarrassing to purchase a condom—mumbling one's request to a pharmacist who invariably asks you to speak up used to be a rite of passage to adulthood—young people do not suffer the same stigma, scrutiny, or self-consciousness teenagers did 30 years ago.

Teenagers can also obtain contraceptives such as pills and diaphragms from family planning clinics free of charge or on a sliding fee scale. In 1992, over 4,000 federally funded clinics served 4.2 million women, some as young as 13. According to AGI, 60 percent of sexually active female teens use clinics to obtain

31. Truth and Consequences

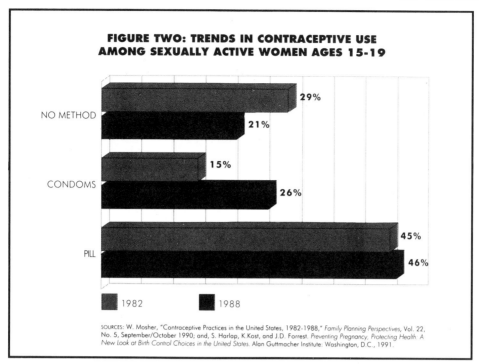

FIGURE TWO: TRENDS IN CONTRACEPTIVE USE AMONG SEXUALLY ACTIVE WOMEN AGES 15-19

NO METHOD: 29% (1982), 21% (1988)
CONDOMS: 15% (1982), 26% (1988)
PILL: 45% (1982), 46% (1988)

SOURCES: W. Mosher, "Contraceptive Practices in the United States, 1982-1988," *Family Planning Perspectives*, Vol. 22, No. 5, September/October 1990; and, S. Harlap, K. Kost, and J.D. Forrest. *Preventing Pregnancy, Protecting Health: A New Look at Birth Control Choices in the United States*. Alan Guttmacher Institute: Washington, D.C., 1991.

contraceptive services, while only 20 percent of women over 30 do. In all states except Utah, teenagers can use clinic services without parental consent. To receive free services under the Medicaid program, however, a teenager must present the family's Medicaid card to prove eligibility.

In 1990, total public expenditures for family planning clinics amounted to $504 million. Adjusted for inflation, however, combined federal and state funding for clinics has declined by about one-third since 1980. But the impact of these cuts is unclear. On the one hand, the U.S. Department of Health and Human Services reports that the number of women using publicly funded clinics actually rose between 1980 and 1990, from 4.0 million to 4.2 million. When William Mosher of the National Center for Health Statistics analyzed the NSFG data, however, he found a slight decline between 1982 and 1988 in the proportion of respondents who had visited a clinic in the 12 months preceding the survey (37 percent versus 35 percent).

Whatever the effect of these cuts, the evidence suggests that as with condoms, teens know how to find a clinic when they want to. When they are younger, they do not feel the need to go to a clinic since condoms tend to be their initial form of contraception.

Susan Davis of Planned Parenthood explains, "The most common reason teenagers come is because they think they are pregnant. They get worried. Or they get vaginal infections. I had a whole slew of girls coming for their first pelvic exam and they all had chlamydia." The median time between a female teenager's first sexual experience and her first visit to a clinic is one year, according to a 1981 survey of 1,200 teenagers using 31 clinics in eight cities conducted by Laurie Zabin of the School of Hygiene and Public Health at the Johns Hopkins University in Baltimore.

The Conception Index

Two pieces of evidence further dispel the notion that lack of availability of contraception is the prime problem. First, reported contraceptive use has increased even more than rates of sexual activity. By 1988, the majority of sexually experienced female teens who were at risk to have an unintended pregnancy were using contraception: 79 percent. (This represents an increase from 71 percent in 1982.) When asked what method they use, 46 percent reported using the pill, 26 percent reported using condoms, and 2 percent reported using foam (see figure 2). In addition, the proportion of teen females who reported using a method of contraception at first intercourse increased from 48 percent in 1982 to 65 percent in 1988.

The second piece of evidence is that as they grow older, teenagers shift the forms of contraception they use. Younger teens tend to rely on condoms, whereas older teens use female-oriented methods, such as a sponge, spermicide, diaphragm, or the pill, reflecting the greater likelihood that an older female will be sexually active.

A major reason for this increase in contraceptive use is the growing number of middle-class youths who are sexually active. But it's more than this. Levels of unprotected first sex have decreased among all socioeconomic groups. Among teens from wealthier families, the proportion who reported using no method at first sex decreased between 1982 and 1988 from 43 percent to 27 percent. During the same period, non-use among teens from poorer families also declined, from 60 percent to 42 percent.

Unprotected first sex also decreased among racial groups. Between 1982 and 1988, the proportion of white females who reported using a method of contraception at first intercourse increased from 55 percent to 69 percent. Among blacks, the increase was from 36 percent to 54 percent.

It's not just that teens are telling interviewers what they want to hear about contraception. Despite large increases in sexual activity, there has not been a corresponding increase in the number of conceptions. Between 1975 and 1988,

5. SEXUALITY THROUGH THE LIFE CYCLE: Youth and Their Sexuality

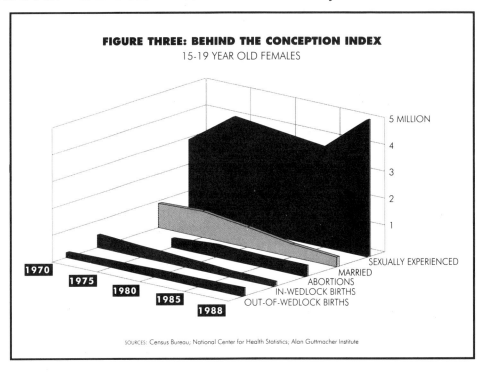

when about 1.3 million more teen females reported engaging in sex (a 39 percent increase), the absolute number of pregnancies increased by less than 21 percent (see figure 3).

In fact, one could create a crude "teen conception index" to measure the changing rate of conception (composed of abortions, miscarriages, and births) among sexually active but unmarried teenagers. If we did so, the 1988 index would stand at .87, representing a decline of 13 percent from 1975 (down from 210 to 182 per 1,000 sexually active, unmarried teens). Most of this decline occurred between 1985 and 1988 as more middle-class teenagers had sex.

The Challenge

Although the conception index among teens is declining, the enormous increase in sexual activity has created a much larger base against which the rate is multiplied. Thus, as we have seen, there have been sharp increases in the rates of abortion, out-of-wedlock births, welfare dependency, and sexually transmitted diseases as measured within the whole teen population.

Teenage sexuality does not have to translate into pregnancy, abortion, out-of-wedlock births, or sexually transmitted diseases. Western Europe, with roughly equivalent rates of teen sexuality, has dramatically lower rates of unwanted pregnancy. According to a 1987 AGI study, the pregnancy rate among American teens (96 per 1,000 women) was twice as high as that in Canada (44), England and Wales (45), and France (43). It was almost three times higher than Sweden's (35) and more than six times higher than in the Netherlands (14). The answer, of course, is effective contraception.

The magnitude of the problem is illustrated by data about reported condom use. Between 1979 and 1988, the reported use of a condom at last intercourse for males ages 17–19 almost tripled, from 21 percent to 58 percent. A decade of heightened concern about AIDS and other sexually transmitted diseases probably explains this tripling. According to Freya Sonenstein and her colleagues at the Urban Institute, over 90 percent of males in their sample knew how AIDS could be transmitted. Eighty-two percent disagreed "a lot" with the statement, "Even though AIDS is a fatal disease, it is so uncommon that it's not a big worry."

As impressive as this progress was, 40 percent did not use a condom at last intercourse. In fact, the 1991 NASM found that there has been no increase in condom use since 1988—even as the threat of AIDS has escalated.

The roots of too-early and too-often unprotected teen sex reach deeply into our society. Robin Williams reportedly asked a girlfriend, "You don't have anything I can take home to my wife, do you?" She said no, so he didn't use a condom. Now both Williams and the girlfriend have herpes, and she's suing him for infecting her. (She claims that he contracted herpes in high school.) When fabulously successful personalities behave this way, should we be surprised to hear about an inner-city youth who refuses his social worker's entreaties to wear a condom when having sex with his AIDS-infected girlfriend?

According to the 1988 NSAM, while 26 percent of white 15-year-old males reported engaging in sex compared to 67 percent of blacks, by age 18 the gap narrowed to 71 percent of whites and 83 percent of blacks.

This is the challenge before us: How to change the behavior of these young men as well as the one in five sexually active female teens who report using no method of contraception. First, all the programs in the world cannot deal with one vital aspect of the problem: many teenagers are simply not ready for sexual relationships. They do not have the requisite emotional and cognitive maturity. Adolescents who cannot remember to hang up their bath towels may be just as unlikely to remember to use contraceptives. Current policies and programs do not sufficiently recognize this fundamental truth.

At the same time, the clock cannot be turned all the way back to the innocent 1950s. Sexual mores have probably been permanently changed, especially for older teens—those who are out of high school, living on their own or off at college. For them, and ultimately all of us, the question is: How to limit the harm being done?

The challenge for public policy is to pursue two simultaneous goals: to lower the rate of sexual activity, especially among young teens, and to raise the level of contraceptive use. Other than abstinence, the best way to prevent pregnancy is to use a contraceptive, and the best way to prevent sexually transmitted diseases is to use a barrier form of contraception. Meeting this challenge will take moral clarity, social honesty, and political courage—three commodities in short supply these days.

SINGLE PARENTS AND DAMAGED CHILDREN

The Fruits of the Sexual Revolution

Lloyd Eby and Charles A. Donovan

Lloyd Eby is assistant senior editor of The World & I. *Charles A. Donovan is senior policy consultant at the Family Research Council, Washington, D.C. Research assistance was provided by Diane Falk, Jayne Turconi, and Mark Petersen.*

Vice President Dan Quayle was right. Murphy Brown—the unmarried TV character who became pregnant on the show—was a bad role model for women, legitimizing and glamorizing single motherhood. But Quayle did not go on to raise or discuss the more difficult problems. Should Murphy Brown have had an abortion? Should she have taken more precautions with contraception so she didn't get pregnant? As she was unmarried, was it wrong for her to have sex? Should she have given up the baby for adoption?

The social science evidence now available shows conclusively that children suffer when they grow up in any family situation other than an intact two-parent family formed by their biological father and mother who are married to each other. As recently as 1960, the biological two-parent family was the norm; in that year, about 75 percent of children in the United States lived with both of their biological parents, who had been married only once, to one another. By 1991 this percentage had declined to about 56 percent. Now, if the darker forecasts are accurate, fewer than 50 percent of children can expect to live continuously throughout their childhood in such families.[1]

The costs of this ever-increasing decline in families and family support of children are huge: to the children, to the larger society, and to the nation. An increasing number of our children, largely from single-parent homes, are unable to participate constructively and ethically in our economic, political, and social life, although many children from single-parent families nonetheless do succeed in life. Costs include immense and ever-increasing welfare rolls; remedial and repeated education; anomie, crime, and lawlessness; high and increasing rates of teen suicide; dealing with unemployable people; and the financial, spiritual, and civic costs of all kinds of social pathologies. All these impose very great financial expenditures as well as enormous psychic and civic burdens. Indeed, it may not be too much to say that family breakdown—with its attendant pathologies and their costs—is our country's most serious social and economic problem, threatening to overwhelm us and even threatening our very democracy and the society on which it rests, unless somehow curbed.

When it was published in 1962, Anthony Burgess' *A Clockwork Orange* seemed overwrought in its depiction of the anomie, violence, pathology, and nihilism of some young people. Today, Burgess' fiction appears to have been remarkably prescient; the murders, rapes, thefts, assaults, burnings and lootings, and other crimes and damages committed by feral and often emotionless youths now surpass his depictions. These developments are very closely linked to the rise in family breakdown and single parenthood.

Today, more than half of all children will live for some period in a single-parent home, either as a result of being born to an unwed mother or as a result of divorce, and the number of such children continues to increase. Some of those children will find themselves having one or more stepparents—even successive series of different stepparents—through the marriage of their previously single parent or parents. But stepfamilies themselves are more prone to divorce, and in general stepparents do not care for or bond as well with children as do biological parents. Increasingly, children are living not with their biological parents but with their grandparents, in foster homes, or in other quasi-family situations. Of course, many children will survive all these troubles and traumas and become fulfilled and productive adults, and many single parents, foster parents, and stepparents cope very well and perform heroically. But those cases increasingly are being overshadowed by the number and severity of other ones. All the children who survive and flourish in these circumstances will do so because they somehow found a way through or around the difficulties, not because of them.

The Murphy Brown character's wealth, prominent social position, education, career success, management skills,

32. Single Parents and Damaged Children

race, and mature age with its attendant relative emotional stability will protect her and her child from some of the worst consequences of unmarried pregnancy and of growing up with a single parent: poverty, lack of educational opportunity and social standing, and the burdens and chaos resulting from having too much responsibility too young. The same holds for the hundreds of real-life actresses, princesses, and other prominent and successful women who have recently—often publicly and defiantly, sometimes quietly or even secretly—borne children out of wedlock. But even in these cases not all is well; children who grow up in single-parent families invariably suffer. The greatest suffering and deprivation, however—for both mothers and children—comes about from unmarried teenage pregnancy.

PREGNANCY OF UNMARRIED TEENAGERS

Today, the United States has a very high and increasing rate of pregnancies to unmarried teenage girls, a much higher rate than any other country in the developed world. In 1950 there were 56,000 births to unmarried teenage girls aged 15 to 19 years, and the birthrate was 12.6 births per thousand such teenagers. In 1960 there were 87,000 such births, and the rate had climbed to 15.3. Between 1961 and 1962 the rate fell slightly, although the number of such births continued to rise. From that date on, the rate has continued to rise every year, and the rate of increase itself has risen—the problem is accelerating. In 1970 there were 190,000 births to unmarried teenage mothers aged 15 to 19, and the rate of such births was 22.4 per thousand unmarried teenagers. In 1980 the figures were 263,000 births and a rate of 27.6. In 1990—the last year for which reliable statistics are available—the rate was 42.5 and the number of births was nearly 350,000—361,000 if we include those children born to girls under 15.

In 1990, 4,158,212 babies were born in the United States to all women. This means that of all births in 1990, about 8.7 percent—or one out of every twelve—was born to an unmarried teenager between 15 and 19 years of age.[2] One birth in twelve may seem relatively insignificant, but the total is for births to unmarried teenagers of all races, compared to all births to all women, of whatever age or race, married or unmarried. If the statistics are broken down by race and restricted to unmarried women, a strong trend appears. Of all births to white women of all ages, the percentage of births to unmarried women in 1990 was 20.35 percent. For all births to women of all races, 28.0 percent were to unmarried women. Of all births to black women of all ages, 66.5 percent were to unmarried women.

The figures for nonmarital births to girls age 15 to 19 are even more bracing. For white teens, 56.4 percent of births were nonmarital in 1990; for black teens, 91.97 percent. Overall, 67.1 percent of teen births in 1990 were nonmarital—a mirror image of the situation as recently as 1970, when 70 percent of *all* teen births were to married women.

PRÉCIS

Attitudes changed in America after the post-1950s social-sexual revolution. Unmarried people came to view having sex as normal and right. Avoidance of divorce for the sake of the children gave way to favoring the happiness of adults as individuals. The taboo against birth out of wedlock eventually disappeared. So ever-more children are being born to unmarried women and growing up in single-parent families. The worst consequences—for both children and parent—occur from births to unmarried teenagers.

The costs—financial and nonfinancial—of the decline of families and family support of children are very high: Ever-increasing welfare rolls. Poor mothers who cannot escape poverty for themselves or their children. An increasing number of our children, largely from single-parent homes, who are lost to our economic, political, and social life. All kinds of social pathologies. Can our society and democracy survive these threats?

Previous attempts to deal with the problem have emphasized technological solutions—various forms of contraception, especially the pill and condoms. But teenagers are poor users of contraception. Rates and virulence of sexually transmitted diseases are rising.

What is to be done about all this? Perhaps we should return to teaching and emphasizing abstinence.

5. SEXUALITY THROUGH THE LIFE CYCLE: Youth and Their Sexuality

If anything, current figures may be worse: More than half the white teens giving birth are unmarried, and among young black mothers fewer than one in ten is married. In short, hardly any births to black teenagers are to married women, and two-thirds of births to all black women are to unmarried women. Each year, one in ten black teenagers will give birth. Nearly half will become unmarried mothers before the end of their teenage years—and many will have more than one child. Another conclusion is that in the United States a large number of children of all races—and the vast majority of black children—are growing up as children of single mothers, that is, as *fatherless* children.

TEENAGE GIRLS AND ABORTION

The figures given above are for live births, not conception rates. To compute conception rates, we need to include the figures for the number of pregnancies terminated through abortion, plus the number of pregnancies that result in miscarriages. The number of miscarriages is unknown but is estimated to be equal to 20 percent of births plus 10 percent of abortions. The number of pregnancies to unmarried teenagers that are terminated through abortion is quite large. Of the 1,590,750 abortions performed in the United States in 1988, 1,314,060, or 82.6 percent, were performed on unmarried women (of all women age 15–44, not just teenagers, and including separated, divorced, and never-married women).[3] Although statistics given by different authorities vary, conservative figures indicate that about 1,033,730 women under 20 became pregnant in 1988, and 40 percent of each age level in that group chose abortion.[4] In any case, we can conclude that a large number of teenage girls are choosing to end their pregnancies through abortion, and that abortion is being used as a last-ditch form of birth control for many teenage girls, establishing a pattern that, after more than two decades of abortion on demand nationwide, is reverberating throughout the cohort of women in their 20s.

Moreover, many counselors and other people concerned with the welfare of teenagers—or, less charitably, with burgeoning welfare rolls—advocate abortion and encourage unmarried pregnant girls to seek abortions.

CAUSES OF TEENAGE PREGNANCY

As Barbara Dafoe Whitehead has noted, a great change in the "social metric" occurred in post-1950s America, from an overarching emphasis on adults sacrificing themselves to achieve child well-being to a concern for adult individual self-fulfillment without regard to whether that is good for children.[5] The attitude that divorce was to be avoided for the sake of the children was exchanged for an attitude that what makes parents happy as individuals is what counts. The ancient taboo against birth out of wedlock was given up. Hugh Hefner's *Playboy* ethic and aesthetic (first issue, December 1953) sanctioned and encouraged young males in their pursuit of unattached sex, and Helen Gurley Brown's *Sex and the Single Girl*, published in 1962, proclaimed loudly that it was not only OK for single women to engage in sex but that they were entitled to it, as an issue of equality. These changes amounted to a great cultural shift, away from the attitude that sex should be restricted to married couples, toward an attitude that proclaimed sex as both good and necessary. The new attitude paid far less attention to marriage, in many cases actually disdaining it. Hollywood and the mass media took up these trends, so that today they are firmly ensconced in American popular culture.

What hardly anyone was willing to see at the time, however, was how children were being affected by these changes. Teenagers—girls especially—became ensnared in a dilemma: Are they adults or children? If they are adults, then they should be able—even encouraged, at least according to one kind of thinking—to participate in all the supposed pleasures of adulthood, including unmarried sex. But if they do engage in sex, then many of the girls will become pregnant. But they are not really adults; they are adolescents, even though their bodies have become sexually mature. The media and the popular culture, however, continually push them toward being sexually active. If the adults who are concerned for their welfare were to claim that it was wrong or misguided for them to be engaging in sex, this would tend to commit those adults to questioning whether the cultural shift is indeed good and beneficial, something very few people who are active in the mass and dominant culture—the universities and the media, including TV, metropolitan newspapers, movies, the magazines, and so on—have been willing to do, until very recently anyway. Those who do question this cultural shift—for example, Dan Quayle—almost always are attacked as meanspirited, stupid, unrealistic religious bigots who want to "turn back the clock" to some supposed past golden era, as old fogies who are unwilling to accept the facts of contemporary life.

One result is widespread and rising amounts of sexual activity among teenagers at younger and younger ages. In 1982, 30.1 percent of women 15–17 years old reported that they were sexually experienced. By 1988 the figure had risen to 37.5 percent. Among unmarried women 18–19 years old, 59.7 percent reported that they were sexually experienced in 1982, and by 1988, 72 percent of such women reported sexual experience. This increased activity is coupled with a rising rate and number of teenage pregnancies. These changes have occurred across all races and social classes, but the most disruptive and devastating effects have been among those who were the poorest and most vulnerable. For them, the allure of sexual freedom broke the threads of cultural and moral cohesion that were the real safety net between temporary poverty and chronic destitution.

THE COSTS OF UNMARRIED PREGNANCY AND FATHERLESSNESS

Teenage pregnancy has costs to the mothers, to the children, and to the larger society and nation. In 1987, more than $19 billion in public funds was spent for income maintenance, health care, and nutrition for support of families begun by teenagers. Babies born to teenagers have a high risk of being born with low birth weight, and low birth weight requires initial hospital care averaging $20,000 per infant. The total lifetime medical costs for each low birth-weight infant average $400,000. For all adolescents (married and unmarried) giving birth, 46 percent go on welfare within four years, and 73 percent of unmarried teenagers giving birth go on welfare within four years.[6] The costs of welfare are extremely high, especially for state budgets. The total state budget for Michigan in 1992, for example, was about $30 billion, and one-third of this—$10 billion—went to the state's social service (welfare) program. Michigan's plight is similar to that of other states—it has neither the lowest nor the highest such expenditure. Moreover, members of these single-parent–headed, welfare-receiving families are at very high risk of remaining poor and ill educated throughout their lives. When married women go on welfare, they tend to get off welfare within a few years. When unmarried women go on welfare, they tend to remain there permanently. We now have the phenomenon in every state of large numbers of families, made up of unmarried women and their children, being on welfare for three or more generations, with no end in sight.

Has anyone ever heard of a child who is happy because he does not know his father? Being a child of a single mother is a handicap, regardless of the wealth, maturity, or social status of that mother.[7]

Numerous studies of child development have shown that growing up as the child of a single parent is linked with lower levels of academic achievement (having to repeat grades in school or receiving lower marks and class standing); increased levels of depression, stress, and aggression; a decrease in some indicators for physical health; higher incidences of needing the services of mental health professionals; and other emotional and behavioral problems.[8] All these effects are linked with lifetime poverty, poor achievement, susceptibility to suicide, likelihood of committing crimes and being arrested, and other pathologies. One such study, based on data from the 1988 National Health Interview Survey, concludes as follows:

> Data ... revealed an excess risk of negative health and performance indicators among children who did not live with both biological parents. These findings are consistent with the hypotheses that children are adversely affected ... by the relative lack of attention, supervision, and opposite-sex role models provided by single parents, regardless of marital status.[9]

THE ATTACK ON FATHERHOOD

While the cultural shift Whitehead describes was occurring, fatherhood was coming under attack. Among other things, these attacks included feminist rhetoric, the claim that fathers are distant and brutal and repressive, based on the observation that some fathers abuse and otherwise mistreat their children and wives, and the observable fact that some children without fathers grow up quite well and become very good and productive adults. "A woman without a man is like a fish without a bicycle," defiantly proclaimed one feminist slogan. But fathers do have a crucial role in rearing children. The small boy with a bicycle wants a father to help him learn to ride it, and both boys and girls usually like their fathers to take them fishing.

Children need two parents, playing into the daily dramas of discipline, self-sacrifice, sincerity, and complementarity. Historically, fathers have given and enforced rules of behavior and provided role models of proper male behavior for both girls and boys. Traditionally, fathers have been very concerned with the sexual virtue of their daughters. Fathers know the attitudes and intentions of teenage boys, having once been teens themselves, and therefore are uniquely able both to guide their daughters and to check out and enforce rules on boyfriends. This does not mean that mothers do not or cannot perform these tasks and roles, but they are handicapped doing it alone. Fathers are vital, and their place cannot be taken by a single mother, however able, resolute, and resourceful she may be. Having fathers as guardians, disciplinarians, and role models is necessary to help teenagers navigate those most difficult experiences and years.

Today, increasing numbers of children, even preteens, are becoming involved in acts of violence and crime, including drug usage and drug dealing, assault, robbery, burglary, theft, carjacking, and shootings and murders, often for seemingly the most trivial of reasons. On the street, a disrespectful look, or an envied pair of sneakers, can provoke a bullet. ("I shot him cause he 'dissed' me.") The costs of this are immense. For all of us—rich and poor, of whatever race—our sense of civic order, safety, and well-being is increasingly threatened, if it has not already collapsed. The monetary

---

It may not be too much to say that family breakdown—with its attendant pathologies and their costs—is our country's most serious social and economic problem.

---

price is enormous; public costs include the expenses of law enforcement, prisons, and other expenditures of crime fighting. Private costs include insurance, security systems, repairing the damages, and the forced exodus of people as they flee our cities in an attempt to find a safe area. A majority of the young people who are responsible for these crimes, with their attendant costs, are products of what once was unashamedly acknowledged as "broken homes."

The criminal and other destructive activities of these teenagers tend to make them into poor or unsuitable prospective marriage partners. Female children of single mothers are more likely to engage in early premarital sex, thus leading to increasing rates of unmarried pregnancy at younger and younger ages. Male children of single mothers are less and less able to become responsible fathers and marriage partners. So we can conclude that unmarried parenthood is feeding on itself, contributing to its own rise.

As already stated, women who are single parents tend to be poorer, more prone to being on welfare, less educationally advantaged, less able to handle careers and work, and more beleaguered in every way than their married sisters. These conclusions hold true for the vast majority of cases, even though there are many instances of such women performing notably and heroically. Single mothers are understandably less likely to be able to accumulate any appreciable amount of savings, purchase homes, afford higher education for themselves and their children, or finance a start-up of any business or profession for themselves or their offspring. In fact, if they are on

5. SEXUALITY THROUGH THE LIFE CYCLE: Youth and Their Sexuality

welfare—Aid to Families with Dependent Children (AFDC)—they are forbidden by the rules to have any significant savings. Some economists have gone so far as to suggest that increasing rates of single motherhood point toward the economic demise of a nation. All these effects are especially pronounced and accentuated for women who become mothers as single teenagers. So the cost of teenage pregnancy to all—to the parents, to the children, to the society, and to the nation—is very high and rising.

PREVENTION OF TEENAGE PREGNANCY

It is estimated that 41 percent of unintended pregnancies among teenagers could be avoided if all sexually active teenagers used contraception. But one-fourth of such teenagers use no contraceptive method or an ineffective one. Half of all teenage pregnancies occur within six months of first sexual intercourse, and more than 20 percent of all initial premarital pregnancies occur in the first month after the initiation of sex. But the use of contraception requires planning, and planned initiation of sexual intercourse among teens is rare. Only 17 percent of women and 25 percent of men report having planned their first intercourse. The contraceptives most widely used by teenagers are the pill and condoms.[10]

Nature equips humans with two differing timetables for maturity; physical and sexual maturity comes first, and emotional and psychological maturity appears later. Teenagers, particularly younger ones, are poorly equipped with the ability to foresee the consequences of their acts and plan accordingly. Teens tend to see themselves as invulnerable to risks. Moreover, this is a time of life when peer pressure and media pressure for engaging in sex are especially acute.

There is reliable but anecdotal evidence that, at least for many inner-city and other poor unmarried teenage girls, their pregnancies are not actually unplanned but actively desired. These studies conclude that the girls are not ignorant about contraception; they do not use it because they actually yearn for babies. Their emotional and psychological immaturity, however, does not allow them to know or understand the real consequences of motherhood, especially teenage motherhood. This is the phenomenon commonly called "babies having babies." Typically, a poor girl who has a baby while unmarried is especially vulnerable to becoming pregnant again while still in her teens.

The primary goal of teenage pregnancy prevention programs since 1970 has been to educate teenagers about the risks of pregnancy and to get them to use contraceptives; this sometimes has been derided as "throwing condoms at the problem." But teenagers typically do not go to see the school nurse or to a health clinic until after they have become sexually active; girls often go for the first time because they think they may be pregnant.

The received approach to the problem of teenage pregnancy has been "technological," in that it has relied on providing teenagers with the technology for avoiding pregnancy, or, once pregnant, with abortions as a technological solution to the pregnancy. But rising rates of teenage pregnancy, abortion, and births to teenage mothers show that these technological solutions have been anything but effective. Advanced as the "realistic" answer to the out-of-wedlock pregnancy problem, these interventions have come athwart the reality of failure statistics. Abortion has reduced the overall adolescent birthrate, but the unmarried adolescent birthrate has gone up dramatically since 1970. Adolescents have become slightly more efficient users of contraception in recent years, but they remain dramatically less so than the adult married population. Moreover, the slight increase in efficiency has been overwhelmed by three factors that are not unrelated to contraceptive availability itself: (a) an increase in the percentage of adolescents in each age cohort having sex; (b) a decrease in the age of the first reported sexual experience; and (c) increases in the frequency of intercourse and the number of sexual partners among adolescents. In this environment, more intense contraceptive use and increased pregnancy rates coexist and may be mutually reinforcing.

All this says nothing about sexually transmitted disease (STD). Increased sexual activity is correlated with rising rates of these diseases in teenagers. Regular, conscientious, and proper use of condoms lowers the incidence of such disease transmission, but we know that teenagers often fail to use them and that, even with conscientious use, condoms sometimes fail. There is no lowering of the risk of sexually transmitted disease through use of oral contraceptives.

This is the fruit of the newfound sexual freedom among adolescents. Not surprisingly, these dismal outcomes are tempting a new generation of advocates to discard freedom when it comes to the latest generation of contraceptive devices.

NORPLANT AND TEENAGE GIRLS

The failure of free-choice use of the pill or condoms to reduce the rates of pregnancy in teenage girls has led to proposals for recommending or requiring Norplant as a contraceptive for teenage girls—particularly for girls who already have a baby. The best technical argument for Norplant is that it removes the diligence factor for those adolescents who receive the implant. Those pregnancies to teenage girls that resulted because they "forgot" to take the pill or because a condom was not readily available will not occur, advocates say, with Norplant. Once inserted in the woman's upper arm, Norplant works automatically, so it would lower the threshold of conscientiousness that adolescents need to practice contraception diligently.

By similar reasoning, however, Norplant will lower adolescent conscientiousness about avoiding sexual encounters on the grounds that pregnancy might result. The conclusion is that, although Norplant arguably would be effective in reducing the number of pregnancies, it may well promote a rise in sexually transmitted diseases.

Norplant advocates recommend continued use of condoms to avert this result, but this raises the diligence problem again. If diligence was not an effective strategy for contraception, will it be so for disease? AIDS infection and syphilis are on the rise, as is antibiotic-resistant gonorrhea. This rise has occurred even in a time when there has been a slight increase in contraceptive efficiency among adolescents. But condom use itself is no guarantee against disease. As one expert notes, "The inescapable fact is that, during one act of intercourse, condoms *may* protect against STD, but for frequent, repeated acts of intercourse over months and years, *they will not*."[11]

Increases also are occurring in other significant venereal diseases, such as chlamydia (which can cause sterility), herpes, and HPV (human papilloma virus). This last is associated with precancerous conditions, from which invasive cancer can develop. So, even if Norplant turns out to reduce the incidence of teenage pregnancy, it may well lead to even more serious problems involving spread of an impressive and growing array of sexually transmissible diseases.

The most recent news on STDs is especially grim. A study released by the Alan Guttmacher Institute on March 21, 1993, reaches the conclusion that 56 million Americans—one in five—are infected with a sexually transmitted viral disease. These diseases can be controlled but not cured. The study estimates that even more Americans are likely to contract an STD during their lifetimes. The greatest effect will be on women and people under the age of 25. According to the study, each year 100,000 to 150,000 women become infertile as a result of STDs. Teenagers and blacks are disproportionately affected by STDs because these people are more likely to be unmarried and thus to have multiple sexual partners. Moreover, teenagers who begin sexual activity earlier are more likely to have more partners. About one in nine women aged 15 to 44 are treated for pelvic inflammatory disease (PID) during their reproductive lifetimes, according to the report, and, if current trends continue, one-half of all women who were 15 in 1970 will have had a PID by the year 2000.[12]

The contraceptive debate of the 1970s occurred in a completely different environment from today's—medically, morally, and socially. Medically, with HIV, the stakes now are significantly higher, even if heterosexual transmission remains relatively uncommon. Morally and socially, the sex education and contraception movements of the 1970s competed against established mores that militated against teen sexual experimentation, but these movements did not compete against an alternative institutional and educational approach, namely abstinence education. Today, the generation that lived through the '70s is debating the policies of the '90s. Experience of what has happened in this field in the ensuing three decades does not lend much respect or hope for any "magic bullet" approach such as is epitomized by free-choice use of Norplant. In today's debate, advocates of the technological approach embodied in Norplant are battling against advocates of an abstinence-based approach who are armed with texts, studies, and curricula of their own. Veterans of the sexual revolution can be found in both camps, leading to a much more realistic—and interesting—public policy debate, with neither side having a monopoly on scientific opinion.

WHAT IS TO BE DONE?

Patterns of teenage pregnancy, abortion, and out-of-wedlock childbearing, although stabilizing somewhat in the 1980s, continue to worsen. After a trial of two decades, national pregnancy policies have failed to reduce pregnancy rates, have succeeded in lowering birthrates only through a sharp concomitant rise in abortions among adolescents, and have coincided with an unprecedented increase in teenage sexual activity. A generation or more of young people—especially inner-city blacks, but others too, of all races—is being lost to productive adulthood and citizenship and is imposing huge and ever-increasing costs—financial, social, and medical—on the larger society and nation. These costs are so great that they threaten to overwhelm us. Can civil liberties, democracy, civil order, and the rule of law survive these present conditions? For the sake of the next generation of American children, it is time for a generous dose of domestic "new thinking" about one of the nation's most intractable social problems.

If all unmarried women of childbearing age for whom it is not medically contraindicated were forced to use it, Norplant would be one solution, but forced imposition is ethically objectionable as well as impossible in a democratic society. Can voluntary use of this method, or any other contraceptive, significantly lower this rising tide of births to single women, with the enormous and ever-rising attendant costs both to children and the nation? So far, voluntary contraceptive methods have failed to curb this problem. Besides, Norplant does not prevent—and may even exacerbate—the spread of venereal disease. Norplant plus condoms has all the problems of teenagers not being sufficiently diligent to practice effective contraception. Perhaps it is time to abandon technological solutions and return to teaching abstinence on moral grounds. Although it sometimes failed, teaching children to abstain was socially, psychologically, and medically far more effective than any of the methods introduced by the sexual revolution—a revolution that was supposed to offer us freedom but that seems instead to have failed us, threatening our livelihoods, our civil order, and perhaps even our liberty itself.

NOTES

1. Deborah A. Dawson, "Family Structure and Children's Health and Well Being: Data from the 1988 National Health Interview Survey on Child Care," *Journal of Marriage and the Family* 53 (August 1991): 573–84.

2. The National Center for Health Statistics, *Vital Statistics of the United States*, annual and *Monthly Vital Statistics Report*, vol. 41, no. 9 supplement, February 25, 1993.

3. Alan Guttmacher Institute, *Abortion Factbook, 1992*.

4. Cited in Center for Population Options, "Adolescents and Abortion Factsheet," February 1993.

5. Barbara Dafoe Whitehead, "Dan Quayle Was Right," *Atlantic Monthly*, April 1993, 47 ff.

6. David A. Hamburg, M.D., *Today's Children: Creating a Future for a Generation in Crisis* (New York: Random House, Times Books, 1992), 198.

7. Margaret Carlson, "Why Quayle Has Half a Point," *Time*, 1 June 1992, 30, 31.

8. Dawson, 573, 574.

9. Dawson, 580.

10. Cited in Center for Population Options, "Adolescent Contraceptive Use Factsheet," June 1990.

11. Joe S. McIlhaney, M.D., *Sexuality and Sexually Transmitted Diseases* (Grand Rapids, Mich.: Baker Book House, 1990), 36. Emphasis in original.

12. Study by Alan Guttmacher Institute, reported in the *New York Times*, 1 April 1993, A1.

Let the Games Begin

Sex and the <u>not</u>-thirtysomethings

SIMON SEBAG MONTEFIORE

In sending out special correspondent Simon Sebag Montefiore to research and report on sex among the thirtysomething set, we realized there existed a new non-sequitur in the language. The term "thirtysomething"—thanks to a moribund and not-quite-defunct TV show (thanks to syndicated reruns)—continues to refer to those lovable, suspender-wearing whiners of the 1980s.

What we needed to do was redefine the term: the "<u>not</u>-thirtysomethings" or, as we refer to them, the "thirtynothings." We don't mean that derogatorily. Rather, that since the focus has shifted with the aging baby boomers, who are now fortysomething, those in their early thirties suddenly find themselves out of the limelight to a certain extent—caught, for the moment at least, in media limbo.

Nevertheless, they are alive and well and living in paradise—enjoying an age of acquisition, feeling their oats and not shy about sowing them.—The Editors

When you begin reading this article, imagine you can hear a clock ticking, like a time bomb. The decade from 30 to 40 years old is the period when everyone is suddenly aware that time is passing faster and faster. I spoke to many people in their thirties about their sexuality, and I don't think there was a single person who did not mention time.

Thirty is the age that dares not speak its name. If sex is a war between male and female, power and trust, orgasm and death, commitment and adventure, then today's early thirtynothings may be living in a kind of no-man's land on the battlefield of sex: Teenagers are in the trenches, twentysomethings are the cavalry, and people in their forties and older are far behind the lines at headquarters. But the young thirties are something of a forgotten race—unsure whether to return to their lines, ride with the cavalry, or simply wait for the war to end.

Unfairly, the sexes are perceived very differently if they are unmarried in their thirties: single men are envied as strutting predatorial bachelors without the slightest pressure to get hitched, while single women are simply "left-on-the-shelf" products whom everyone tries to introduce to any single man they know.

In America, the country that invented feminism, there is an "age-ism"—which is really sexism—that implies that real beauty rests in teenage looks. (Ironically, in Europe, the traditional home of chauvinism, there has always been a cult of the earthy, knowing, sensual, and utterly attractive thirtynothing woman, which alters the way men and women of that age are treated in real life.) But it turns out that—despite the media images of youth as "Adonis culture" (hard waists; big breasts and biceps), the risk of AIDS, the miseries of divorce, and the fear of being "on the shelf"—today's thirtynothings, especially women, are enjoying a concealed festival of sexuality. While the MTV culture condemns them to grim silence (relegated to "soft-rock" radio stations and VH1), the thirtynothings are having the time of their sexual lives.

My mission was to discover whether thirtynothing sex was different from twentysomething sex. In addition, to answer these important questions: Does marriage improve sex, end it, or lead to extramarital affairs? Has AIDS affected the love life of the thirtynothings in the same way it has the twentysomethings? Are the thirties the woman's decade? Do divorcées in their thirties face celibate loneliness or the ultimate liberation? For this I canvassed a random sampling of 30 to 40 people from around the country, in different careers and from different backgrounds, married and single, religious and not.

33. Let the Games Begin

"It's far too late to start that!" was the usual response from men, as if safe sex was like a healthy diet.

Roberta is an attractive pediatrician from Philadelphia, 34, who loves her career, works very hard, possesses a bustling cheerful energy that is as sexy as it is wholesome. She finished an eight-year relationship with an older man two years ago and laughs when I ask if she wishes she were married.

Roberta: "Not exactly. You want to know the truth about sex in the thirties? It's the best. I dreaded the end with Eric. I mourned for weeks. Of course I'd have married him. I was *dying* to get married. I yearn to get married now. But I have my own career. I don't need a man in the traditional way. But lately, I've been dating the most *in*-appropriate guys. It's my right—guys without jobs, younger guys, even really dumb buys with great bodies who were 21 years old. I mean I haven't had sex like that since *I* was 21! It's difficult to suddenly go back to a nice marriageable guy of my age after that. I guess folks in their thirties just say: Let the games begin! And boy, they do!"

Scott, 37, a New York advertising executive who got married three years ago, puts it another way: "I know what I like now. That makes the sex you have in your thirties the best, whether you're married or not. And women my age know what they like. *That's* sexy. It's as if there's no hanging around anymore—they're less inhibited than at any other age. But this goes for both sexes; we're all scared of time ticking, so why beat around the bush? Sex now is the best. This is the icing on the cake."

More Sex, Less Safe

Do thirtynothings practice safe sex in "this day and age"—that euphemism for the AIDS era that I discovered in my previous article on sex amongst the twentysomethings? Since they are older and (supposedly) more responsible, I expected to find that singles in their thirties practice safer sex than any other group. While the twentysomethings are a generation formed by fear, who feel deeply guilty when they do not use condoms (which is often), the thirtynothings are even more oblivious to the danger. The twentysomethings at least have the shame (if we can call it that) to lie to their parents, lovers, and friends about condom use; the thirtynothings apparently do not lie at all.

"It's far too late to start *that*!" was the usual response from men, as if safe sex was like a healthy diet. The women were even less aware of it. Their general response was: "I hate those things. Besides, he was younger." As if their lover's youth were some sort of shield. If their boyfriend was divorced, the general comment was: "Look, Jim's fine. He's been married for five of the last ten years!" If they are divorced themselves, the women answer: "After what I've been through, don't expect me to live in a nunnery."

The whole culture which came of adolescence in the '70s and '80s was built around promiscuity, from James Bond to the Rolling Stones, from Jack Kennedy to Warren Beatty. It is difficult to change in midstream, and the thirtynothings were just getting a taste for the apple when it rotted in their hands.

This attitude is partly the result of the fact that the safe-sex campaign is aimed at teenagers and twentysomethings, presuming (wrongly, as it turns out) that thirtynothings are more responsible and less promiscuous. On the contrary, thirtynothings (single ones) told me that they have far more sex now than they did as teenagers. Why?

I talked to Jane, 37, a consumer-affairs reporter in Chicago, who divorced her husband, Ron, in 1992: "I guess we're more desperate. I could kid you with stuff about women's later sexual peak—I know that's true. But I think it's just because time is ticking and I've got to take advantage of it while I've got it. I don't have one relationship at the moment. I've done that and I'm having a great time, better than ever before."

Jane's answer to the condom question? "I usually insist but the guys never want to."

Martin, 39, is a graphic designer in Miami who has never been married, always has a girlfriend, and thinks it may be time to get married ("I'd like kids"): "Look, I should probably use a condom, but it's just too late now. I'm not even sure I could do it with a condom. I'm stuck in a time warp—I was at Studio 54, I had a wild time in the '80s and I figure, if I've got it, I've got it; and if I *do* have it, I don't even want to know!"

Of course, the ones who are most afraid of AIDS and are fully aware of the importance of safe sex are those who are least at risk—the happily married couples who are faithful and sit tight in the warm castle of marriage, which they imagine is surrounded by all manner of threats, seducers, and deadly diseases. Josh, 36, an investment banker, cuddles his pregnant wife and shivers as he looks back at his single days: "I'm so glad I'm married and safe. I was scared of getting AIDS when I was single because I enjoyed the scene, the bars. One thing terrified me more than anything—not the actual dying, but the thought of dying foolishly from an unsatisfying, silly encounter with a woman whom I hated the next morning."

Only the Good Look Young

Three traditions are increasingly extinct amongst the thirtynothings: dating, foreplay, and the outdated need to marry in order to have children. "I don't date," says Martin. "Everyone's so obsessed with diet, weight, what health club you're a member of, whether you eat red meat, and whether you've seen the latest TV show. Thirtynothings just end up telling each other how young they are. It makes me sick."

Most of the single thirtynothings I interviewed had given up the good old all-American date—that familiar tradition designed for two people to investigate one another before going further. The date has died of the "youth pantomime," which is the sad and comical result of the Adonis cult and is practiced by many segments of our society; although the thirtynothings appear to be its most avid experts.

Roberta: "So many women in their thirties have been so intimidated by advertising that they are desperate to look

5. SEXUALITY THROUGH THE LIFE CYCLE: Sexuality in the Adult Years

and sound young. That's why I gave up dates. What's the point of sitting there with one person saying, 'I'm young,' and then the other one saying, 'I'm young, too.' It's humiliating."

As for motherhood, Jennifer, 38, sums up her feelings: "What women in their thirties need these days is a kid, not a man. I've had what you might call an active love life since I was 15, and I still do. I know who the father is, but I'm with a different guy now, so why should I parade around his name as if I'm scared without it? I'd have given my right hand to fall in love and marry a guy and have kids, but it didn't happen and I'm almost 40 so I got pregnant. It's the best thing I ever did."

Sex and Marriage

The majority of people in their thirties are married or have been married, but since divorce is more common than ever before, there are more single thirtynothings today than ever before. Many of the married couples I spoke to have fantastic sex lives—even if none of them have sex as much as they did when they were single—and are blissfully happy. Amongst married couples, the biggest lie I found, passed man to man and woman to woman, is that they are all having lots of sex. Several married couples admitted that they imply to their friends that they have far more sex than they actually do.

Brent and Lisa, both 31, have been married for four years, have a daughter, and live in Detroit. He works in the auto industry and she works in hotels. Brent: "When you're married, there's no pressure to do it when you don't want to or are tired. We have great fun when we do. When I go out, sure, I look, but it's great not having to chase girls anymore."

Lisa: "When we were living together, we had sex every night. Now it varies, but usually it's a treat on Sunday, once a week. We know each other so well it's always great. I don't lie to my friends about it, but I do sort of imply that we do it a bit more than we do. I don't know why, because we're probably all lying to one another."

The big question is: Can a marriage be truly happy without sex?

Stephanie, a 34-year-old public-relations powerhouse who is itching to get married (the guys always seem to run away), sees boring sex as the price of happy marriage: "If I'm going out with a guy, I won't stand for bad sex. If the guy needs guidance, it's curtains. The only circumstance in which I'd put up with bad sex is in a relationship that'll end in marriage. If I was married, I'd get used to bad sex."

Pat, who married at 23 and is still happily married 10 years later, is an investment banker, outgoing, a sportsman, who has always hinted at his vigorous sex life with his wife, Kelly. Pat does not usually confide about sex, but when I told him I was writing this article, he dropped this bombshell: "You won't believe this, but we simply don't have sex anymore. On holidays we do, but really only to convince each other. I mean, we *talk* as if we had good sex. I guess we're happy, but it suddenly occurred to me that a chunk of my life that used to be real important to me is dead. Maybe it's the cost of being happy."

Kelly: "After about five years, we stopped doing it. Maybe we're just not sexual people. Some people are, some people aren't; but I feel I could be more. We go skiing in Vail every year and share a house with our best friends. The walls there are paper thin and, as we lie there, we can hear them making love—real noisy, shouting and grunting. If we're at home, Pat is as quiet as a mouse when we make love. Not a whisper. But when we're in Vail, he grunts and screams which is great. I guess it turns him on hearing them. Or else he's into the competition thing."

The Fornication Express

The blonde does not look like a married woman. That is the first rule of being married, being in your 30s, and living in the '90s: never *look* married. Married is dull even if marriage is healthy. Linda, 35 (but looks 28) is a tanned Californian visiting New York City.

Brad, 31, single, met her at an Irish bar with a group of friends from work. He persuaded her to go dancing. At the end of the evening, outside his apartment, they kissed on the stairwell.

Linda: "I haven't kissed anyone since I was married, two years ago. We fooled around. I let him touch me because I figured that's okay. That's not sex. Then I went home. But he made me promise to have tea with him at home next day. I guess I shouldn't have gone."

Brad: "Why else would she agree to come unless she wanted to have sex? The moment she arrived we started fooling around, and spent the whole day in bed. We did everything *but*. She was very cute, and it was a big turn-on that she belonged to another man, that it was illicit. She really believed she had not been unfaithful."

Linda: "I really want the marriage to work, but we were having financial problems and job problems and I just had to get away, so a month in New York was good for me. Brad helped a lot. I don't know if the marriage will last but, you know, I didn't give myself a hundred percent. I mean, not really. So I didn't have to lie to my husband."

The story is typical of the infidelities of a thirtynothing couple because it contains: a) a rocky marriage haunted by money worries, job worries, and angst about aging; b) a transcontinental trip to think things through that has become the staple of the '90s relationship (one married woman told me rather proudly that she called the flight from L.A. to New York "The Fornication Express"); c) condoms were not even mentioned; and d) Linda denied the affair had really happened, while shamelessly enjoying its benefits.

A Thousand Days of Desire

Barbet Schroeder, the film director, said that, even in marriage, sexual attraction only lasts a thousand days, but that love can last forever. Of course, the figure is absurd, but, whatever the life span of desire, it often ends in the 30s. What happens when it ends?

When I visited an upscale strip bar in New York, in which thirtynothing men provide the overwhelming percentage of the clientele, I discovered there are harmless, if extremely bland, ways to find some fun without breaking the marital vows. I talked to Kevin, an attorney with a small firm, who is about 35 years old and happily married. His wife knows he goes to strip joints after work: "She thinks it's a boys' thing. For me, it's a relaxing way to have a beer, and the girls are cute as hell. She knows we guys need some entertainment every now and then. Even if I could, I wouldn't touch 'em."

The clubs exist on the premise that nothing illicit happens there. Hence, the very '90s concept of "lap-dancing," wherein a stripper sits in a man's lap and shakes her breasts in his face, while he enjoys the gyrations without touching. Yet while the implant-enhanced women appeared quite bizarre and borderline attractive, the bigger and more unnatural the breasts appeared, the louder the men shout and yahoo.

Ladies, Start Your Engines

The biggest difference in sexual practices between the twenties and thirties lies in the behavior of the women, not the men. Partly, I believe this is the result of the sexual revolution that made it completely acceptable for women to pursue their own desires.

33. Let the Games Begin

The thirties is the decade when the office becomes a major component in the lives of most women and men these days, especially if they are married. Often the marriage itself becomes a threesome: husband, wife, office. Of course, the office has always provided a wealth of secret opportunities for men, hence that very thirtynothing alibi for infidelity—"working late at the office." Ironically, many men and women (who would never truly be unfaithful) become "married to the office" (another thirtynothing cliché), because they have no great desire to return home to their spouses.

The change in the '90s is not that sex is happening in the office, but that nowadays it is *women* in positions of power who are acting as the predators. This is the first decade where there are scores of women in the top echelons of most companies. Like men, the women begin to reach senior positions in their thirties, offering plenty of opportunities for a new sexual formula. (How long before a man brings a case of sexual harassment against his female boss?)

Geoff is a 27-year-old associate in a big Wall Street law firm. His boss, Kim, is a tall, forbidding 35-year-old, a cruel taskmaster engaged to a partner in his forties. Geoff always enjoyed following her along the corridors in her tight-fitting, short-skirted business suits, but was terrified of her too.

Geoff: "At the party the senior associates give for the first years, we went dancing and she didn't dance with anyone except me. I barely dared speak because she's the terror of the firm. Finally, she just says, 'Let's go somewhere else!' So we go to another club and dance until we're sweaty and drunk. Then she pulls me over and says, 'Kiss me.' So I'm wondering if she's gonna fire me if I try to go any further. I decide it's better not to so she makes every move. She makes me take her home. She gives all the orders. We make love for hours. She says she has to be hard at work because she's determined to be a partner and a woman has to be that much better than a man to make it. But, she joshes, I am her reward, her prerogative. Next day, she says hello with a nice smile, but it's all business. We never mentioned it again."

Shelly, 38, is a fit, long-legged real estate broker with an eternal tan who loves her husband and children. She relays a common frustration among women: "I can't believe how long I spent being a little lady, waiting for guys to hit on me, never doing a thing until we'd had at least four dates. I was like that until I turned 30 and thought, why shouldn't I live like men always live? It takes you until you're thirtyish to be free about what you want."

But there is a balance between the sudden celebration of sexual jubilation of the Kims and the Shellys of this world and the miserable, lonely feeling of creeping age coupled with a culture that so elevates every part of young erotic energy. The books of Madonna, the videos on MTV, the photos of Cindy Crawford and models in *Vogue* and on television have all taken their toll.

Cornelia, 39, talked to me the week her divorce from her husband, Harry, came through about what she calls "the erotic dictatorship." She learned suddenly that he had a young girlfriend and was leaving her for this nubile: "My nightmare rival came to life in the young girls in "Beverly Hills 90210." I was jealous of her. I knew she was 25 and I just felt and still feel so inadequate, so…*old*. It's so unfair. How can I compete? I hate all these videos. They make watching television agony."

> The sound of passing time is so deafening to the thirtynothings that they are more afraid of age than they are of death from AIDS.

Tick…Tick…Tick…

The other difficult part of the thirtynothing experience is finding out who you really are before it is too late. Again, everyone I spoke to mentioned a sense of time ticking. Just as the women I've quoted discovered their true sexual calling, so Julia recently discovered her husband's calling was towards his own sex.

Julia's husband, Ted, is the most conventionally straight guy imaginable. At the time they were married, it never occurred to her that Ted was gay. Maybe it never occurred to Ted either. He says simply: "I'd always had feelings, but my parents wanted me to marry, my colleagues at the firm expected me to marry, and I wanted to marry too, to have kids. In fact, I wanted to stay married. But as my thirties went on, I felt that it wasn't the real me and that I was wasting my youth on a false premise. I couldn't tell Julia but I really wanted her to know."

Julia: "I felt de-womanized by his leaving me for another man. So here I am, married for life, but now suddenly single with a big problem: I just don't know how to trust men after Ted. I can't really forgive him. I'm 38 and I don't want to be alone."

Time is still ticking. Time is passing. The sound of passing time is so deafening to the thirtynothings that they are more afraid of age than they are of death from AIDS. Hence, they will be the last generation who play on, condomless, at their game of sexual roulette while younger generations fear death itself. But the thirtynothings, especially women, are also the first generation to really enjoy the benefits of the feminist and sexual revolutions together, so they can climb up corporate ladders, hunt in the office, have marital affairs—just as men always have. But if this piece has a theme, it is that despite the Adonis cult that dominates U. S. culture today, the 30s is the golden age of sexuality, the feast at which both sexes—having learned the mistakes of the past and knowing what pleases them now—collect the glittering prizes before they are gone forever. Or as Roberta puts in biblically, "It's the *knowing* that's sexy about being in your thirties. We're like Adam and Eve—we've eaten of the fruit of the knowledge of good and evil, and we like both."

Article 34

25 ways to make your MARRIAGE SEXIER

What does it take to keep a relationship exciting and passionate? We asked top marriage and sex experts—including America's sexiest wives, who answered LHJ's own survey. Here's their advice.

Lynn Harris

Call to say "I love you." Just one quick romantic phone call during the day, even a message on his voice mail, is "all it takes to keep your sex life simmering," says Judy Seifer, Ph.D., R.N., a sex educator and therapist, and spokesperson for the Chicago-based American Association of Sex Educators, Counselors and Therapists. "If you connect like that during the day, you don't feel disconnected when you get home together," she says. "Those little things can go a long way to make our whole lives sexier."

Listen to one another. You may be an expert at the fine art of lovemaking, but the most important skill is the ability to *listen* to your partner. "Everything else is secondary," says Andrew Stanway, M.D., a psychosexual and marital physician, and author of the just-published book *The Art of Sexual Intimacy* (Carroll & Graf). "If you are able to truly listen to each other—really putting yourself aside and listening with your heart—great sex will follow."

Take bubble baths—together. You don't need a heart-shaped bathtub in a honeymoon suite in order to share some sex suds. According to psychologist Dr. Joyce Brothers, a bubble bath may be a new, exciting experience for your husband no matter where you take the plunge. "He may have been parasailing and skydiving, but he's probably never taken a bubble bath," she says. "Women tend to associate bubble baths with pampering, relaxation and taking care of themselves. Our society doesn't really give men permission to do these things—but a bubble bath does. It gives men a chance to express something that's not ordinarily in their lexicon." And besides, what could be sexier—and more fun—than the two of you in the tub?

Take a walk down memory lane. Remember how hot and heavy your love life used to be? If you bring back some of the things you used to do, your relationship can be steamy once again. "Think back to all your flirtatious, impulsive courtship behavior, and return to what worked for you before," says Helen Singer Kaplan, M.D., Ph.D., director of the Human Sexuality Program at New York Hospital–Cornell Medical Center. If you still live in the same area, seduce each other at an old haunt, or steam up the car windows while parking on the old lovers' lane. You'll find all those sparks are still smoldering—and you may kindle some new ones.

Be mischievous. There's nothing wrong with sex as usual, but why not try something different every so often? "Create an exotic world for yourselves now and then," suggests Pepper Schwartz, Ph.D., president of the Society for the Scientific Study of Sex and professor of sociology at the University of Washington, in Seattle. But that doesn't mean you have to pack your bags and head off to the tropics. For busy couples, "exotic" may simply mean a few hours away from work, kids and telephones. "Allow yourself to be mischievous," Schwartz advises. Take a long lunch and meet your husband for a quickie, or borrow the keys to a friend's cabin for an afternoon of romance. Trying these kinds of sexy escapades every now and then will add adventure to your at-home lovemaking.

Indulge in fantasy. "A lot of women don't realize that fantasizing about or during sex is perfectly okay," says Janet Reibstein, Ph.D., a psychotherapist and co-author of the soon-to-be-published *Sexual Arrangements* (Scribners). Fantasizing doesn't mean that something is wrong with your sex life. On the contrary, "It's having a sense of freedom in your relationship that allows fantasy to awaken," she says. "You can use it to introduce novelty into your sex life"—and that's often what keeps things steamy. There's no need for complex plots—"a fantasy can be nothing more than a fleeting idea," says Reibstein. "Simply let your imagination run free. If you decide to share your fantasy with your partner, you'll find that just talking about it can be a turn-on." Be sure to give him the chance to tell you about one of his fantasies, too.

34. Make Your Marriage Sexier

Try a new technique. Just like experimenting with when or where you make love, experimenting with *how* you do it can rescue your sex life from the routine. "Have sex in a position you've never tried before," suggests sex therapist Dr. Ruth Westheimer. "Thinking of new positions together means that you're thinking about one another's pleasure, and that you're taking responsibility for making sex as sexy as it can be." You can't go wrong: While you might discover that variety is the spice of your sex life, you may also rediscover how much you enjoy it the old way.

Make love with your eyes open. You're supposed to close your eyes when you're transported by passion, right? Not necessarily. Keeping your eyes open during sex creates an intense connection between you and your partner, according to David Schnarch, Ph.D., sex therapist and author of *Constructing the Sexual Crucible: An Integration of Sex and Marital Therapy* (Norton, 1991). "Opening your eyes gives you a real jolt—you'll realize that sex can be more intimate than you ever thought."

Remember that getting there is more than half the fun. "Many people think of sex like a staircase, with

---

Take him by surprise.

Maintaining a spirit of surprise and inventiveness is the key to sensational sex, says Brenda Venus, author of the just-published *Secrets of Seduction: How to Be the Best Lover Your Woman Ever Had* (Dutton). Her favorite surprise scenario: "Show up at his desk, whisper that you're not wearing any underwear, kiss him passionately, say, 'See you tonight,' and leave." He'll definitely be ready for love when he comes home—or maybe you'll both end up taking the rest of the day off! Outrageous, perhaps, but, says Venus, "You can never go too far with the one you love."

Reverse your roles. Does one of you usually take charge during lovemaking? If so, then try doing things the other way around, suggests New York clinical psychologist and sex therapist Janet Wolfe, Ph.D., author of *What to Do When He Has a Headache* (Penguin, 1993). "Changing roles expands your repertoire," she says. So if you tend to be more passive, here's your chance to initiate sex, decide on positions and discover new ways of arousing your partner. If you're usually the aggressive one, allow yourself to relax and let go. Chances are, both of you will find the switch exciting—and erotic.

Take your time. As you initiate lovemaking, take a few extra minutes to wind down from your day—and to wind up for each other. When you make that transition slowly, "You have a chance to relax and start thinking sexy thoughts," says Maggie Scarf, author of *Intimate Partners: Patterns in Love and Marriage* (Ballantine, 1988). "Otherwise, you're still worrying about getting that report done and reminding yourself to call the window washer," she says. What should the two of you do during that transition time? "Forget about genitals for a while—practice the art of the tender caress," says Scarf. "Remember that every inch of our skin is erotic."

each step leading to the next. But unfortunately, they're unhappy if they don't reach the top, and then they forget about all the other steps along the way," says Beverly Whipple, Ph.D., R.N., a sex researcher and co-author of *Safe Encounters* (McGraw-Hill, 1989), and *The G-Spot* (Holt, 1982). "How about thinking of it as a circle, where each point along the circumference is an end in itself?" she suggests. That way, kissing, cuddling and other foreplay will feel just as important as intercourse.

Stop worrying about your flaws. "What gets in the way of good sex is that we're so hung up on our bodies, we're busy trying to hide instead of turning each other on," says Patti Putnicki, author of the forthcoming *101 Things Not to Say During Sex* (Warner). "Kiss his bald spot, touch his potbelly—celebrate the very things that he probably thinks are his worst flaws." Transforming trouble spots into erogenous zones will make you more comfortable and able to focus on pleasing one another.

Don't expect him to read your mind—or to read his. "There's not a man born who knows how to touch a woman until she tells him how," says Bill Young, director of the Masters & Johnson Institute, in St. Louis. "If you don't know how to say it, tell him that you don't know how—and then say it anyway. He'd rather have your guidance than worry about looking inept." While you're at it, ask him what *he* likes. Even if you've asked each other this question before, ask again. Sexual tastes can change with your mood, the time of day, even the weather. "Check it out with each other constantly," recommends Young.

Be playful. "Just have lots of fun together," says Bernie Zilbergeld, Ph.D., sex and marriage therapist and author of *The New Male Sexuality: The Truth About Men, Sex and Pleasure* (Bantam, 1992). Sheer enjoyment is what gives you the energy that fuels hot sex. Dancing is a great way to generate some of that kind of electricity. Says Zilbergeld, "It's not that far from the dance floor to the bedroom floor."

Talk to each other during lovemaking. Your sex life may be active, but is it all action and no talk? "Many people say, 'Was it good?' or 'All done? Okay, thanks,' but there's so much more to be said," says Samuel Janus, Ph.D., co-author with his wife, Cynthia Janus, M.D., of the recently released *The Janus Report on Sexual Behavior* (Wiley). Instead, try a little body language: "During sex, describe what you're about to do, what you're doing, what you'd like your partner to do, how good it feels." This openness, says Janus, "can add an extra dimension of sharing and satisfaction to your sex life."

Give yourself pleasure. Touching yourself will light up lovemaking for both of you, says Susan Crain Bakos, author of *Sexual Pleasures* (St. Martin's, 1992) and *What Men Really Want* (St. Martin's, 1990). Your own pleasure will skyrocket—and what could be sexier than this expression of total freedom and intimacy? He'll be turned on, too: "There's nothing more exciting for a man than an aroused woman," says Bakos.

Build anticipation. When you were first together, the thrill of anticipating sex practically outdid the act itself. That passion is something you need never outgrow. "Prolong [lovemaking] as long as possible," says Shirley Zussman, Ed.D., a sex and marital therapist in New York City. "You can start arousing yourself even before he's there." Take a warm bath, put on some perfume, conjure up your favorite fantasy. When he joins you, don't take things too fast. Instead, build up to lovemaking: Undress one another garment by garment, indulge in some slow, luxurious foreplay. Then, says Zussman, "when you feel you're nearing orgasm, slow down and bring yourself back again to where you were to prolong the pleasurable experience."

5. SEXUALITY THROUGH THE LIFE CYCLE: Sexuality in the Adult Years

Advice from America's sexiest wives

Last February, when we published the results of our landmark survey "The love life of the American wife," the response was tremendous. Suddenly everyone from Jay Leno on the *Tonight Show* to morning talk-show hosts all over the country were commenting on our survey results.

What we found was that women today are more honest and adventurous in the bedroom than ever before. And through in-depth statistical analysis and cross-tabulation, we discovered a group of highly eroticized wives who reported making love more often and with greater skill than our other respondents. We decided to look more closely at this group to try to find out what their secrets are. Below, the characteristics they share and what we can all learn from them:

Be affectionate. Our respondents report that their marriages are still romantic, and that sexuality is not something they save only for the bedroom. Instead, they eroticize their lives with affection. To let their husband know how much they love him, the sexiest wives cook his favorite foods, call him during the day to tell him they love him and write him love notes. Happily, husbands reciprocate with romantic gestures of their own: They bring their wife flowers, call her to say "I love you" and write her love letters. In fact, according to our findings, men are getting *more* romantic today. In our previous sex survey ten years ago, just 35 percent of women reported that their husband sent them flowers and wrote them love notes; today, 56 percent do. These gestures make both partners feel loved and loving, and keep the spark in their marriage. Says one wife, "Because we are kind and thoughtful of each other at all times, it spills over to make our lovemaking indescribably beautiful."

Every couple can benefit from this kind of nurturing, according to Evelyn Moschetta, D.S.W., a marriage counselor in New York City. "Sex begins *outside* the bedroom," she says. "Physical and verbal affection, helping each other out, coming through for each other—that's what activates the warm, close, loving feelings you take into the bedroom." Telling him how much you love him is especially important, adds New York–based family therapist Bonnie Eaker Weil, Ph.D., author of the forthcoming book, *Adultery: The Forgivable Sin* (Birch Lane Press). "You and your partner should say loving and endearing things to one another for at least thirty seconds twice a day, once right when you wake up and again before you go to sleep," she says.

Make time for love. Busy as they may be, sexy wives make lovemaking a top priority in their relationships. "We put each other first and foremost," says one wife. She, like our other respondents, knows that lovemaking is a habit: The more you do it, the more you like it, and the more you like it, the more you do it. And like it they do: These wives make love with their husbands an average of three to five times a week.

How can you increase the frequency of lovemaking? Schedule time for [it], says Lonnie Barbach, Ph.D., co-author of *Going the Distance: Finding and Keeping Lifelong Love* (Plume, 1993). "Set aside one whole evening a week that's yours alone; don't let anything get in the way. Find a baby-sitter, dress up, go out to eat or get take-out, flirt with each other—even try candles, lingerie, exotic videos—whatever will make it a wonderful event for both of you to look forward to."

Be good to each other in bed. Erotic wives are self-confident—and that's one of the greatest sexual turn-ons. If a woman thinks she's good in bed, then she *is* good in bed. She also knows that it's important to be just as loving to her partner, telling him he's sexy and exciting. And chances are, he is. Our respondents rate themselves and their husbands as good or excellent lovers. "Making love is a special gift we give to each other," says one very satisfied wife who's been married for seventeen years. "We're just positive that we've got a corner on the market and no one in the world could possibly make love like we do!"

Try a little spontaneity. Sexy wives break the rules. Instead of making love only at night after the kids are in bed, they seduce their husbands (or let him seduce them) when the urge strikes. Some women say they've called in late for work so they can spend a passionate morning with their partners; others have left a party early to go home and make love. These sexy couples know that the thrill of giving in to desire—when they should be at work or at a social function—keeps their relationship fresh and passionate.

"Where is it written that sex happens *after* dinner and a movie?" asks psychologist Carol Cassell, Ph.D., author of the about-to-be-published *Tender Bargaining: Negotiating an Equal Partnership with the Man You Love* (Lowell House). "What about right now, when you're feeling what you're feeling, and what about right there on the kitchen counter?" She says couples should ask themselves: What patterns have we established? What can we do to jolt them? Explore changes—even small ones—that both of you feel comfortable with.

Be adventurous. The bedroom isn't the only place where the most erotic wives rendezvous with their husbands. In fact, they've learned that getting out of the bedroom is one of the ways married couples can be more innovative. Instead of safe sex behind a locked bedroom door, they enjoy being daring and making love in unexpected places: the tub or shower, the living room, outdoors, even in the car.

"For many of us, the naughtiest and most thrilling sex there is is sex that flirts with public discovery," says sex therapist Dagmar O'Connor, author of *How to Put the Love Back into Making Love* (Doubleday, 1989) and *How to Make Love to the Same Person for the Rest of Your Life* (Bantam, 1986). "It's the secretiveness that makes these public games exciting and that can make us feel closer to one another." She suggests starting with something tame, like a public kiss. If you're ready for something a little more risqué, try advanced under-the-table footsie—the more formal the dinner affair, the better!

Be willing to experiment. The sexiest couples are open-minded. They don't censor sexual experiences or tell each other what they should and shouldn't like. The sexiest wives in our study report using erotic materials—videos, sex toys and sexy magazines—with their husbands. And 54 percent report that *both* partners find it exciting.

Learn what pleases you the most. Sexy wives almost always reach orgasm—usually through intercourse or oral sex. What's the key? They're familiar with their bodies, they know what pleases them, and they focus on their own sensuality. And when a wife gets turned on, so does her husband. The more excited she gets, the more excited he gets and the more exciting their relationship is. "We both want to please each other," says one wife. "And our sex life just keeps getting better than ever!"

Beyond Betrayal: Life After Infidelity

Frank Pittman III, M.D.

Hour after hour, day after day in my office I see men and women who have been screwing around. They lead secret lives, as they hide themselves from their marriages. They go through wrenching divorces, inflicting pain on their children and their children's children. Or they make desperate, tearful, sweaty efforts at holding on to the shreds of a life they've betrayed. They tell me they have gone through all of this for a quick thrill or a furtive moment of romance. Sometimes they tell me they don't remember making the decision that tore apart their life: "It just happened." Sometimes they don't even know they are being unfaithful. (I tell them: "If you don't know whether what you are doing is an infidelity or not, ask your spouse.") From the outside looking in, it is insane. How could anyone risk everything in life on the turn of a screw? Infidelity was not something people did much in my family, so I always found it strange and noteworthy when people did it in my practice. After almost 30 years of cleaning up the mess after other people's affairs, I wrote a book describing everything about infidelity I'd seen in my practice. The book was *Private Lies: Infidelity and the Betrayal of Intimacy* (Norton). I thought it might help. Even if the tragedy of AIDS and the humiliation of prominent politicians hadn't stopped it, surely people could not continue screwing around after reading about the absurd destructiveness of it. As you know, people have *not* stopped having affairs. But many of them feel the need to write or call or drop by and talk to me about it. When I wrote *Private Lies*, I thought I knew everything there was to know about infidelity. But I know now that there is even more.

Accidental Infidelity

All affairs are not alike. The thousands of affairs I've seen seem to fall into four broad categories. Most first affairs are cases of *accidental infidelity*, unintended and uncharacteristic acts of carelessness that really did "just happen." Someone will get drunk, will get caught up in the moment, will just be having a bad day. It can happen to anyone, though some people are more accident prone than others, and some situations are accident zones.

Many a young man has started his career as a philanderer quite accidentally when he is traveling out of town on a new job with a philandering boss who chooses one of a pair of women and expects the young fellow to entertain the other. The most startling dynamic behind accidental infidelity is misplaced politeness, the feeling that it would be rude to turn down a needy friend's sexual advances. In the debonair gallantry of the moment, the brazen discourtesy to the marriage partner is overlooked altogether.

Both men and women can slip up and have accidental affairs, though the most accident-prone are those who drink, those who travel, those who don't get asked much, those who don't feel very tightly married, those whose running buddies screw around, and those who are afraid to run from a challenge. Most are men.

After an accidental infidelity, there is clearly the sense that one's life and marriage have changed. The choices are:

1. To decide that infidelity was a stupid thing to do, to confess it or not to do so, but to resolve to take better precautions in the future;

2. To decide you wouldn't have done such a thing unless your husband or wife had let you down, put the blame on your mate, and go home and pick your marriage to death;

3. To notice that lightning did not strike you dead, decide this would be a safe and inexpensive hobby to take up, and do it some more;

4. To decide that you would not have done such a thing if you were married to the right person, determine that this was "meant to be," and declare yourself in love with the stranger in the bed.

Romantic Infidelity

Surely the craziest and most destructive form of infidelity is the temporary insanity of *falling in love*. You do this, not when you meet somebody wonderful (wonderful people don't screw around with married people) but when you are going through a crisis in your own life, can't continuing living your life, and aren't quite ready for suicide yet. An affair with someone grossly inappropriate—someone decades younger or older, someone dependent or dominating, someone with problems even bigger than your own—is so crazily stimulating that it's like a drug that can lift you out of your depression and enable you to feel things again. Of course, between moments of ecstasy, you are more depressed, increasingly alone and alienated in your life, and increasingly hooked on the affair partner. Ideal romance partners are damsels or "dumsels" in distress, people without a life but with a lot of problems, people with bad reality testing and little concern with understanding reality better.

Romantic affairs lead to a great many divorces, suicides, homicides, heart attacks, and strokes, but not to very many successful remarriages. No matter how many sacrifices you make to keep the love alive, no matter how many sacrifices your family and children make for this crazy relationship, it will gradually burn itself out when there is nothing more to sacrifice to it. Then you must face not only the wreckage of several lives, but the original depression from which the affair was an insane flight into escape.

People are most likely to get into these romantic affairs at the turning points of life: when their parents die or their children grow up; when they suffer health crises or are under pressure to give up an addiction; when they achieve an unexpected level of job success or job failure; or when their first child is born—any situation in which they must face a lot of reality and grow up. The better the marriage, the saner and more sensible the spouse, the more alienated the romantic is likely to feel. Romantic affairs happen in good marriages even more often than in bad ones.

MYTHS OF INFIDELITY

The people who are running from bed to bed creating disasters for themselves and everyone else don't seem to know what they are doing. They just don't get it. But why should they? There is a mythology about infidelity that shows up in the popular press and even in the mental health literature that is guaranteed to mislead people and make dangerous situations even worse. Some of these myths are:

1. Everybody is unfaithful; it is normal, expectable behavior. Mozart, in his comic opera *Cosi Fan Tutti*, insisted that women all do it, but a far more common belief is that men all do it: "Higgamous, hoggamous, woman's monogamous; hoggamous, higgamous, man is polygamous." In Nora Ephron's movie, *Heartburn*, Meryl Streep's husband has left her for another woman. She turns to her father for solace, but he dismisses her complaint as the way of all male flesh: "If you want monogamy, marry a swan."

We don't know how many people are unfaithful; if people will lie to their own husband or wife, they surely aren't going to be honest with poll takers. We can guess that one-half of married men and one-third of married women have dropped their drawers away from home at least once. That's a lot of infidelity.

Still, most people are faithful most of the time. Without the expectation of fidelity, intimacy becomes awkward and marriage adversarial. People who expect their partner to betray them are likely to beat them to the draw, and to make both of them miserable in the meantime.

Most species of birds and animals in which the male serves some useful function other than sperm donation are inherently monogamous. Humans, like other nest builders, are monogamous by nature, but imperfectly so. We can be trained out of it, though even in polygamous and promiscuous cultures people show their true colors when they fall blindly and crazily in love. And we have an escape clause: nature mercifully permits us to survive our mates and mate again. But if we slip up and take a new mate while the old mate is still alive, it is likely to destroy the pair bonding with our previous mate and create great instinctual disorientation—which is part of the tragedy of infidelity.

2. Affairs are good for you; an affair may even revive a dull marriage. Back at the height of the sexual revolution, the *Playboy* philosophy and its *Cosmopolitan* counterpart urged infidelity as a way to keep men manly, women womanly, and marriage vital. Lately, in such books as Annette Lawson's *Adultery* and Dalma Heyn's *The Erotic Silence of the American Wife,* women have been encouraged to act out their sexual fantasies as a blow for equal rights.

It is true that if an affair is blatant enough and if all hell breaks loose, the crisis of infidelity can shake up the most petrified marriage. Of course, any crisis can serve the same detonation function, and burning the house down might be a safer, cheaper, and more readily forgivable attention-getter.

However utopian the theories, the reality is that infidelity, whether it is furtive or blatant, will blow hell out of a marriage. In 30 odd years of practice, I have encountered only a handful of established first marriages that ended in divorce without someone being unfaithful, often with the infidelity kept secret throughout the divorce process and even for years afterwards. Infidelity is the *sine qua non* of divorce.

3. People have affairs because they aren't in love with their marriage partner. People tell me this, and they even remember it this way. But on closer examination it routinely turns out that the marriage was fine before the affair happened, and the decision that they were not in love with their marriage partner was an effort to explain and justify the affair.

Being in love does not protect people from lust. Screwing around on your loved one is not a very loving thing to do, and it may be downright hostile. Every marriage is a thick stew of emotions ranging from lust to disgust, desperate love to homicidal rage. It would be idiotic to reduce such a wonderfully rich emotional diet to a question ("love me?" or "love me not?") so simplistic that it is best asked of the petals of daisies. Nonetheless, people do ask themselves such questions, and they answer them.

Falling out of love is no reason to betray your mate. If people are experiencing a deficiency in their ability to love their partner, it is not clear how something so hateful as betraying him or her would restore it.

4. People have affairs because they are oversexed. Affairs are about secrets. The infidelity is not necessarily in the sex, but in the dishonesty.

Swingers have sex openly, without dishonesty and therefore without betrayal (though with a lot of scary bugs.) More cautious infidels might have chaste but furtive lunches and secret telephone calls with ex-spouses or former affair partners—nothing to sate the sexual tension, but just enough to prevent a marital reconciliation or intimacy in the marriage.

Affairs generally involve sex, at least enough to create a secret that seals the conspiratorial alliance of the affair, and makes the relationship tense, dangerous, and thus exciting. Most affairs consist of a little bad sex and hours on the telephone. I once saw a case in which the couple had attempted sex once 30 years before and had limited the intimacy in their respective marriages while they maintained their sad, secret love with quiet lunches, pondering the crucial question of whether or not he had gotten it all the way in on that immortal autumn evening in 1958.

> **Every marriage is a thick stew of emotions ranging from lust to disgust.**

Both genders seem equally capable of falling into the temporary insanity of romantic affairs, though women are more likely to reframe anything they do as having been done for love. Women in love are far more aware of what they are doing and what the dangers might be. Men in love can be extraordinarily incautious and willing to give up everything. Men in love lose their heads—at least for a while.

MARITAL ARRANGEMENTS

All marriages are imperfect, and probably a disappointment in one way or another, which is a piece of reality, not a license to mess around with the neighbors. There are some marriages that fail to provide a modicum of warmth, sex, sanity, companionship, money. There are awful marriages people can't get all the way into and can't get all the way out of, divorces people won't call off and can't go through, marriages that won't die and won't recover. Often people in such marriages make a *marital arrangement* by calling in marital aides to keep them company while they avoid living their life. Such practical affairs help them keep the marriage steady but

In general, monogamous couples have a lot more sex than the people who are screwing around.

5. Affairs are ultimately the fault of the cuckold. Patriarchal custom assumes that when a man screws around it must be because of his wife's aesthetic, sexual, or emotional deficiencies. She failed him in some way. And feminist theory has assured us that if a wife screws around it must be because men are such assholes. Many people believe that screwing around is a normal response to an imperfect marriage and is, by definition, the marriage partner's fault. Friends and relatives, bartenders, therapists, and hairdressers, often reveal their own gender prejudices and distrust of marriage, monogamy, intimacy, and honesty, when they encourage the infidel to put the blame on the cuckold rather than on him- or herself.

One trick for avoiding personal blame and responsibility is to blame the marriage itself (too early, too late, too soon after some event) or some unchangeable characteristic of the partner (too old, too tall, too ethnic, too smart, too experienced, too inexperienced.) This is both a cop-out and a dead end.

One marriage partner can make the other miserable, but can't make the other unfaithful. (The cuckold is usually not even there when the affair is taking place). Civilization and marriage require that people behave appropriately however they feel, and that they take full responsibility for their actions. "My wife drove me to it with her nagging"; "I can't help what I do because of what my father did to me"; "She came on to me and her skirt was very short"; "I must be a sex addict"; et cetera. Baloney! If people really can't control their sexual behavior, they should not be permitted to run around loose.

There is no point in holding the cuckold responsible for the infidel's sexual behavior unless the cuckold has total control over the sexual equipment that has run off the road. Only the driver is responsible.

6. It is best to pretend not to know. There are people who avoid unpleasantness and would rather watch the house burn down than bother anyone by yelling "Fire!" Silence fuels the affair, which can thrive only in secrecy. Adulterous marriages begin their repair only when the secret is out in the open, and the infidel does not need to hide any longer. Of course, it also helps to end the affair.

A corollary is the belief that infidels must deny their affairs interminably and do all that is possible to drive cuckolds to such disorientation that they will doubt their own sanity rather than doubt their partner's fidelity. In actuality, the continued lying and denial is usually the most unforgivable aspect of the infidelity.

One man was in the habit of jogging each evening, but his wife noticed that his running clothes had stopped stinking. Suspicious, she followed him—to his secretary's apartment. She burst in and confronted her husband who was standing naked in the secretary's closet. She demanded: "What are you doing here?" He responded: "You do not see me here. You have gone crazy and are imagining this." She almost believed him, and remains to this day angrier about *that* than about the affair itself. Once an affair is known or even suspected, there is no safety in denial, but there is hope in admission.

I recently treated a woman whose physician husband divorced her 20 years ago after a few years of marriage, telling her that she had an odor that was making him sick, and he had developed an allergy to her. She felt so bad about herself she never remarried.

I suspected there was more to the story, and sent her back to ask him whether he had been unfaithful to her. He confessed that he had been, but had tried to shield her from hurt by convincing her that he had been faithful and true but that she was repulsive. She feels much worse about him now, but much better about herself. She now feels free to date.

7. After an affair, divorce is inevitable. Essentially all first-time divorces occur in the wake of an affair. With therapy though, most adulterous marriages can be saved, and may even be stronger and more intimate than they were before the crisis. I have rarely seen a cuckold go all the way through with a divorce after a first affair that is now over. Of course, each subsequent affair lowers the odds drastically.

It doesn't happen the way it does in the movies. The indignant cuckold does scream and yell and carry on and threaten all manner of awful things—which should not be surprising since his or her life has just been torn asunder. But he or she quickly calms down and begins the effort to salvage the marriage, to pull the errant infidel from the arms of the dreaded affairee.

When a divorce occurs, it is because the infidel can not escape the affair in time or cannot face going back into a marriage in which he or she is now known and understood and can no longer pose as the chaste virgin or white knight spotless and beyond criticism. A recent *New Yorker* cartoon showed a forlorn man at a bar complaining: "My wife understands me."

Appropriate guilt is always helpful, though it must come from inside rather than from a raging, nasty spouse; anger is a lousy seduction technique for anyone except terminal weirdos. Guilt is good for you. Shame, however, makes people run away and hide.

The prognosis after an affair is not grim, and those who have strayed have not lost all their value. The sadder but wiser infidel may be both more careful and more grateful in the future.

5. SEXUALITY THROUGH THE LIFE CYCLE: Sexuality in the Adult Years

distant. They thus encapsulate the marital deficiency, so the infidel can neither establish a life without the problems nor solve them. Affairs can wreck a good marriage, but can help stabilize a bad one.

People who get into marital arrangements are not necessarily the innocent victims of defective relationships. Some set out to keep their marriages defective and distant. I have seen men who have kept the same mistress through several marriages, arranging their marriages to serve some practical purpose while keeping their romance safely encapsulated elsewhere. The men considered it a victory over marriage; the exploited wives were outraged.

I encountered one woman who had long been involved with a married man. She got tired of waiting for him to get a divorce and married someone else. She didn't tell her husband about her affair, and she didn't tell her affairee about her marriage. She somehow thought they would never find out about one another. After a few exhausting and confusing weeks, the men met and confronted her. She cheerfully told them she loved them both and the arrangement seemed the sensible way to have her cake and eat it too. She couldn't understand why both the men felt cheated and deprived by her efforts to sacrifice their lives to satisfy her skittishness about total commitment.

Some of these arrangements can get quite complicated. One woman supported her house-husband and their kids by living as the mistress of an older married man, who spent his afternoons and weekend days with her and his evenings at home with his own children and his sexually boring wife. People averse to conflict might prefer such arrangements to therapy, or any other effort to actually solve the problems of the marriage.

Unhappily married people of either gender can establish marital arrangements to help them through the night. But men are more likely to focus on the practicality of the arrangement and diminish awareness of any threat to the stability of the marriage, while women are more likely to romanticize the arrangement and convince themselves it is leading toward an eventual union with the romantic partner. Networks of couples may spend their lives halfway through someone's divorce, usually with a guilt-ridden man reluctant to completely leave a marriage he has betrayed and even deserted, and a woman, no matter how hard she protests to the contrary, eternally hopeful for a wedding in the future.

Philandering

Philandering is a predominantly male activity. Philanderers take up infidelity as a hobby. Philanderers are likely to have a rigid and concrete concept of gender; they worship masculinity, and while they may be greatly attracted to women, they are mostly interested in having the woman affirm their masculinity. They don't really like women, and they certainly don't want an equal, intimate relationship with a member of the gender they insist is inferior, but far too powerful. They see women as dangerous, since women have the ability to assess a man's worth, to measure him and find him wanting, to determine whether he is man enough.

These men may or may not like sex, but they use it compulsively to affirm their masculinity and overcome both their homophobia and their fear of women. They can be cruel, abusive, and even violent to women who try to get control of them and stop the philandering they consider crucial to their masculinity. Their life is centered around displays of masculinity, however they define it, trying to impress women with their physical strength, competitive victories, seductive skills, mastery of all situations, power, wealth, and, if necessary, violence. Some of them are quite charming and have no trouble finding women eager to be abused by them.

Gay men can philander too, and the dynamics are the same for gay philanderers as for straight ones: the obvious avoidance of female sexual control, but also the preoccupation with masculinity and the use of rampant sexuality for both reassurance and the measurement of manhood. When men have paid such an enormous social and interpersonal price for their preferred sexuality, they are likely to wrap an enormous amount of their identity around their sexuality and express that sexuality extensively.

Philanderers may be the sons of philanderers, or they may have learned their ideas about marriage and gender from their ethnic group or inadvertently from their religion. Somewhere they have gotten the idea that their masculinity is their most valuable attribute and it requires them to protect themselves from coming under female control. These guys may consider themselves quite principled and honorable, and they may follow the rules to the letter in their dealings with other men. But in their world women have no rights.

To men they may seem normal, but women experience them as narcissistic or even sociopathic. They think they are normal, that they are doing what every other real man would do if he weren't such a wimp. The notions of marital fidelity, of gender equality, of honesty and intimacy between husbands and wives seem quite foreign from what they learned growing up. The gender equality of monogamy may not feel compatible to men steeped in patriarchal beliefs in men being gods and women being ribs. Monogamous sexuality is difficult for men who worship Madonnas for their sexlessness and berate Eves for their seductiveness.

Philanderers' sexuality is fueled by anger and fear, and while they may be considered "sex addicts" they are really "gender compulsives," desperately doing whatever they think will make them look and feel most masculine. They put notches on their belts in hopes it will make their penises grow bigger. If they can get a woman to die for them, like opera composer Giacomo Puccini did in real life and in most of his operas, they feel like a real man.

Female Philanderers

There are female philanderers too, and they too are usually the daughters or ex-wives of philanderers. They are angry at men, because they believe all men screw around as their father or ex-husband did. A female philanderer is not likely to stay married for very long, since that would require her to make peace with a man, and as a woman to carry more than her share of the burden of marriage. Marriage grounds people in reality rather than transporting them into fantasy, so marriage is too loving, too demanding, too realistic, and not romantic enough for them.

I hear stories of female philanderers, such as Maria Riva's description of her mother, Marlene Dietrich. They appear to have insatiable sexual appetites but, on closer examination, they don't like sex much, they do like power over men, and underneath the philandering anger, they are plaintively seeking love.

Straying wives are rarely philanderers, but single women who mess around with married men are quite likely to be. Female philanderers prefer to raid other people's marriages, breaking up relationships, doing as much damage as possible, and then dancing off reaffirmed. Like male philanderers, female philanderers put their vic-

Spider Woman

There are women who, by nature romantics, don't quite want to escape their own life and die for love. Instead they'd rather have some guy wreck his life for them. These women have been so recently betrayed by unfaithful men that the wound is still raw and they are out for revenge. A woman who angrily pursues married men is a "spider woman"—she requires human sacrifice to restore her sense of power.

When she is sucking the blood from other people's marriages, she feels some relief from the pain of having her own marriage betrayed. She simply requires that a man love her enough to sacrifice his life for her. She may be particularly attracted to happy marriages, clearly envious of the woman whose husband is faithful and loving to her. Sometimes it isn't clear whether she wants to replace the happy wife or just make her miserable.

The women who are least squeamish and most likely to wreak havoc on other people's marriages are victims of some sort of abuse, so angry that they don't feel bound by the usual rules or obligations, so desperate that they cling to any source of security, and so miserable that they don't bother to think a bit of the end of it.

Josephine Hart's novel *Damage*, and the recent Louis Malle film version of it, describe such a woman. She seduces her fiancee's depressed father, and after the fiancee discovers the affair and kills himself, she waltzes off from the wreckage of all the lives. She explains that her father disappeared long ago, her mother had been married four or five times, and her brother committed suicide when she left his bed and began to date other boys. She described herself as damaged, and says, "Damaged people are dangerous. They know they can survive."

Bette was a spider woman. She came to see me only once, with her married affair partner Alvin, a man I had been seeing with his wife Agnes. But I kept up with her through the many people whose lives she touched. Bette's father had run off and left her and her mother when she was just a child, and her stepfather had exposed himself to her. Most recently Bette's manic husband Burt had run off with a stripper, Claudia, and had briefly married her before he crashed and went into a psychiatric hospital.

While Burt was with Claudia, the enraged Bette promptly latched on to Alvin, a laid-back philanderer who had been married to Agnes for decades and had been screwing around casually most of that time. Bette was determined that Alvin was going to divorce Agnes and marry her, desert his children, and raise her now-fatherless kids. The normally cheerful Alvin, who had done a good job for a lifetime of pleasing every woman he met and avoiding getting trapped by any of them, couldn't seem to escape Bette, but he certainly had no desire to leave Agnes. He grew increasingly depressed and suicidal. He felt better after he told the long-suffering Agnes, but he still couldn't move in any direction. Over the next couple of years, Bette and Alvin took turns threatening suicide, while Agnes tended her garden, raised her children, ran her business, and waited for the increasingly disoriented and pathetic Alvin to come to his senses.

Agnes finally became sufficiently alarmed about her husband's deterioration that she decided the only way she could save his life was to divorce him. She did, and Alvin promptly dumped Bette. He could not forgive her for what she had made him do to dear, sweet Agnes. He lost no time in taking up with Darlene, with whom he had been flirting for some time, but who wouldn't go out with a married man. Agnes felt relief, and the comfort of a good settlement, but Bette was once again abandoned and desperate.

She called Alvin hourly, alternately threatening suicide, reciting erotic poetry, and offering to fix him dinner. She phoned bomb threats to Darlene's office. Bette called me to tell me what a sociopathic jerk Alvin was to betray her with another woman after all she had done in helping him through his divorce. She wrote sisterly notes to Agnes, offering the comfort of friendship to help one another through the awful experience of being betrayed by this terrible man. At no point did Bette consider that she had done anything wrong. She was now, as she had been all her life, a victim of men, who not only use and abuse women, but won't lay down their lives to rescue them on cue.

EMOTIONALLY RETARDED MEN IN LOVE

About the only people more dangerous than philandering men going through life with an open fly and romantic damsels going through life in perennial distress, are emotionally retarded men in love. When such men go through a difficult transition in life, they hunker down and ignore all emotions. Their brain chemistry gets depressed, but they don't know how to feel it as depression. Their loved ones try to keep from bothering them, try to keep things calm and serene—and isolate them further.

An emotionally retarded man may go for a time without feeling pleasure, pain, or anything else, until a strange woman jerks him back into awareness of something intense enough for him to feel it—perhaps sexual fireworks, or the boyish heroics of rescuing her, or perhaps just fascination with her constantly changing moods and never-ending emotional crises.

With her, he can pull out of his depression briefly, but he sinks back even deeper into it when he is not with her. He is getting addicted to her, but he doesn't know that. He only feels the absence of joy and love and life with his serenely cautious wife

> **All marriages are imperfect, a disappointment in one way or another.**

> **Once an affair is known or even suspected, there's no safety in denial but there is hope in admission.**

5. SEXUALITY THROUGH THE LIFE CYCLE: Sexuality in the Adult Years

and kids, and the awareness of life with this new woman. It doesn't work for him to leave home to be with her, as she too would grow stale and irritating if she were around full time.

What he needs is not a crazier woman to sacrifice his life for, but treatment for his depression. However, since the best home remedies for depression are sex, exercise, joy, and triumph, the dangerous damsel may be providing one or more of them in a big enough dose to make him feel a lot better. He may feel pretty good until he gets the bill, and sees how much of his life and the lives of his loved ones this treatment is costing. Marriages that start this way, stepping over the bodies of loved ones as the giddy couple walks down the aisle, are not likely to last long.

Howard had been faithful to Harriett for 16 years. He had been happy with her. She made him feel loved, which no one else had ever tried to do. Howard devoted himself to doing the right thing. He always did what he was supposed to do and he never complained. In fact he said very little at all.

Howard worked at Harriett's father's store, a stylish and expensive men's clothiers. He had worked there in high school and returned after college. He'd never had another job. He had felt like a son to his father-in-law. But when the old man retired, he bypassed the stalwart, loyal Howard and made his own wastrel son manager.

Howard also took care of his own elderly parents who lived next door. His father died, and left a nice little estate to his mother, who then gave much of it to his younger brother, who had gotten into trouble with gambling and extravagance.

Howard felt betrayed, and sank into a depression. He talked of quitting his job and moving away. Harriett pointed out the impracticality of that for the kids. She reminded him of all the good qualities of his mother and her father.

Howard didn't bring it up again. Instead, he began to talk to Maxine, one of the tailors at the store, a tired middle-aged woman who shared Howard's disillusionment with the world. One day, Maxine called frightened because she smelled gas in her trailer and her third ex-husband had threatened to hurt her. She needed for Howard to come out and see if he could smell anything dangerous. He did, and somehow ended up in bed with Maxine. He felt in love. He knew it was crazy but he couldn't get along without her. He bailed her out of the frequent disasters in her life. They began to plot their getaway, which consumed his attention for months.

Harriett noticed the change in Howard, but thought he was just mourning his father's death. They continued to get along well, sex was as good as ever, and they enjoyed the same things they had always enjoyed. It was a shock to her when he told her he was moving out, that he didn't love her anymore, and that it had nothing whatever to do with Maxine, who would be leaving with him.

Harriett went into a rage and hit him. The children went berserk. The younger daughter cried inconsolably, the older one

Most affairs consist of a little bad sex and a lot of telephoning.

became bulimic, the son quit school and refused to leave his room. I saw the family a few times, but Howard would not turn back. He left with Maxine, and would not return my phone calls. The kids were carrying on so on the telephone, Howard stopped calling them for a few months, not wanting to upset them. Meanwhile he and Maxine, who had left her kids behind as well, borrowed some money from his mother and moved to the coast where they bought into a marina—the only thing they had in common was the pleasure of fishing.

A year later, Harriett and the kids were still in therapy but they were getting along pretty well without him. Harriett was running the clothing store. Howard decided he missed his children and invited them to go fishing with him and Maxine. It surprised him when they still refused to speak to him. He called me and complained to me that his depression was a great deal worse. The marina was doing badly. He and Maxine weren't getting along very well. He missed his children and cried a lot, and she told him his preoccupation with his children was a betrayal of her. He blamed Harriett for fussing at him when she found out about Maxine. He believed she turned the children against him. He couldn't understand why anyone would be mad with him; he couldn't help who he loves and who he doesn't love.

MEN AND WOMEN WHO CHEAT

Howard's failure to understand the complex emotional consequences of his affair is typically male, just as Bette's insistence that her affair partner live up to her romantic fantasies is typically female. Any gender-based generalization is both irritating and inaccurate, but some behaviors are typical. Men tend to attach too little significance to affairs, ignoring their horrifying power to disorient and disrupt lives, while women tend to attach too much significance, assuming that the emotions are so powerful they must be "real" and therefore concrete, permanent, and stable enough to risk a life for.

A man, especially a philandering man, may feel comfortable having sex with a woman if it is clear that he is not in love with her. Even when a man understands that a rule has been broken and he expects consequences of some sort, he routinely underestimates the extent and range and duration of the reactions to his betrayal. Men may agree that the sex is wrong, but may believe that the lying is a noble effort to protect the family. A man may reason that outside sex is wrong because there is a rule against it, without understanding that his lying establishes an adversarial relationship with his mate and is the greater offense. Men are often surprised at the intensity of their betrayed mate's anger, and then even more surprised when she is willing to take him back. Men rarely appreciate the devastating long-range impact of their infidelities, or even their divorces, on their children.

Routinely, a man will tell me that he assured himself that he loved his wife before he hopped into a strange bed, that the women there with him means nothing, that it is just a meaningless roll in the hay. A woman is more likely to tell me that at the sound of the zipper she quickly ascertained that she was not as much in love with her husband as she should have been, and the man there in bed with her was the true love of her life.

A woman seems likely to be less concerned with the letter of the law than with the emotional coherence of her life. It may be okay to screw a man if she "loves" him, whatever the status of his or her marriage, and it is certainly appropriate to lie to a man who believes he has a claim on you, but whom you don't love.

Women may be more concerned with the impact of their affairs on their children than they are with the effect on their mate, whom they have already devalued and discounted in anticipation of the affair. Of course, a woman is likely to feel the children would be in support of her affair, and thus may involve them in relaying her messages, keeping her secrets, and telling her lies. This can be mind-blowingly seductive and confusing to the kids. Sharing the secret of one parent's affair, and hiding it from the other parent, has essentially the same emotional impact as incest.

Some conventional wisdom about gender differences in infidelity is true.

More men than women do have affairs, but it seemed to me that before the AIDS epidemic, the rate for men was dropping (philandering has not been considered cute since the Kennedy's went out of power) and the rate for women was rising (women who assumed that all men were screwing around saw their own screwing around as a blow for equal rights.) In recent years, promiscuity seems suicidal so only the suicidal—that is, the romantics—are on the streets after dark.

Men are able to approach sex more casually than women, a factor not only of the patriarchal double standard but also of the difference between having genitals on the outside and having them on the inside. Getting laid for all the wrong reasons is a lot less dangerous than falling in love with all the wrong people.

Men who get caught screwing around are more likely to be honest about the sex than women. Men will confess the full sexual details, even if they are vague about the emotions. Women on the other hand will confess to total consuming love and suicidal desire to die with some man, while insisting no sex ever took place. I would believe that if I'd ever seen a man describe the affair as so consumingly intense from the waist up and so chaste from the waist down. I assume these women are lying to me about what they know they did or did not do, while I assume that the men really are honest about the genital ups and downs—and honestly confused about the emotional ones.

Women are more likely to discuss their love affairs with their women friends. Philandering men may turn their sex lives into a spectator sport but romantic men tend to keep their love life private from their men friends, and often just withdraw from their friends during the romance.

On the other hand, women are not more romantic than men. Men in love are every bit as foolish and a lot more naive than women in love. They go crazier and risk more. They are far more likely to sacrifice or abandon their children to prove their love to some recent affairee. They are more likely to isolate themselves from everyone except their affair partner, and turn their thinking and feeling over to her, applying her romantic ways of thinking (or not thinking) to the dilemmas of his increasingly chaotic life.

Men are just as forgiving as women of their mates' affairs. They might claim ahead of time that they would never tolerate it, but when push comes to shove, cuckolded men are every bit as likely as cuckolded women to fight like tigers to hold on to a marriage that has been betrayed. Cuckolded men may react violently at first, though cuckolded women do so as well, and I've seen more cases of women who shot and wounded or killed errant husbands. (The shootings occur not when the affair is stopped and confessed, but when it is continued and denied.)

Betrayed men, like betrayed women, hunker down and do whatever they have to do to hold their marriage together. A few men and women go into a rage and refuse to turn back, and then spend a lifetime nursing the narcissistic injury, but that unusual occurrence is no more common for men than for women. Marriage can survive either a husband's infidelity or a wife's, if it is stopped, brought into the open, and dealt with.

I have cleaned up from more affairs than a squad of motel chambermaids. Infidelity is a very messy hobby. It is not an effective way to find a new mate or a new life.

It is not a safe treatment for depression, boredom, imperfect marriage, or inadequate gender splendor. And it certainly does not impress the rest of us. It does not work for women any better than it does for men. It does excite the senses and the imaginations of those who merely hear the tales of lives and deaths for love, who melt at the sound of liebestods or country songs of love gone wrong.

I think I've gotten more from infidelity as an observer than all the participants I've seen. Infidelity is a spectator sport like shark feeding or bull fighting—that is, great for those innocent bystanders who are careful not to get their feet, or whatever, wet. For the greatest enjoyment of infidelity, I recommend you observe from a safe physical and emotional distance and avoid any suicidal impulse to become a participant.

What Doctors and Others Need to Know

Six Facts on Human Sexuality and Aging

Richard J. Cross, MD

Certified Specialist in Internal Medicine, Professor Emeritus at the Robert Wood Johnson Medical School, NJ

Most of us find that our definition of old age changes as we mature. To a child, anyone over forty seems ancient. Sixty-five and older is the common governmental definition of a senior citizen, and it is the definition that I will follow here, although the author (who is in his late 70s) long ago began to find it hard to accept. There is, of course, no specific turning point, but rather a series of gradual physical and emotional changes, some in response to societal rules about retirement and entitlement to particular benefits.

Demographically, the elderly are a rapidly growing segment of the population. In 1900, there were about three million older Americans; by the year 2000, there will be close to 31 million older Americans. Because of high male mortality rates, older women outnumber men 1.5 to 1, and since most are paired off, single women outnumber single men by about 4 to 1. By definition, the elderly were born in the pre–World War I era. Most were thoroughly indoctrinated in the restrictive attitudes toward sex that characterized these times.

In my opinion, the care of the elderly could be significantly improved if doctors and other health workers would remember the following six, simple facts.

it is a carryover of the victorian belief that sex is dangerous and evil, though necessary for reproduction, and that sex for recreational purposes is improper and disgusting. Second is what Mary S. Calderone, SIECUS co-founder, has called a tendency for society to castrate its dependent members: to deny the sexuality of the disabled, of prisoners, and of the elderly. This perhaps reflects a subconscious desire to dehumanize those who we believe to be less fortunate than ourselves in order to assuage our guilt feelings. Third, Freud and many others have pointed out that most of us have a hard time thinking of our parents as being sexually active, and we tend to identify all older people with our parents and grandparents.

For whatever reason, it is unfortunate that young people so often deny the sexuality of those who are older. It is even more tragic when older people themselves believe the myth and then are tortured by guilt when they experience normal, healthy sexual feelings. Doctors and other health workers need to identify and alleviate such feelings of guilt.

How many people are sexually active? It is generally agreed upon by experts that the proportion of both males and females who are sexually active declines, decade by decade, ranging according to one study from 98% of married men in their 50s to 50% for unmarried women aged 70 and over.[1] At each decade, there are also some people who are inactive. It is important to accept abstinence as a valid lifestyle as well—at any age—as long as it is freely chosen.

Fact #1: All Older People Are Sexual

Older people are not all sexually active, as is also true of the young, but they all have sexual beliefs, values, memories, and feelings. To deny this sexuality is to exclude a significant part of the lives of older people. In recent decades, this simple truth has been repeatedly stated by almost every authority who has written about sexuality, but somehow the myth persists that the elderly have lost all competence, desire, and interest in sexuality, and that those who remain sexual, particularly if sexually active, are regarded as abnormal and, by some, even perverted. This myth would seem to have at least three components. First,

Fact #2: Many Older People Have a Need for a Good Sexual Relationship

To a varying extent, the elderly experience and must adapt to gradual physical and mental changes. They may find themselves no longer easily able to do the enjoyable things they used to do; their future may seem fearful; retirement and an "empty nest" may leave many with reduced incomes and no clear goals in life; friends and/or a lifetime partner may move away, become ill, or die; and the threat of loneliness may be a major concern. Fortunately, many older people are not infirm, frustrated, fearful, bored, or lonely; nonetheless, some of these elements

36. What Doctors and Others Need to Know

may be affecting their lives. An excellent antidote for all this is the warmth, intimacy, and security of a good sexual relationship.

Fact #3: Sexual Physiology Changes with Age

In general, physiological changes are gradual and are easily compensated for, if one knows how. But when they sneak up on an unsuspecting, unknowledgeable individual, they can be disastrous. Health workers need to be familiar with these changes and with how they can help patients adapt to them.

Older men commonly find that their erections are less frequent, take longer to achieve, are less firm, and are more easily lost. Ejaculation takes longer, is less forceful, and produces a smaller amount of semen. The refractory period (the interval between ejaculation and another erection) is often prolonged to many hours or even days. The slowing down of the sexual response cycle can be compensated for simply by taking more time, a step usually gratifying to one's partner, especially if he or she is elderly. But in our society many men grow up believing that their manliness, their power, and their competence depend on their ability to "get it up, keep it up, and get it off." For such an individual, slowing of the cycle may induce performance anxiety, complete impotence, and panic. Good counseling about the many advantages of a leisurely approach can make a world of difference for such an individual.

The prolonged refractory period may prevent a man from having sexual relations as often as he formerly did, but only if he requires that the sexual act build up to his ejaculation. If he can learn that good, soul-satisfying sexual activity is possible without male ejaculation, then he can participate as often as he and his partner wish. Finally, men (and sometimes their partners) need to learn that wonderful sex is possible without an erect penis. Tongues, fingers, vibrators, and many other gadgets can make wonderful stimulators and can alleviate performance anxiety.

Some women find the arrival of menopause disturbing; others feel liberated. If one has grown up in a society that believes that the major role for women is bearing children, then the loss of that ability may make one feel no longer a "real woman." The most common sexual problem of older women, however, is vaginal dryness which can make sexual intercourse painful, particularly if her partner is wearing an unlubricated condom. The obvious solution is to use one of the many water-soluble lubricants available in drug stores. Saliva is a fairly good lubricant and it does have four advantages over commercial products: 1) it is readily available wherever one may be; 2) it is free; 3) it is at the right temperature; and 4) its application is more intimate than something from a tube.

An alternative approach attacks the root of the problem. Vaginal drying results from a decrease in estrogen and can be reversed with estrogen replacement which also prevents other consequences of menopause like hot flashes and loss of calcium from the skeleton. But estrogen administration may increase the risk of uterine cancer; therefore, each woman and her doctor will need to balance out the risks and benefits in her particular situation.

Aging inevitably changes physical appearance and, in our youth-oriented culture, this can have a profound impact on sexuality. It is not easy to reverse the influence of many decades of advertisements for cosmetics and clothes, but doctors can at least try to avoid adding to the problem. Many medical procedures—particularly mastectomy, amputations, chemotherapy, and ostomies—have a profound impact on body image. It is of utmost importance to discuss this impact before surgery and to be fully aware of the patient's need to readjust during the postoperative period. When possible, involvement of the patient's sexual partner in these discussions can be very helpful.

Fact #4: Social Attitudes Are Often Frustrating

As indicated above, society tends to deny the sexuality of the aged, and in so doing creates complications in their already difficult lives. Laws and customs restrict the sexual behavior of older people in many ways. This is particularly true for women, since they have traditionally enjoyed less freedom and because, demographically, there are fewer potential partners for heterosexual, single women, and many of the few men that are available are pursuing women half their age.

Some professionals have suggested that women explore sexual behaviors with other women. However, we know that sexual orientation, although potentially fluid throughout a life-span, is more complicated than the suggestion implies. While some women discover lesbian sexuality at an older age, it is rarely the result of a decrease in the availability of male partners. When doctors see an older woman as a patient, they can, at least, inquire into sexual satisfaction. If sexual frustration is expressed, they can be understanding. Some women can be encouraged to try masturbating, and some will find a vibrator a delightful way to achieve orgasm.

Older people are living in a variety of retirement communities and nursing homes. This brings potential sexual partners together, but tends to exaggerate the gender imbalance. In retirement homes, single women often outnumber single men, eight or ten to one. Furthermore, rules, customs, and lack of privacy severely inhibit the establishment of intimate relationships at these sites. Administrators of such homes are often blamed for this phenomenon. Some are, indeed, unsympathetic, but we must also consider the attitudes of the trustees, the neighbors, and the legislators who oversee the operation, and particularly the attitudes of the family members. If two residents establish a sexual relationship, it is often followed by a son or daughter pounding the administrator's desk and angrily shouting, "That's not what I put Mom (Dad) here for!"

Fact #5: Use It Or Lose It

Sexual activity is not a commodity that can be stored and saved for a rainy day. Rather, it is a physiologic function that tends to deteriorate if not exercised, and it is particularly fragile in the elderly. If interrupted, it may be difficult (though not impossible) to reinvigorate. Doctors should work with the patient and partner on reestablishing the ability if desired.

5. SEXUALITY THROUGH THE LIFE CYCLE: Sexuality in the Adult Years

Research Note

Andrew Greeley, priest, author, and sociologist at the University of Chicago analyzed national-poll data of 6,000 respondents and found that sexual activity is plentiful, even after the age of 60. He reported in 1992 that 37 percent of married people over 60 have sexual relations at least once a week—and one in six respondents had sexual relations more often. Greeley concluded that sexually active married men are happier with their spouses at 60 than 20-year-old single males who have many sexual partners. His report, "Sex After Sixty: A Report," based on surveys by the Gallup Organization and the National Opinion Research Center included the following results:

	Married Men and Women: in their 20s	in their 60s
Who have sexual relations outdoors	55%	20%
Who have sexual relations once a week	80%	37%
Who undress each other	70%	27%

Fact #6: Older Folks Do It Better

This may seem like an arrogant statement to some, but much depends on what is meant by "better." If the basis is how hard the penis is, how moist the vagina, how many strokes per minute, then the young will win out, but if the measure is satisfaction achieved, the elderly can enjoy several advantages. First, they have usually had considerable experience, not necessarily with many different partners. One can become very experienced with a single partner. Second, they often have more time, and a good sexual relationship takes a lot of time. The young are often pressured by studies, jobs, hobbies, etc., and squeeze their sexual activities into a very full schedule. Older folks can be more leisurely and relaxed. Finally, attitudes often improve with aging. The young are frequently insecure, playing games, and acting out traditional roles because they have not explored other options. Some older folks have mellowed and learned to roll with the punches. They no longer need to prove themselves and can settle down to relating with their partner and meeting his or her needs. Obviously one does not have to be old to gain experience, to set aside time, or to develop sound attitudes. Perhaps the next generation of Americans will discover how to learn these simple things without wasting thirty or forty years of their lives playing silly games. One hopes so.

Conclusion

In summary, older people are sexual, often urgently need sexual contact, and yet encounter many obstacles to enjoying its pleasures, some medical, most societal. Doctors and other healthcare providers need to be aware of these problems and need to help those who are aging cope with them.

Dr. Richard J. Cross originally wrote this article for the SIECUS Report in 1988.

Author's References

1. Brecher EM. *Love, Sex, and Aging.* New York: Little, Brown and Company, 1984

Sexuality and Aging

What it means to be sixty or seventy or eighty in the '90s

"It's the awfulness of it," Harry said when asked how he was getting along after his wife died. It was the way he said the "awe"—with a stunned sound, as if he hadn't expected the blow to be so crushing.

Pounder of pianos, designer of great, black locomotives, father of five, this new fragility was a surprise. But it wasn't his last surprise.

After a year of bridge parties with old friends, Harry and his new fiancée turned up at a family dinner.

Their only worry was that the children would think Martha too young for him. She was 69. He was 78.

When they left for their honeymoon, the family still had questions. No one, including Harry and Martha, knew quite what to expect.

Three trends, longer life expectancy, early retirement and better health, are stretching the time between retirement and old age. These trends are redefining our image of aging for the 31 million Americans older than age 65. If you are in your 60s or 70s, you are probably more active and healthy than your parents were at a similar age. Many people are retiring earlier. This opens a whole new segment of your life.

Like everyone of every age, you probably want to continue sharing your life with others in fulfilling relationships. And, you may want to include sex in an intimate relationship with someone you love.

WHAT IS SEXUALITY?

The sexual drive draws humans together for biological reproduction, but it goes beyond this. Your sexuality influences your behavior, speech, appearance; indeed, many aspects of your life.

You might express your sexuality by buying an attractive blouse, playing a particular song or holding hands. Some people express their sexuality through shared interests and companionship. A more physical expression of sexuality is intimate contact, such as sexual intercourse.

Sexuality brings people together to give and receive physical affection. Although it's an important form of intimacy, sexual intimacy isn't the only one. For many people, sexual intimacy isn't an available or desired form of closeness. A close friendship or a loving grandparent-grandchild relationship, for example, can provide rewarding opportunities for non-sexual intimacy.

For some older people, though, sexual intimacy remains important. Despite this importance, sexuality in people after age 60 or 70 is not openly acknowledged.

MYTHS AND REALITIES

The widespread perception in America is that older people are not sexually active. Try to remember the last time the media portrayed two seniors in a passionate embrace. In America, sex is considered the exclusive territory of the young.

Comedian Sam Levinson expressed it well when he quipped, "My parents would never do such a thing; well, my father—maybe. But my mother—NEVER!"

This is a myth.

Realities

The reality is that many older people enjoy an active sex life that often is better than their sex life in early adulthood. The idea that your sexual drive dissolves sometime after middle age is nonsense. It's comparable to thinking your ability to enjoy good food or beautiful scenery would also disappear at a certain point.

In now famous studies, Dr. Alfred C. Kinsey collected information on sexual behavior in the 1940s. Drs. W. B. Masters and V. E. Johnson continued this research in the 1970s. Little of their research looked at people over 60. But in the last decade, a few telling studies show the stark difference between myth and reality.

In a 1992 University of Chicago study, Father Andrew Greeley, author and professor of sociology, released "Sex After Sixty: A Report." According to Greeley, "The happiest men and women in America are married people who continue to have sex frequently after they are 60. They are also most likely to report that they are living exciting lives."

Greeley's report, an analysis of two previous surveys involving 5,738 people, showed 37 percent of married people over 60 have sex once a week or more, and 16 percent have sex several times a week.

A survey of 4,245 seniors done by Consumers Union (*Love, Sex, and Aging,* 1984), concludes that, "The panorama of love, sex and aging is far richer and more diverse than the stereotype of life after 50. Both the quality and quantity of sexual activity reported can be properly defined as astonishing."

These surveys are helping today's seniors feel more comfortable acknowledging their sexuality. A 67-year-old consultant to the Consumers Union report wrote:

"Having successfully pretended for decades that we are nonsexual, my generation is now having second thoughts. We are increasingly realizing that denying our sexuality means denying an essential aspect of our common humanity. It cuts us off from communication with our children, our grandchildren and our peers on a subject of great interest to us all—sexuality."

5. SEXUALITY THROUGH THE LIFE CYCLE: Sexuality in the Adult Years

Health and sexuality

Sex, like walking, doesn't require the stamina of a marathoner. It does require reasonably good health. Here are some guidelines:

- *Use it or lose it* — Though the reason is unclear, prolonged abstinence from sex can cause impotence. Women who are sexually active after menopause have better vaginal lubrication and elasticity of vaginal tissues.

- *Eat healthfully* — Follow a balanced, low-fat diet and exercise regularly. Fitness enhances your self-image.

- *Don't smoke* — Men who smoke heavily are more likely to be impotent than men who don't smoke. Smokers are at an increased risk of hardening of the arteries, which can cause impotence (see page 4). Similar studies for women are needed.

- *Control your weight* — Moderate weight loss can sometimes reverse impotence.

- *Limit alcohol* — Chronic alcohol and drug abuse causes psychological and neurological problems related to impotence.

- *Moderate coffee drinking may keep sex perking* — A recent study reported that elderly people who drank at least one cup of coffee a day were more likely to be sexually active than those who didn't. The reason for this association is unknown; further studies are needed.

- *Protect against AIDS and other STDs* — The best protection against AIDS and other sexually transmitted diseases (STDs) is a long-standing, monogamous relationship. Next best: Use a condom.

SEX AFTER SIXTY: WHAT CAN YOU EXPECT?

Once you've reshaped your idea of what society should expect of you, you're faced with the sometimes more worrisome obstacle of what you can expect of yourself. Sex, something you've taken for granted most of your life, may suddenly be "iffy" at sixty.

Changes in women

Many women experience changes in sexual function in the years immediately before and after menopause. Contrary to myth, though, menopause does not mark the end of sexuality.

Generally, if you were interested in sex and enjoyed it as a younger woman, you probably will feel the same way after menopause. Yet menopause does bring changes:

- *Desire*—The effects of age on your sexual desire are the most variable of your sexual responses. Although your sex drive is largely determined by emotional and social factors, hormones like estrogen and testosterone do play a role.

Estrogen is made in your ovaries; testosterone, in your adrenal glands. Surprisingly, sexual desire is affected mainly by testosterone, not estrogen. At menopause, your ovaries stop producing estrogen, but most women produce enough testosterone to preserve their interest in sex.

- *Vaginal changes*—After menopause, estrogen deficiency may lead to changes in the appearance of your genitals and how you respond sexually.

The folds of skin that cover your genital region shrink and become thinner, exposing more of the clitoris. This increased exposure may reduce your sensitivity or cause an unpleasant tingling or prickling sensation when touched.

The opening to your vagina becomes narrower, particularly if you are not sexually active. Natural swelling and lubrication of your vagina occur more slowly during arousal. Even when you feel excited, your vagina may stay somewhat tight and dry. These factors can lead to difficult or painful intercourse (dyspareunia = DYS - pa - ROO - nee - ah).

- *Orgasm*—Because sexual arousal begins in your brain, you can have an orgasm during sexual stimulation throughout your life. You may have diminished or slower response. Women in their 60s and 70s have a greater incidence of painful uterine contractions during orgasm.

Changes in men

Physical changes in a middle-aged man's sexual response parallel those seen in a postmenopausal woman.

- *Desire*—Although feelings of desire originate in your brain, you need a minimum amount of the hormone testosterone to put these feelings into action. The great majority of aging men produce well above the minimum amount of testosterone needed to maintain interest in sex into advanced age.

- *Excitement*—By age 60, you may require more stimulation to get and maintain an erection, and the erection will be less firm. Yet a man with good blood circulation to the penis can attain erections adequate for intercourse until the end of life.

- *Orgasm*—Aging increases the length of time that must pass after an ejaculation and before stimulation to another climax. This interval may lengthen from just a few minutes at the age of 17, to as much as 48 hours by age 70.

Changes due to illness or disability

Whether you're healthy, ill or disabled, you have your own sexual identity and desires for sexual expression. Yet illness or disability can interfere with how you respond sexually to another person. Here's a closer look at how some medical problems can affect sexual expression:

- *Heart attack*—Chest pain, shortness of breath or the fear of a recurring heart attack can have an impact on your sexual behavior. But a heart attack will rarely turn you into a "cardiac cripple." If you were sexually active before your heart attack, you can probably be again. If you have symptoms of angina, your doctor may recommend nitroglycerine before intercourse. Most people who have heart disease are capable of a full, active sex life (see "Sex after a heart attack: Is it safe?").

Even though pulse rates, respiratory rates and blood pressure rise during intercourse, after intercourse they return to normal within minutes. Sudden death during sex is rare.

- *Prostate surgery*—For a benign condition, such as an enlarged prostate, surgery rarely causes impotence. Prostate surgery for cancer causes impotence 50

37. Sexuality and Aging

to 60 percent of the time. However, this type of impotence can be treated (see next page).

• *Hysterectomy*—This is surgery to remove the uterus and cervix, and in some cases, the fallopian tubes, ovaries and lymph nodes. A hysterectomy, by itself, doesn't interfere with your physical ability to have intercourse or experience orgasm once you've recovered from the surgery. Removing the ovaries, however, creates an instant menopause and accelerates the physical and emotional aspects of the natural condition.

When cancer is not involved, be sure you understand why you need a hysterectomy and how it will help your symptoms. Ask your doctor what you can expect after the operation. Reassure yourself that a hysterectomy generally doesn't affect sexual pleasure and that hormone therapy should prevent physical and emotional changes from interfering.

• *Drugs*—Some commonly used medicines can interfere with sexual function. Drugs that control high blood pressure, such as thiazide diuretics and beta blockers, can reduce desire and impair erection in men and lubrication in women. In contrast, calcium channel blockers and angiotensin converting enzyme (ACE) inhibitors have little known effect on sexual function.

Other drugs that affect sexual function include antihistamines, drugs used to treat depression and drugs that block secretion of stomach acid. If you take one of these drugs and are experiencing side effects, ask your doctor if there is an equally effective medication that doesn't cause the side effects. Alcohol also may adversely affect sexual function.

• *Hardening of the arteries and heart disease*—About half of all impotence in men past age 50 is caused by damage to nerves or blood vessels to the penis. Hardening of the arteries (atherosclerosis) can damage small vessels and restrict blood flow to the genitals. This can interfere with erection in men and swelling of vaginal tissues in women.

• *Diabetes*—Diabetes can increase the collection of fatty deposits (plaque) in blood vessels. Such deposits restrict the flow of blood to the penis. About half of men with diabetes become impotent. Their risk of impotence increases with age. Men who've had diabetes for many years and who also have nerve damage are more likely to become impotent.

If you are a woman with diabetes, you may suffer dryness and painful inter-

Sex and illness

Changes in your body due to illness or surgery can affect your physical response to sex. They also can affect your self-image and ultimately limit your interest in sex. Here are tips to help you maintain confidence in your sexuality:

■ *Know what to expect* — Talk to your doctor about the usual effects your treatment has on sexual function.

■ *Talk about sex* — If you feel weak or tired and want your partner to take a more active role, say so. If some part of your body is sore, guide your mate's caresses to create pleasure and avoid pain.

■ *Plan for sex* — Find a time when you're rested and relaxed. Taking a warm bath first or having sex in the morning may help. If you take a pain reliever, such as for arthritis, time the dose so that its effect will occur during sexual activity.

■ *Prepare with exercise* — If you have arthritis or another disability, ask your doctor or therapist for range-of-motion exercises to help relax your joints before sex.

■ *Find pleasure in touch* — It's a good alternative to sexual intercourse. Touching can simply mean holding each other. Men and women can sometimes reach orgasm with the right kind of touching.

If you have no partner, touching yourself for sexual pleasure may help you reaffirm your own sexuality. It can also help you make the transition to intercourse after an illness or surgery.

course that reduce the frequency of orgasm. You may have more frequent vaginal and urinary tract infections.

• *Arthritis*—Although arthritis does not affect your sex organs, the pain and stiffness of osteoarthritis or rheumatoid arthritis can make sex difficult to enjoy. If you have arthritis, discuss your capabilities and your desires openly with your partner. As long as you and your partner keep communications open, you can have a satisfying sexual relationship.

• *Cancer*—Some forms of cancer cause anemia, loss of appetite, muscle wasting or neurologic impairment that leads to weakness. Surgery can alter your physical appearance. These problems can decrease your sexual desire or pleasure.

Cancer may also cause direct damage to your sexual organs or to their nerve and blood supplies; treatment can produce side effects that may interfere with sexual function, desire or pleasure. Discuss possible effects of your treatment with your doctor. If cancer has disrupted your usual sexual activity, seek other ways of expression. Sometimes cuddling or self-stimulation can be enough.

TO REMAIN SEXUALLY ACTIVE, WHAT CAN A WOMAN DO?

Long-term estrogen replacement therapy (ERT) can not only prevent osteoporosis (bone thinning) and heart disease, it can help prevent changes in vaginal tissue, lubrication and desire as well.

Testosterone enhances sexual desire in women. But, it also can produce unwanted, sometimes irreversible side effects such as deepening of the voice and increased facial hair.

Your doctor may prescribe estrogen cream which, applied to your genital area, can prevent dryness and thinning of vaginal tissue. You can also use over-the-counter lubricants just before sexual activity. It's best to use a water-based lubricant, such as K-Y jelly, rather than oil-based mineral oil or petroleum jelly.

If you have problems reaching orgasm, talk to your physician. Your doctor might adjust your medications or offer other options, including counseling, if the problem is non-medical; or, your doctor may refer you to a specialist.

What can a man do?

Only a few years ago doctors generally thought that about 90 percent of

5. SEXUALITY THROUGH THE LIFE CYCLE: Sexuality in the Adult Years

Sex after a heart attack: Is it safe?

If you can climb a flight of stairs without symptoms, you can usually resume sexual activity. Ask your doctor for specific advice. Here are some guidelines:

- *Wait after eating* — Wait three or four hours after eating a large meal or drinking alcohol before intercourse. Digestion puts extra demands on your heart.
- *Rest* — Make sure you are well rested before you have intercourse, and rest after.
- *Find comfortable positions* — Positions such as side-by-side, or your partner on top, are less strenuous.

---

impotence was psychological. Now they realize that 50 to 75 percent of impotence is caused by physical problems. There is a wide range of treatments. Keep in mind that the success of any treatment depends, in part, on open communication between partners in a close, supportive relationship. Here are some treatment options:

- *Psychological therapy*—Many impotence problems can be solved simply by you and your partner understanding the normal changes of aging and adapting to them. For help in this process, your doctor may recommend counseling by a qualified psychiatrist, psychologist or therapist who specializes in the treatment of sexual problems.
- *Hormone adjustment*—Is testosterone a magic potion for impotence? No. Although testosterone supplementation is used in rare instances, its effectiveness for aging men experiencing a normal, gradual decline in testosterone is doubtful.
- *Vascular surgery*—Doctors sometimes can surgically correct impotence caused by an obstruction of blood flow to the penis. However, this bypass procedure is appropriate in only a small number, less than 2 percent, of young men who have impotence problems. The long-term success of this surgery is too often disappointing.
- *Vacuum device*—Currently one of the most common treatments for impotence, this device consists of a hollow, plastic cylinder that fits over your flaccid penis. With the device in place, you attach a hand pump to draw air out of the cylinder. The vacuum created draws blood into your penis, creating an erection.

Once your penis erect, you slip an elastic ring over the cylinder onto the base of your penis. For intercourse, you remove the cylinder from your penis. The ring maintains your erection by reducing blood flow out of your penis. Because side effects of improper use can damage the penis, you should use this device under your doctor's care.

- *Self-injection*—Penile injection therapy is another option. It involves injecting a medication directly into your penis.

After age 60, intercourse may require some planning

Problems	Solutions	
Decreased desire	■ Use mood enhancers (candlelight, music, romantic thoughts). ■ Hormone replacement therapy (estrogen or testosterone).	■ Treatment for depression. ■ Treatment for drug abuse (alcohol). ■ Behavioral counseling.
Vaginal dryness; Vagina expands less in length and width	■ Use a lubricant. ■ Consider estrogen replacement therapy.	■ Have intercourse regularly. ■ Pelvic exercises prescribed by your doctor.
Softer erections; More physical and mental stimulation to get and maintain erection	■ Use a position that makes it easy to insert the penis into the vagina. ■ Accept softer erections as a normal part of aging.	■ Don't use a condom if disease transmission is not possible. ■ Tell your partner what is most stimulating to you.
Erection lost more quickly; Takes longer to get another	■ Have intercourse less frequently. ■ Emphasize quality, not quantity.	■ Emphasize comfortable sexual activities that don't require an erection.

One or more drugs (papaverine, phentolamine and prostaglandin-E1) are used. The injection is nearly painless and produces a more natural erection than a vacuum device or an implant.

- *Penile implants*—If other treatments fail or are unsatisfactory, a surgical implant is an alternative. Implants consist of one or two silicone or polyurethane cylinders that are surgically placed inside your penis. Implants are not the perfect solution. Mayo experts say there is a 10 to 15 percent chance an implant will malfunction within five years, but the problem almost always can be corrected. Many men still find the procedure worthwhile.

There are two major types of implants: one uses malleable rods and the other uses inflatable cylinders. Malleable rods remain erect, although they can be bent close to your body for concealment. Because there are no working parts, malfunctions are rare.

Inflatable devices consist of one or two inflatable cylinders, a finger-activated pump and an internal reservoir, which stores the fluid used to inflate the tubes. All components—the cylinders, pump and reservoir—are implanted within your penis, scrotum and lower abdomen. These devices produce more "natural" erections.

- *Medications*—Neurotransmitters are chemicals in your brain and nerves that help relay messages. Nitric oxide is now recognized as one of the most important of these chemicals for stimulating an erection. Unfortunately, there is as yet no practical way to administer nitric oxide for treatment of impotence. Other drugs have not proven effective.

SEX IN SYNC

You might wonder how sex can survive amidst tubes, pumps and lubes that can make you feel more like a mechanic than a romantic.

Actually, many people discover that late-life sexuality survives in an increased diversity of expression, sometimes slow, tender and affectionate, and sometimes more intense and spontaneous.

In some ways, middle- and late-life sex are better than the more frantic pace of your younger years. Biology finally puts sex in sync. As a young man, you were probably more driven by hormones and societal pressure . You may find that now desire, arousal and orgasm take longer and aren't always a sure thing. You may find setting and mood more important. Touch and extended foreplay may become as satisfying as more urgent needs for arousal and release.

As a woman, you were probably more dependent on setting and mood when you were younger. You may feel more relaxed and less inhibited in later life. You may be more confident to assert your sexual desires openly.

COMMUNICATION

It can be difficult to talk to another person about sex—doctor, counselor or even a lifetime lover. But, good communication is essential in adapting your sex life to changes caused by aging. Here are three cornerstones of good communication:

- *Be informed*—To start the process, know the facts. Gather as much reliable information as you can about sex and aging and share the facts with your partner.
- *Be open*—If there are unresolved problems in your relationship, sex won't solve them. Be sensitive to the views and feelings of your partner. Work out the differences that are inevitable - before you go to bed. Appropriate sexual counseling can help you and your partner work out problems and enhance your relationship.
- *Be warned*—Most likely, your physician will be willing to discuss questions concerning sexuality with you, but it would be unusual if he or she were a specialist in treating sexual problems. Ask your doctor to refer you to a specialist in this area.

ADAPTING TO CHANGES

As your sexual function changes, you may need to adapt not only to physical changes but to emotional changes as well, perhaps even to changes in your living arrangements.

Lovemaking can lose its spontaneity. Adapting may mean finding the courage to experiment with new styles of making love with the same partner. It may mean trying alternatives to intercourse. You may feel self-conscious about suggesting new ways to find pleasure. But, by changing the focus of sex, you minimize occasional erectile failures that occur.

And, adapting may mean having the flexibility to seek a new partner if you're single. Because women outlive men an average of seven years, women past 50 outnumber unmarried men almost three to one. Older women have less opportunity to remarry.

Families also need to be flexible. Children may need to deal with issues such as inheritance and acceptance of the new spouse.

Another factor that often limits sexual activity is the status of your living arrangements. If you live in a nursing home, you may face an additional problem. Although there are a few nursing homes that offer the privacy of apartment-style living, most do not. Fortunately, this problem is becoming more widely recognized. In the future, nursing homes may offer more privacy.

If you live independently, getting out and around may be a chore. Yet, many older men and women do find new partners. And, they report the rewards of sharing your life with someone you care for may be well-worth the extra effort.

THE NEED FOR INTIMACY IS AGELESS

It takes determination to resist the "over-the-hill" mentality espoused by society today. Age brings changes at 70 just as it does at 17. But you never outgrow your need for intimate love and affection. Whether you seek intimacy through non-sexual touching and companionship or through sexual activity, you and your partner can overcome obstacles. The keys are caring, adapting and communicating.

Old/New Sexual Concerns

- Sexual Hygiene (Articles 38–41)

- Sexual Abuse and Violence (Articles 42–44)

- Males/Females—War and Peace (Articles 45–49)

This final unit deals with several topics that are of interest or concern for different reasons. In one respect, however, these topics have a common denominator—they have all recently emerged in the public's awareness as social issues. Unfortunately, public awareness of issues is often a fertile ground for misinformation and misconceptions. In recognition of this, it is the overall goal of this section to provide some objective insights into pressing sexual concerns.

Health consciousness has increased through the 1980s and 1990s. In the past decade, the term wellness has been coined to represent healthiness encompassing physical and emotional factors. Sexual health or wellness is often incorporated into this sought-after ideal. In order to accomplish sexual health, we must learn about sexual hygiene, normative sexual processes, and the effects of diseases on our sexual functioning. Of particular concern in this area are diseases misappropriately labeled "social diseases." This stigmatization may reflect a still-prevalent negative aura that surrounds sexuality. We can achieve sexual health only when we can learn about and be responsible for our sexual selves in a positive and accepting way.

The Sexual Hygiene subsection of this unit strives to help readers achieve sexual health by providing articles on sexually transmitted diseases, including both those presently getting media coverage and public attention (such as HIV/AIDS) and others that receive less attention, but should be of concern to sexually active persons (such as chlamydia, syphilis, gonorrhea, genital warts, and herpes). Each of the four articles in this subsection provides fact-filled objectivity with strong words of warning designed to get readers to sit up and take notice. Several of the articles confront people's failure to discuss disease-prevention methods with partners as a big part of the problem.

Sexual abuse and violence are especially pernicious when the abuser is an acquaintance, family member, or someone who has some kind of power over the victim (or potential victim), such as an employer, professor, or landlord. In all of these abuse scenarios, victims feel a violation of trust and of personal and sexual integrity, and the relationships involved are likely to be damaged beyond repair. The healing process may take many years.

Two other kinds of sexual abuse or misuse are now gathering increasing public and media attention: sexual harassment and date rape. Like other kinds of sexual abuse and violence, these problems have existed for some time prior to gaining media attention. Public attention was fueled by two media events: the televising of Professor Anita Hill's sexual harassment allegations about Supreme Court nominee Clarence Thomas, and media coverage of the date-rape trial of the nephew of Senator Edward Kennedy. Sexual harassment and date rape are surrounded by myths, misinformation, tendencies to "blame the victim," and emotional and legal controversies.

The three articles in the Sexual Abuse and Violence subsection address some issues in abuse that are just receiving attention by experts and the media. "A Question of Abuse" addresses the controversy surrounding recovered memories, the term for sexual abuse survivors remembering of a long-buried experience. "The Sexually Abused Boy" details the special needs and difficulties of male child-abuse victims, both as children and as they grow to manhood. Because acknowledgement and support for abuse victims has focused more on the female victim and because we confuse male perpetrator-male victim with homosexuality, these boys and men carry especially heavy psychic burdens. The final article of the subsection, "Who's to Blame for Sexual Violence?" includes some very interesting, yet troubling, accounts of convicted rapists' perceptions of responsibility and blame for their actions. It challenges readers to reassign blame in a way that will facilitate less abuse, violence, and coercion in sexual interaction.

The final subsection, Males/Females—War and Peace, focuses on the age-old battle of the sexes, which many today report is taking on a more overt, adversarial, and

Unit 6

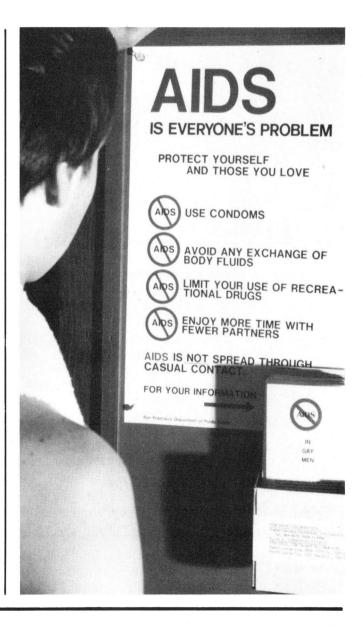

louder tone. While examining the likelihood that biological, genetic, even brain differences contribute to male and female similarities and differences, the articles also illustrate the social, interpersonal, and cultural factors involved in gender roles, behaviors, and conflicts. It is likely that some of the articles will kindle a fire, or at least an emotional response, in many readers of both genders. It is hoped, however, that they also provide insights, strategies, and impetus for ending the battle between the sexes. For only by seeing through our differences to our similarities, and stopping the blame game can we truly find the intimacy for which we yearn.

Looking Ahead: Challenge Questions

How knowledgeable are you about sexual health issues and sexually transmitted diseases? What keeps you from being more informed and involved in your own sexual wellness?

How concerned do you perceive your classmates and friends are about contracting AIDS? Are you more or less concerned than they? Why?

What does your college/university do with respect to HIV/AIDS? Do you support HIV/AIDS education efforts? Condom distribution? Have you considered the likelihood that there are HIV positive (and possibly full-blown AIDS) students in some of your classes? What do you feel and think about your and your school's role in the HIV/AIDS crisis?

Is date or acquaintance rape a problem for many college women? What are some of its causes, and what relationship changes may prevent it?

Have you or someone close to you been affected by rape? What was your response and how has it affected you and your actions?

How do you feel about male/female differences? Have you ever wished that the opposite sex could be more like you? When and why?

THE FUTURE OF AIDS

Dr. William A. Haseltine

William A. Haseltine, Ph.D., is Chief of the Division of Human Retrovirology, Dana-Farber Cancer Institute; Professor, Department of Pathology Harvard Medical School; and Professor, Department of Cancer Biology, Harvard School of Public Health.

AIDS is caused by infection with the human immunodeficiency virus. It is estimated that ten to twenty million people are now infected. In the absence of an effective vaccine or other effective means to control the spread of the disease, a reasonable estimate is that 100 million people will be infected with the virus by the year 2000, given the explosive growth of the epidemic in Sub-Saharan Africa, in the Indian subcontinent, in South America and in Asia. In Bombay, the number of people infected rose from a few hundred to more than half a million over the past five years. In northern Thailand, the fraction of 18-year-old men inducted into the army infected by the AIDS virus rose from an almost undetectable level to about 20 percent in the same period.

New epidemics will also arise from the ever expanding population of patients with depleted immune function. The population of people with AIDS serves as a reservoir and breeding ground for deadly diseases. For example, the new worldwide epidemic of tuberculosis arises directly from the population of people with AIDS. Unlike AIDS, some of these diseases will not require sexual contact for transmission. As the number of immune suppressed people grows, it is likely that the world population will suffer multiple, concurrent, lethal epidemics consequent to the AIDS epidemic.

The future of AIDS is indeed the future of humanity.

SCIENTIFIC AND MEDICAL PROGRESS

The collective ability of humans to analyze and to comprehend natural phenomena is formidable. Our power to rearrange nature to our own liking is limited. The AIDS epidemic illustrates human strengths and weaknesses.

AIDS was first identified as a new, progressive disease of the immune system in 1981. By the end of 1982, it was known that the AIDS virus was caused by an infectious agent that could be transmitted by sexual contact and by blood. The AIDS virus was first observed in 1983. In 1984, a test for the AIDS virus was developed. The test permits identification of those people who are infected and who are capable of transmitting the infection to others. This is remarkable progress by any standard.

Given what we know today, it cannot be predicted when, or even if, effective treatments and vaccines for AIDS will be developed.

AIDS research has grown into a formidable scientific and medical endeavor world-wide. AIDS research is at the cutting edge of discovery in many fields, including structural biology, drug design and vaccine development. The effort is characterized by close, collaborative, collegial relations.

Why then, does the future look so bleak? The answer, simply put, is that given what we know today, it cannot be predicted when, or even if, effective treatments and vaccines for AIDS will be developed.

TREATMENT

Medical interventions can be designed either to cure or to treat a disease. The intent of curative therapy is to eliminate the cause of the disease. Alternatively, treatment aims to control disease without necessarily eliminating the underlying cause.

The nature of the AIDS virus demands that medical intervention be designed to treat rather than to cure the disease. Upon infection, the genetic information of

---

This article contains excerpts from a speech Dr. Haseltine delivered to the French Academy of Science in Paris, November 16, 1992. [Priorities] invites responses to this provocative text.

the AIDS virus is inserted into the genetic material—the DNA—of the host. Infection by AIDS involves most of the major organ systems including blood, lymph nodes, bone marrow, brain, skin, intestines, liver and heart. Infection cannot be eliminated from these tissues or organs without destroying the infected cells. In many of these tissues the viral information remains silent, undetectable by any means for months or years. Eradication of the infection is beyond our present comprehension. For this reason, work is directed toward discovery of a treatment, not a cure.

The collective ability of humans to analyze and to comprehend natural phenomena is formidable.

Our power to rearrange nature to our own liking is limited.

The AIDS epidemic illustrates human strengths and weaknesses.

A consequence of life-long infection by the AIDS virus is the requirement for lifelong therapy. Therefore, the cumulative effects of treatment over months and years must be non-toxic.

Drugs have been developed that slow the progress of the virus. Typically, such drugs inhibit the function of critical components of the virus. To date such drugs have had, at best, only limited effect on the disease for two reasons—virus resistance to the drugs develops, and the drugs are toxic. New methods of treatment are now being developed, specifically immune therapy—a treatment that alters the ability of the immune system to respond to the virus after infection and gene therapy—the insertion of novel genes into normal cells to decrease growth of the virus. Unfortunately, it is likely that the AIDS virus will also develop resistance to both new types of treatment.

Progress toward developing an effective treatment is also slowed by the lengthy course of the disease. A minimum of two to three years is required to evaluate the effectiveness of each new treatment. The treatments now being developed are both labor intensive and expensive. Such treatments will not be available to most infected people for both economic and social reasons.

For these reasons, an AIDS cure is unlikely. Moreover, it cannot be predicted when, or even if, effective AIDS treatment will be developed in the foreseeable future.

VACCINES

The greatest hope for a medical solution to the AIDS epidemic is an effective vaccine. Vaccines educate the immune system and prevent or limit the consequences of infection. There are multiple obstacles for the creation of an effective AIDS vaccine. The AIDS virus changes as it grows. The rate of change is so great that no one virus is identical to any other. Vaccines rely on prior recognition of similarity. The development of a single or even a limited set of vaccines that prevent infection by the virus is not likely.

Most vaccines work by preparing the immune system for early elimination of the virus once infection is initiated. Such vaccines will probably not work for AIDS. Once infection is established, the AIDS virus is not naturally eliminated by the immune system. Central features of the lifestyle of the AIDS virus, including the ability of the virus to change within a single individual, its ability to establish a silent state of infection that is invisible to the immune system and its ability to infect immunologically privileged organs such as lymph node and brain, permit the virus to evade immune surveillance.

For these reasons, it is likely that to be effective, an AIDS vaccine must prevent the establishment of primary infection altogether, a requirement not met by any existing vaccine.

To be effective, an AIDS vaccine must prevent the establishment of primary infection altogether, a requirement not met by any existing vaccine.

The most common route of infection, exposure of sexual membranes to virus and virus infected cells in seminal and vaginal fluid, poses an additional difficulty for vaccine development. Immune cells at the surface of mucous membranes can be infected by the AIDS virus. These cells naturally migrate from the surface to the interior of the body where they come in contact with other cells of the immune system. By this means the virus infection spreads from the initial point of contact at the surface of the mucous membrane throughout the body. It has proved difficult to induce long lasting immune protection at the surface of mucous membranes to any microorganism.

For these reasons, it is not possible at present to predict when, or even if, a vaccine will be developed.

Consideration of the difficulties inherent in the discovery of an AIDS cure and AIDS vaccine does not mean that the efforts should be abandoned. To the contrary, the peril to human existence of the AIDS epidemic is so great the efforts to discover both a cure and a vaccine must be redoubled.

6. OLD/NEW SEXUAL CONCERNS: Sexual Hygiene

INDIVIDUAL AND COLLECTIVE BEHAVIOR

Society cannot rely on a medical miracle—a cure or vaccine—for salvation from the AIDS epidemic. Despite our best efforts such a miracle may never occur. Scientists around the world are working night and day in an unprecedented effort to control this disease. Each day there is a new hope. On a weekly basis, the scientific and lay press describe new insights regarding the AIDS virus, progress towards treatment and progress in vaccine development. This cumulative, intense effort may bring about the hoped for miracle, but society should not count on it. If such a miracle does occur it may be too late for many millions of people. Until that day, it is the responsibility of individuals, singly and collectively, to save themselves.

There is very little evidence to suggest that sexual behavior in any part of the world has changed significantly in response to the AIDS epidemic. The changes in sexual behavior that have been reported to have occurred in some groups have proved, for the most part, to be transient. For example, bath houses and sex clubs in many cities have either reopened or were never closed. The incidence of other sexually transmitted diseases such as syphilis and gonorrhea, as well as the incidence of unwanted pregnancies, both indirect measures of risky sexual behavior, have not declined significantly. In some parts of the world the incidence of sexually transmitted diseases has risen sharply over the past five years.

The dance of sex and death is not new. The lessons of the past, most recently the experience in the nineteenth century with syphilis, are not comforting. Until individuals recognize the risk to themselves and to their families of unprotected sex with multiple partners, and until they learn to modify their behavior accordingly, the AIDS epidemic will continue.

It is our collective responsibility to provide education to all so that each person has clear, unambiguous knowledge of how to avoid infection.

There are two known means to reduce the risk of infection in addition to abstinence and limiting the number of sexual partners. Condoms have been shown to reduce the risk of infection by about 90 percent. For this reason, condoms should be available to everyone world-wide, at a cost that is affordable. Although condoms do not provide absolute protection from infection, universal use of condoms would dramatically slow the progress of the epidemic.

The risk of infection can also be reduced by testing a potential partner for evidence of infection by the AIDS virus. The AIDS test detects more than 95 percent of those infected. The current AIDS test fails to detect infection for only a very brief period, the first four to six weeks immediately following primary exposure. Simple, reliable tests that require only five to ten minutes to complete, that are easy to use, that require no equipment and that require only a drop of blood or a spot of saliva are now available. Such AIDS tests should be made widely available at affordable cost. Prior testing of each new sexual partner substantially reduces the risk of infection. If used routinely, such tests could dramatically slow the progress of the epidemic.

Richer countries should provide resources to poorer countries for education, condom distribution and AIDS testing. Richer countries should also provide medical treatment for highly infectious diseases that arise from the AIDS infected population, including tuberculosis. The world is truly united by this global epidemic. The health of all nations is intertwined.

This epidemic makes it clear that we are all our brother's keeper. We must be compassionate for those who are ill; we must provide care and support for those afflicted; and we must work together to find both social and medical solutions.

Campuses Confront AIDS:

Tapping the Vitality of Caring and Community

Richard P. Keeling, M.D.

Richard P. Keeling, M.D., is director of the University of Wisconsin-Madison Health Center.

The Human Immunodeficiency Virus (HIV) causes a chronic, progressive immunodeficiency disorder called HIV disease, which is manifest in a spectrum of conditions that occur in a sequence of phases; AIDS is the most advanced phase of HIV disease. HIV is responsible for significant premature mortality among both male and female adolescents and young adults in the United States.[1] Many college and university communities are now all too familiar with the enormous cost of the epidemic in human terms: the losses of potential, creativity, contributions, physical and psychological health, and life itself. In response to the needs created by the epidemic, higher education institutions have struggled to develop or redesign effective policies and services and to promote behavioral changes that can help prevent the spread of this disease.

Epidemiology of HIV on campus

After three years of nationwide studies on a variety of campuses, the American College Health Association (ACHA) and the Centers for Disease Control and Prevention (CDC) estimate that approximately 0.2 percent of college and university students (25,000–30,000 individuals) have HIV.[2] The frequency of infection is higher at colleges located in metropolitan areas with a greater prevalence of AIDS; on some such campuses, as many as 0.9 to 1.0 percent of students have HIV.[3] The collective campus experience is that the overwhelming majority of students with HIV acquired their infection through unprotected sexual intercourse. Research observations also confirm that the majority of students with HIV disease in any of its stages are men who have sex with men. The number of women on campus with HIV has been considerably smaller, but with shifts in the patterns of HIV transmission, it is likely that in the future, college communities will have to address the needs of more HIV-positive women.

Future trends in the tempo and extent of the HIV epidemic on campus are difficult to predict. One major influence will be the patterns of sexual and needle-sharing behavior among students. The existence of a reservoir of HIV among students (and among their partners off campus) makes it likely that unprotected intercourse will continue to spread HIV gradually in student communities. Another important factor is the increasing occurrence of HIV disease among adolescents: especially among disadvantaged teenagers, HIV has become a critical health problem.[4] Whereas higher education institutions have been concerned chiefly with the possibility that students would become infected with HIV while on campus, it is likely that in the future, urban and commuter colleges will admit more students who already have HIV.

Teenagers have been more likely than college students to con-

6. OLD/NEW SEXUAL CONCERNS: Sexual Hygiene

tact HIV through heterosexual contact,[5] but heterosexual transmission of HIV is becoming more common in all age groups;[6] at the same time, the proportion of cases of AIDS resulting from unprotected intercourse between men has been declining.[7] Thus, to the extent that campus trends reflect national patterns, in coming years it is less likely that the average student with HIV will be a gay or bisexual man. Currently, however, there are no reliable longitudinal studies with which to define year-to-year trends and variations in the frequency or patterns of HIV transmission among students. A balanced view is that campuses will witness a progressive but gradual expansion of the population of students with HIV and that many, but not all, students who need care and services in the next five years will be gay or bisexual men.

Responding to HIV in campus communities

Reacting to the AIDS Issue
College and university communities shared an unusual early experience with HIV disease: campuses initially reacted to possibilities, rather than to actual occurrences. Concern about HIV first appeared less often in the context of caring for people who were sick than as a reaction to risks — the possibility that someday the campus would have to deal with someone who had HIV, or that behavior in the community (usually assumed to be exclusively, or at least especially, among students) might result in someone's acquiring HIV. Thus, for many schools, HIV and AIDS were theoretical concerns; AIDS was an issue, not a person.

Early AIDS policies were administratively protective: salvaging the institution's reputation, preventing legal liability, rendering enrollment safe and healthy, and providing safety for persons who interacted with someone with HIV. Many of these policies affirmed the institution's commitment to avoid discriminating against people with HIV or AIDS in its academic, recreational, and residential programs, but most did not construct mechanisms of investigation or redress to employ in the event of discrimination. Although some policies recognized the need to provide care and assistance for students or employees who already had HIV, few

More than ten years into the HIV epidemic, HIV remains a theoretical problem on a great many campuses; the fear of what might happen still drives discussions and molds policy.

specifically assigned responsibility for services or defined the nature and extent of the care to be offered. Similarly, while many policies addressed the value and importance of educational programs in preventing HIV infection, only a few required specific kinds of activities or demanded assessments and evaluations to guide those programs.

Services and education: informal networks
Effective responses to HIV and AIDS as educational or humane concerns for a campus community typically resulted from highly personal events and connections: students, faculty, or staff knew someone who had HIV. HIV was no longer just an issue or an external hazard, but a personal challenge that created real-life problems. Fearing that disclosing the existence or identity of a student or employee with HIV might result in protest, violence, embarrassment, or sanctions, people who knew of the presence or needs of persons with HIV frequently remained officially silent. Simple, unstructured referral networks emerged on hundreds of campuses; it was not unusual for employees or students to provide support or services in a quiet, private way (often unknown, in detail, to campus administrators). In many ways, this careful though informal system evolved not only because of the social stigmas and fears associated with HIV, but also because of the connection between HIV and gay men. Because support for gay, lesbian, or bisexual students, faculty, and staff was (and, unfortunately, sometimes still is) hard to sustain in the official policies, procedures, programs, and agencies of many higher education institutions, informal service and education networks already existed around campus gay communities. It was almost inevitable that similar structures, commonly developed by the same people, arose to deal with HIV.

Health promotion and education programs designed to reduce the risk of HIV infection initially evolved informally, as well. Worried clinicians, counselors, and educators whose personal experience with HIV (or with other health issues associated with sexual behavior on campus) organized to write brochures, conduct programs, and begin training peer educators. Details of the methods and content of educational activities were commonly protected from ordinary administrative oversight; educators often felt that their superiors might restrict the kind of programming they could provide and thereby limit its efficacy. Controversies over the appropriateness of distributing condoms, talking explicitly about sexual acts, or developing specific prevention programs for gay students embroiled educators and administrators in disputes that commonly became vividly public.

Evolution of campus responses

Interestingly, the development of an official administrative process and the rise of an informal network of services and education often proceeded in tandem and even shared some people and resources, though the two typically operated with different assumptions and values. The breadth and depth of the difference between them varied, depending on the particular community s values and attitudes. On some campuses, the influence of exemplary leadership and a spirit of caring and community has served to merge the informal network into the official system without damaging the humane qualities of the former. But on other campuses, only the presence of a few committed individuals preserves reasonable educational programs or effective services. In some places, attitudes remain so negative and fear so strong that students concerned about HIV seek services off campus. More than ten years into the HIV epidemic, HIV remains a theoretical problem on a great many campuses; the fear of what might happen still drives discussions and molds policy.

An effective campus response

Policies

Many campus AIDS policies are now obsolete; few institutions have used, reviewed, or revised them. More recent policies and revisions reflect the enlightened institution s humane desire to make affirmative statements about its recognition of the importance of HIV disease as a campus concern; its knowledge of the contributions people with HIV have made, and will make, to the life of its community; its commitment to developing effective educational programs; and its intent to care gently and reasonably for people with HIV disease. On campuses where fear and moralism still reign, such affirmative statements may have both particular preciousness and tremendous potential to generate controversy and dismay. On any campus, writing down in the language of policy the basics of what the institution believes about HIV and how it wants to relate to people with HIV remains helpful in clarifying values and creating a humane atmosphere.

One area currently demanding attention in institutional policies and procedures is occupational risk. The CDC and the Occupational Safety and Health Administration (OSHA) have issued guidelines and rules for reducing the likelihood of HIV transmission in the workplace, and especially in health care settings.[8] These guidelines and standards also apply to teaching and training sites; likewise, they apply to athletic training rooms, physical therapy units, and teaching and research laboratories. Wise and ethical occupational health practices for both students and employees require that institutions adopt, monitor, and enforce policies and procedures that enhance safety in accordance with these agency mandates.

Services

The spectrum of services required by students with HIV is determined by the medical and psychological history of HIV disease, the age of the student population, the institution's commitment to student services, and the availability and accessibility of care in the community. Because of the prolonged asymptomatic phase of HIV disease, people who have symptomatic or advanced disease are likely to be older than traditional undergraduates – graduate and professional students, nontraditional or returning learners, and members of the faculty or staff. Comprehensive universities and urban or commuter campuses are, accordingly, more likely to encounter students with symptoms, and therefore may need to provide (or refer for) more complex or specialized services.

Most students with HIV, though, have relatively early-stage disease, and many are unaware of their infection. Available and accessible HIV antibody testing programs; counseling services for students concerned about risks of HIV; psychological support (including employee assistance programs) for people who are trying to manage having HIV; and clinical evaluation and management of early-stage disease are, therefore, appropriate interventions for a campus community. The success of currently available medical therapies in improving both the quality and length of life for people with HIV disease demands that everyone who thinks they might have HIV have barrier-free access to counseling and antibody testing programs, and that individuals shown to have HIV be referred for competent, comprehensive medical case management. Being a college student should not be a barrier to safe testing or to receiving appropriate medical and psychological services for established HIV disease.

Different institutions have developed different patterns of providing, or referring for, those services; in general, the level of care an institution offers students

6. OLD/NEW SEXUAL CONCERNS: Sexual Hygiene

with HIV should be parallel to the level of services it provides for students with other chronic illnesses. The presence (or absence) of a dedicated student health service does not of itself determine an institution's need (or capacity) to respond; for example, every college and university should know where to refer students for safe, accessible HIV antibody testing. Some institutions have integrated HIV-related services effectively into their counseling and health programs; health services on some campuses provide exemplary primary care for students with HIV; and many college counseling services have developed special programs to address the individual and group needs of students who have HIV, feel concerned about the possibility of HIV, or are dealing with a loved one's HIV disease. Providing effective medical and psychological services for people with HIV often requires professional development programs for clinicians and counselors. Because many people with HIV are gay or bisexual men, it is essential that providers of medical or psychological services be knowledgeable, competent, and comfortable in understanding and addressing the concerns of this population.

Education and prevention
The greatest challenge on most campuses is to imagine, design, and implement effective prevention programs: strategies that will reliably and durably help students change risky behavior. Both qualitative and quantitative research studies repeatedly confirm what campus educators and clinicians regularly observe in their work: while students have learned a lot about HIV, they continue to put themselves at risk.[9] Because HIV is a particularly serious health issue, and because what we are learning about behavior in regard to HIV is deeply relevant to risks of other sexually transmitted diseases, unwanted pregnancies, sexual assault, regretted intercourse, and substance abuse, addressing the inconsistencies of knowledge and behavior is critically important.

The evolution of college and university HIV prevention programs away from information-based strategies and toward a more holistic recognition of the multiple determinants of behavior reflects the synthesis of developmental theory with educational practice.[10] The resulting second generation of activities designed to promote sexual health in higher education is diverse in character and content, but the activities do share certain consistent key features:

- **A clear, comprehensive focus on behavior, rather than on information.** The most important strategy in most programs is skills building. Since skills in decision making, intimacy, communication, and assertiveness are adaptable to a great many other health decisions, HIV prevention programs are now closely integrated with activities dealing with other sexual health issues, campus violence, and substance abuse. Decisions about alcohol, for example, are often closely connected to choices (or the abrogation of choices) about sex. Thus, interventions that enhance the possibility of coherent decision making about alcohol also reduce the probability of unwanted pregnancy and HIV infection.

- **Highly specific, targeted programs for each student's needs.** Using peer education as a dominant model, campuses now use diverse approaches in addressing the specific needs of different constituencies (such as people of color; women; gay, lesbian, and bisexual students; athletes; and faculty and staff). Programs now reflect different learning styles and can be adapted to different patterns of campus life and different levels of presence and involvement in activities.

- **Comprehensive inclusion of students.** Involving students at all levels of decision making (from planning to evaluation) has helped college and university health promotion programs to be more effective and more respectful of students in their content, format, and tone; in respectful programs, health education messages do not preach, demean, belittle, frighten, blame, or infantilize students.

- **A focus on self, identity, and competency as deep determinants of behavior.** New strategies recognize the critical influence of the sense of self on motivation, competency, and resistance — all fundamental factors in making and sustaining changes in behavior. Confronting the enormous problems of self-esteem, identity, and meaning among college and university students is the major challenge of health promotion; through exciting interdisciplinary programs, campus clinicians, counselors, and educators have begun to test and evaluate innovative strategies that help students regain a centered, self-conscious approach to understanding risks, making choices, and assuming responsibility for themselves.

Understanding and addressing social and cultural norms. Community standards, values, and norms have enormous power to shape the context within which individual decisions about health behavior are made.[11] Managing individual risks and decisions therefore requires understanding and managing the social and cultural context that both mirrors and determines them. Changing the social norms that influence behavior is, of course, a gargantuan task in a complex culture; the conflicting messages of media, advertising, music, and entertainment create a seductive, visual environment in which the advice and precautions of health promotion may seem sterile and uninteresting. In the future, health educators at all levels will be agents of deep community change, and the highest and best use of peer educators probably will be to exert a kind of growth pressure on community values.

Building caring and community. Experienced as a sense of caring and connectedness, the spirit of community on campus promotes safety and health among students.[12] A broader and deeper sense of community on campus transcends health promotion, but is absolutely integral to the success of such programs. For example, a spirit of community promotes both a broad sense of justice and respect for difference and a specific commitment to caring about the consequences of unprotected intercourse. This high-level work is just beginning on many campuses; it represents only the latest in a series of cycles by which the wisdom of a community is incorporated into the community's systems.

Conclusion

The experience of responding to HIV in campus communities has taught us much. As students, faculty, and staff, we have come to understand human differences more deeply and to appreciate the richness of our communities in a more authentic way. We have confronted uncertainty, dealt with fear, and found new dimensions in love and caring. In the harsh clarity fostered by this epidemic, we have come to a more lively, but more careful, understanding of sexuality, intimacy, risk, and the influence of social and cultural forces on human behavior. We have felt the power and vitality of caring and community. Campus communities have found ways to be more caring toward people with HIV and more responsive to their needs. Thus, in becoming more connected to their experience, we have, in fact, strengthened the spirit of community that unites us all.

Notes

[1] Buehler, J.W., O.J. Devine, R.L. Berkelman, and F.M. Chevarley. "Impact of the Human Immunodeficiency Virus Epidemic on Mortality Trends in Young Men," *American Journal of Public Health*, 80: 1080-1086; Chu, S.Y., J.W. Buehler, and R.L. Berkelman. "Impact of the Human Immunodeficiency Virus Epidemic on Mortality in Women of Reproductive Age," *JAMA: The Journal of the American Medical Association*, 264: 225–229.

[2] Gayle, H.D., R.P. Keeling, M. Garcia-Tunon, et al. "Prevalence of HIV Infection Among College and University Students," *New England Journal of Medicine*, 393: 1526–1531.

[3] Brian Edlin and Richard P. Keeling, Principal Investigators, personal communication.

[4] St. Louis, M.E., G.A. Conway, C.R. Hayman, et al. "Human Immunodeficiency Virus Infection in Disadvantaged Adolescents. Findings from the U.S. Job Corps," *JAMA: The Journal of the American Medical Association*, 266: 2387–2391.

[5] Ibid.

[6] Centers for Disease Control. "HIV/AIDS Surveillance – United States. AIDS Cases Reported Through December, 1991." Washington, DC: U.S. Department of Health and Human Services, 1992; Holmes, K.K., J.M. Karon, and J. Kreiss. "The Increasing Frequency of Heterosexually Acquired AIDS in the United States," *American Journal of Public Health*, 80: 858–863.

[7] Brookmeyer, R. "Reconstruction and Future Trends of the AIDS Epidemic in the United States," *Science*, 253: 37–42.

[8] Centers for Disease Control. "Recommendations for Prevention of HIV Transmission in Health Care Settings," *Morbidity and Mortality Weekly Report*, 36(suppl 2): 1S–18S.

[9] Carroll, L. "Gender, Knowledge About AIDS, Reported Sexual Behavioral Change, and the Sexual Behavior of College Students," *Journal of American College Health*, 40: 5–12; Fennell, R. "Knowledge, Attitudes, and Beliefs of Students Regarding AIDS: A Review," *Health Education*, 21: 260–267; Keeling, R.P. "Time to Move Forward: An Agenda for Campus Sexual Health Promotion in the Next Decade," *Journal of American College Health*, 40: 51–55.

[10] Gould, J. and R.P. Keeling. "Principles of Effective Sexual Health Promotion: Theory to

6. OLD/NEW SEXUAL CONCERNS: Sexual Hygiene

Practice," in Richard P. Keeling, M.D., ed., *Effective AIDS Education on Campus*. San Francisco: Jossey-Bass, 1992.

[11] Ekstrand, M.L. and T.J. Coates, "Maintenance of Safer Sexual Behaviors and Predictors of Risky Sex: The San Francisco Men's Health Study," *American Journal of Public Health*, 80: 973–977; Fisher, J.D., "Possible Effects of Reference-Group Based Social Influence on AIDS Risk Behavior and AIDS Prevention", *American Psychologist*, 43: 914–920; Keeling, R.P. and E.L. Engstrom, "Building Community for Effective Health Promotion," in Keeling, R.P., ed., *Effective AIDS Education on Campus*. San Francisco: Jossey-Bass, 1992.

[12] Burns, W.D. and M. Klawunn. "The Web of Caring: An Approach to Accountability in Alcohol Policy." Unpublished manuscript. Office of the Assistant Vice President for Student Life Policy and Services, Rutgers, the State University of New Jersey, New Brunswick; Keeling and Engstrom, ibid.

Article 40

Preventing STDs

This article is part of a series with important health information for teenagers. Unlike previous articles, however, it contains sexually explicit material in an effort to reduce the incidence of STDs among teens. Parents and teachers may want to review the article before giving it to teenagers.

Judith Levine Willis

Judith Levine Willis is editor of FDA Consumer.

It's important to read the information printed on the package to make sure a condom's made of latex and labeled for disease prevention. The label may also give an expiration date and tell you if there is added spermicide or lubricant.

You don't have to be a genius to figure out that the only sure way to avoid getting sexually transmitted diseases (STDs) is to not have sex.

But in today's age of AIDS, it's smart to also know ways to lower the risk of getting STDs, including HIV, the virus that causes AIDS.

Infection with HIV, which stands for human immunodeficiency virus, is spreading among teenagers. From 1990 to 1992, the number of teens diagnosed with AIDS nearly doubled, according to the national Centers for Disease Control and Prevention. Today, people in their 20s account for 1 out of every 5 AIDS cases in the United States. Because HIV infection can take many years to develop into AIDS, many of these people were infected when they were teenagers.

You may have heard that birth control can also help prevent AIDS and other STDs. This is only partly true. The whole story is that *only one form of birth control—latex condoms* (thin rubber sheaths used to cover the penis)—is highly effective in reducing the transmission (spread) of HIV and many other STDs.

(When this *FDA Consumer* went to press, the Food and Drug Administration was preparing to approve Reality Female Condom, a form of birth control made of polyurethane. It may give limited protection against STDs, but it is not as effective as male latex condoms.)

So people who use other kinds of birth control, such as the pill, sponge, diaphragm, Norplant, Depo-Provera, cervical cap, or IUD, also need to use condoms to help prevent STDs.

Here's why: Latex condoms work against STDs by keeping blood, a man's semen, and a woman's vaginal fluids—all of which can carry bacteria and viruses—from passing from one person to another. For many years, scientists have known that male condoms (also called safes, rubbers, or prophylactics) can help prevent STDs transmitted by bacteria, such as syphilis and gonorrhea, because the bacteria can't get through the condom. More recently, researchers discovered that latex condoms can also reduce

6. OLD/NEW SEXUAL CONCERNS: Sexual Hygiene

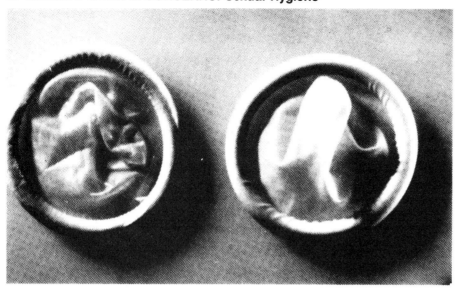

If a condom is sticking to itself, as is the one on the left, it's damaged and should not be used. The one on the right is undamaged and okay to use.

the risk of getting STDs caused by viruses, such as HIV, herpes, and hepatitis B, even though viruses are much smaller than bacteria or sperm.

After this discovery, FDA, which regulates condoms as medical devices, worked with manufacturers to develop labeling for latex condoms. The labeling tells consumers that although latex condoms cannot entirely eliminate the risk of STDs, when used properly and consistently they are highly effective in preventing STDs. FDA also provided a sample set of instructions and requested that all condoms include adequate instructions.

Make Sure It's Latex

Male condoms sold in the United States are made either of latex (rubber) or natural membrane, commonly called "lambskin" (but actually made of sheep intestine). Scientists found that natural skin condoms are not as effective as latex condoms in reducing the risk of STDs because natural skin condoms have naturally occurring tiny holes or pores that viruses may be able to get through. Only latex condoms labeled for protection against STDs should be used for disease protection.

Some condoms have lubricants added and some have spermicide (a chemical that kills sperm) added. The package labeling tells whether either of these has been added to the condom.

Lubricants may help prevent condoms from breaking and may help prevent irritation. But lubricants do not give any added disease protection. If an unlubricated condom is used, a water-based lubricant

New Information on Labels

Information about whether a birth control product also helps protect against sexually transmitted diseases (STDs), including HIV infection, is being given added emphasis on the labeling of these products.

"In spite of educational efforts, many adolescents and young adults, in particular, are continuing to engage in high-risk sexual behavior," said FDA Commissioner David A. Kessler, M.D., in announcing the label strengthening last April. "A product that is highly effective in preventing pregnancy will not necessarily protect against sexually transmitted diseases."

Labels on birth control pills, implants such as Norplant, injectable contraceptives such as Depo Provera, intrauterine devices (IUDs), and natural skin condoms will state that the products are intended to prevent pregnancy and do not protect against STDs, including HIV infection (which leads to AIDS). Labeling of natural skin condoms will also state that consumers should use a latex condom to help reduce risk of many STDs, including HIV infection.

Labeling for latex condoms, the only product currently allowed to make a claim of effectiveness against STDs, will state that if used properly, latex condoms help reduce risk of HIV transmission and many other STDs. This statement, a modification from previous labeling, will now appear on individual condom wrappers, on the box, and in consumer information.

Besides highlighting statements concerning sexually transmitted diseases and AIDS on the consumer packaging, manufacturers will add a similar statement to patient and physician leaflets provided with the products.

Consumers can expect to see the new labels by next fall. Some products already include this information in their labeling voluntarily. FDA may take action against any products that don't carry the new information.

FDA is currently reviewing whether similar action is necessary for the labeling of spermicide, cervical caps, diaphragms, and the Today brand contraceptive sponge.

Looking at a Condom Label

Like other drugs and medical devices, FDA requires condom packages to contain certain labeling information. When buying condoms, look on the package label to make sure the condoms are:
- made of latex
- labeled for disease prevention
- not past their expiration date (EXP followed by the date).

40. Preventing STDs

(such as K-Y Jelly), available over-the-counter (without prescription) in drugstores, can be used but is not required for the proper use of the condom. Do *not* use petroleum-based jelly (such as Vaseline), baby oil, lotions, cooking oils, or cold creams because these products can weaken latex and cause the condom to tear easily.

Condoms with added spermicide give added birth control protection. An active chemical in spermicides, nonoxynol-9, kills sperm. Although it has not been scientifically proven, it's possible that spermicides may reduce the transmission of HIV and other STDs. But spermicides alone (as sold in creams and jellies over-the-counter in drugstores) and spermicides used with the diaphragm or cervical cap do not give adequate protection against AIDS and other STDs. For the best disease protection, a latex condom should be used from start to finish every time a person has sex.

FDA requires condoms with spermicide to be labeled with an expiration date. Some condoms have an expiration date even though they don't contain spermicide. Condoms should not be used after the expiration date, usually abbreviated EXP and followed by the date.

Condoms are available in almost all drugstores, many supermarkets, and other stores. They are also available from vending machines. When purchasing condoms from vending machines, as from any source, be sure they are latex, labeled for disease prevention, and are not past their expiration date. Don't buy a condom from a vending machine located where it may be exposed to extreme heat or cold or to direct sunlight.

Condoms should be stored in a cool, dry place out of direct sunlight. Closets and drawers usually make good storage places. Because of possible exposure to extreme heat and cold, glove compartments of cars are *not* a good place to store condoms. For the same reason, condoms shouldn't be kept in a pocket, wallet or purse for more than a few hours at a time.

How to Use a Condom
• Use a new condom for every act of vaginal, anal and oral (penis-mouth contact) sex. Do not unroll the condom before placing it on the penis.

STD Facts

• Sexually transmitted diseases affect more than 12 million Americans each year, many of whom are teenagers or young adults.
• Using drugs and alcohol increases your chances of getting STDs because these substances can interfere with your judgment and your ability to use a condom properly.
• Intravenous drug use puts a person at higher risk for HIV and hepatitis B because IV drug users usually share needles.
• The more partners you have, the higher your chance of being exposed to HIV or other STDs. This is because it is difficult to know whether a person is infected, or has had sex with people who are more likely to be infected due to intravenous drug use or other risk factors.
• Sometimes, early in infection, there may be no symptoms, or symptoms may be confused with other illnesses.
• You cannot tell by looking at someone whether he or she is infected with HIV or another STD.

STDs can cause:
• pelvic inflammatory disease (PID), which can damage a woman's fallopian tubes and result in pelvic pain and sterility
• tubal pregnancies (where the fetus grows in the fallopian tube instead of the womb), sometimes fatal to the mother and always fatal to the fetus
• cancer of the cervix in women
• sterility—the inability to have children—in both men and women
• damage to major organs, such as the heart, kidney and brain, if STDs go untreated
• death, especially with HIV infection.

See a doctor if you have any of these STD symptoms:
• discharge from vagina, penis or rectum
• pain or burning during urination or intercourse
• pain in the abdomen (women), testicles (men), or buttocks and legs (both)
• blisters, open sores, warts, rash, or swelling in the genital or anal areas or mouth
• persistent flu-like symptoms—including fever, headache, aching muscles, or swollen glands—which may precede STD symptoms.

• Put the condom on after the penis is erect and before *any* contact is made between the penis and any part of the partner's body.

• If the condom does not have a reservoir top, pinch the tip enough to leave a half-inch space for semen to collect. Always make sure to eliminate any air in the tip to help keep the condom from breaking.

• Holding the condom rim (and pinching a half inch space if necessary), place the condom on the top of the penis. Then, continuing to hold it by the rim, unroll it all the way to the base of the penis. If you are also using water-based lubricant, you can put more on the outside of the condom.

• If you feel the condom break, stop immediately, withdraw, and put on a new condom.

• After ejaculation and before the penis gets soft, grip the rim of the condom and carefully withdraw.

• To remove the condom, gently pull it off the penis, being careful that semen doesn't spill out.

• Wrap the condom in a tissue and throw it in the trash where others won't handle it. (Don't flush condoms down the toilet because they may cause sewer problems.) Afterwards, wash your hands with soap and water.

Latex condoms are the only form of contraception now available that human studies have shown to be highly effective in protecting against the transmission of HIV and other STDs. They give good disease protection for vaginal sex and should also reduce the risk of disease transmission in oral and anal sex. But latex condoms may not be 100 percent effective, and a lot depends on knowing the right way to buy, store and use them.

Syphilis in the '90s

The "great imposter" is taking on new manifestations in HIV-infected patients and is staging a major comeback in inner-city populations in the wake of poverty and crack addiction.

JANE R. SCHWEBKE, MD

Jane R. Schwebke is Medical Director, Seattle–King County/Harborview Sexually Transmitted Disease Clinic, and Acting Instructor in Infectious Diseases, University of Washington School of Medicine, Seattle.

In the late 1970s and early 1980s, the incidence of syphilis rose in this country, with especially dramatic increases noted among homosexual and bisexual men. Then, from 1982 to 1985, the overall incidence declined, with much of the decrease seen in male homosexuals, probably stemming from changes in sexual behavior in response to the AIDS epidemic.

Now, the overall incidence of syphilis is again on the rise, increasing 61% from 1985 to 1989 (see Figure 1 for a breakdown by sex). During those five years, the incidence soared from 11.4 cases per 100,000 population to 18.4 (see Figure 2 for geographic distribution).[1,2]

The startling increase in syphilis among inner-city ethnic groups of low socioeconomic status has been a major factor in this overall jump. From 1985 to 1989, syphilis incidence rose by 132% among African-Americans, with a larger increase among women than men. This increase in women has led to a similar rise in cases of congenital syphilis (see Figure 1). And although the disease continues to decline in white men, there are signs that the safer sex message may be losing its urgency as gonorrhea cases start to climb again in certain urban homosexual populations.[3]

The epidemiology of both syphilis and gonorrhea (see "Gonorrhea in the '90s," *Medical Aspects of Human Sexuality*, March 1991) appears to be closely linked to illicit drug use and prostitution.[4] Syphilis rates began to skyrocket among inner-city populations at the same

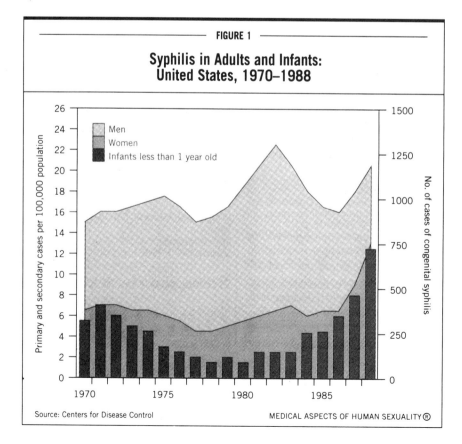

FIGURE 1

Syphilis in Adults and Infants: United States, 1970–1988

Source: Centers for Disease Control

time as use of crack-cocaine reached epidemic proportions. The trading of sex for drugs at crack houses encourages frequent sexual encounters with anonymous partners. This phenomenon not only increases a person's risk for infection but also renders traditional methods of contact tracing much less effective.[5]

Syphilis has been called the "great imposter," since it has many different clinical presentations and may be easily misdiagnosed. With the advent of penicillin therapy, syphilis cases declined and many physicians-in-training in the late 1970s and early 1980s never saw a case of early syphilis. Now in HIV-infected patients, the disease is presenting with unusual manifestations and serology that may confound diagnosis (see box, "Syphilis in the HIV-Infected Patient").

The recent increased incidence of syphilis, along with evidence that the genital ulcers of syphilis may facilitate transmission of HIV, has spurred renewed interest in its diagnosis and treatment.

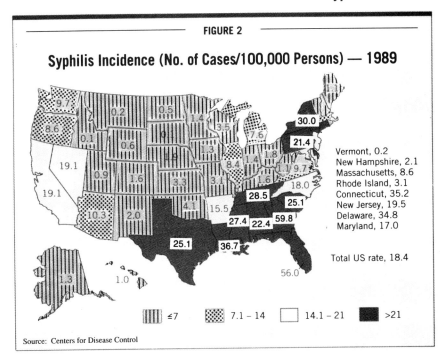

FIGURE 2

Syphilis Incidence (No. of Cases/100,000 Persons) — 1989

Vermont, 0.2
New Hampshire, 2.1
Massachusetts, 8.6
Rhode Island, 3.1
Connecticut, 35.2
New Jersey, 19.5
Delaware, 34.8
Maryland, 17.0

Total US rate, 18.4

≤7   7.1 – 14   14.1 – 21   >21

Source: Centers for Disease Control

> The incidence of syphilis rose by 61% in the United States from 1985 to 1989.

The Three Stages

Syphilis is caused by the spirochete, *Treponema pallidum*. Clinically it is divided into three stages—primary, secondary, and tertiary. The primary stage occurs 10 to 90 days after infection and usually presents as a painless, solitary, indurated genital ulcer (see Figure 3.)[6] Darkfield examination of the ulcer should reveal treponemes unless the patient has been using oral or topical antibiotics. Serologic tests for syphilis may or may not be positive.

Left untreated, the chancre will heal in two to six weeks, and the syphilis will enter the secondary stage, or disseminated phase of infection, which includes invasion of the central nervous system (CNS). Patients may present with any combination of the following signs: generalized skin eruption (see Figure 4), palmar/plantar lesions, widespread adenopathy, oral mucous patches, and alopecia. Results of serology tests are almost always reactive.

If left untreated, the manifestations of secondary syphilis usually heal within two to six week. The organism then becomes latent, and the patient enters the tertiary stage. In this stage, the patient is usually asymptomatic but may develop gummas and/or cardiac and CNS involvement years to decades later.

Diagnosis

Syphilis is usually easily diagnosed in the patient who presents with signs and symptoms of early disease. In general, syphilis should be suspected with any genital lesion that is not clearly herpetic. As noted above, darkfield examinations may be falsely negative due to prior usage of antibiotics, too few organisms, or a healing lesion. If the patient has a

---

Syphilis in the HIV-Infected Patient

Syphilis patients who are also infected with HIV may have unusual presentations. There is some evidence to suggest that syphilis serologies may be unreliable in these patients,[7,8] and they may have an increased incidence of syphilitic meningitis.[9] Thus, it is advisable to screen for HIV infection in the patient with syphilis and vice versa.

Although *T pallidum* frequently invades the CNS in its secondary stage, most normal hosts are able to contain the infection. Patients with HIV infection, however, are less able to do so. Numerous cases have been reported in which HIV-infected patients treated for early syphilis with benzathine penicillin later developed neurosyphilis.[8] The combination of benzathine penicillin and high-dose amoxicillin/probenecid is currently being studied as a possible means of decreasing the incidence of CNS relapse in this group.

suspicious lesion that is darkfield negative, an inguinal node may be aspirated and examined by darkfield as well. Also, many outlying clinics may not have access to a darkfield microscope or have sufficient expertise to perform the examination adequately. Reliance on serology alone to diagnose primary syphilis may miss a substantial number of cases.

The rash of secondary syphilis is frequently mistaken for other dermatologic conditions, and, unless a high degree of suspicion is maintained, the diagnosis may be overlooked.

Syphilis in Pregnancy

All pregnant women should be screened for syphilis, and those with reactive serology results should be treated promptly. Women who are at high risk for STDs should be rescreened in the third trimester. Most cases of congenital syphilis occur in infants of mothers who did not seek prenatal care.

> Relying on serology to diagnose primary syphilis may lead to a substantial number of missed cases.

Syphilis is diagnosed in the infant by demonstrating live treponemes in lesions or nasal discharge, or by documenting a reactive serology in infants with clinical signs that suggest syphilis. It is important to remember that many infants with active infection may be asymptomatic at birth.

If the mother was treated for syphilis during pregnancy (discussed below), the infant's blood may be reactive due to passive transfer of antibody. Serial serologies may be necessary in these infants to assure that their titers are declining.

Neurosyphilis

Neurosyphilis can occur at any stage of the disease but may be difficult to diagnose. When performed on spinal fluid, the VDRL test has a sensitivity of only 50%, although specificity is 100%.[10] Spinal fluid FTA-ABS (fluorescent treponemal antibody absorption) is fraught with technical difficulties and frequently yields false-positive results. Thus, the diagnosis often relies upon pleocytosis and/or elevated protein levels. However, in HIV-infected patients these nonspecific findings often occur for other reasons.

The Centers for Disease Control (CDC) recommends examination of the spinal fluid in all persons with late latent syphilis of greater than one year's duration if possible, and specifically in such patients who have treatment failure (ie, rising VDRL titer or failure of titer to decline), any evidence of active syphilis (iritis, neurologic signs or symptoms, aortitis, gummas), a positive HIV test, or who were treated with an antibiotic other than penicillin.[11]

Examination of the spinal fluid in patients infected with HIV is controversial. Some authorities recommend a lumbar puncture only if the patient has late latent syphilis,[11]

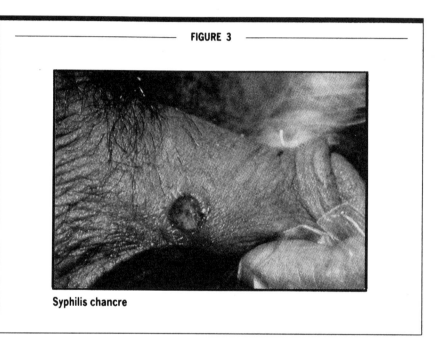

FIGURE 3

Syphilis chancre

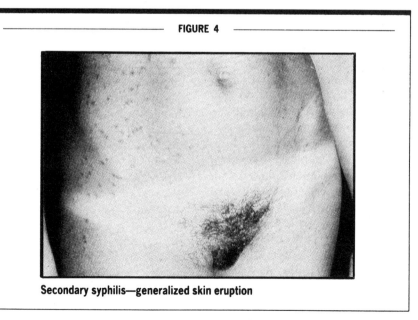

FIGURE 4

Secondary syphilis—generalized skin eruption

while others recommend it for any stage.[9]

Treatment

Over the years, *T pallidum* has remained exquisitely sensitive to penicillin. Benzathine penicillin, which is absorbed slowly, is used because of the organism's long replication time. For early syphilis, give 2.4 million units IM in one dose. For syphilis of greater than one year's duration or of unknown duration, the recommended regimen is 2.4 million units IM each week for three weeks.

Doxycycline, tetracycline, or erythromycin may be used in the penicillin-allergic patient, but they require compliance since they must be taken for 15 to 30 days, depending on the stage of disease.[11]

In pregnancy. Pregnant patients must be treated with penicillin since there have been reports of babies born with congenital syphilis following treatment of the mother with erythromycin. If the pregnant woman is allergic to penicillin, she must be desensitized.[11]

Neurosyphilis. None of the abovementioned antibiotics cross the blood-brain barrier adequately to treat neurosyphilis. Thus, patients with neurosyphilis must receive high-dose aqueous crystalline penicillin G IV for 10 to 14 days. If inpatient therapy is not possible, an alternative regimen is procaine penicillin, 2.4 million units IM daily, plus probenecid, 500 mg PO four times daily, both for 10 to 14 days.[11] Studies are currently underway to assess the efficacy of ceftriaxone in the treatment of neurosyphilis, and early data suggest that it may be a possible alternative.[9]

Patients with neurosyphilis should have a repeat cerebrospinal fluid examination in three to six months to assess response. All patients with syphilis should be followed with serial VDRL titers, which should decline significantly with treatment.

Screening

In addition to the aforementioned groups (HIV-infected individuals and pregnant women), screening for syphilis should be routine for clients attending STD clinics and for any patient with multiple sexual partners. Depending on the level of risk, screening should be performed at one- to three-month intervals. Premarital screening is still required in many states. However, routine screening of hospital admissions is no longer generally performed.

Unusual Presentation in an HIV-Positive Man

Tim, a 26-year-old homosexual man, presented with a generalized, mildly pruritic skin rash. He said that his last sexual encounter had been approximately six weeks earlier with a casual partner. On examination, he had a generalized maculopapular rash, which also affected his palms and soles. There were no penile or rectal ulcers, and his oral mucosa appeared normal.

A stat RPR was strongly reactive. Subsequent testing revealed a titer of 1:128 and a reactive FTA-ABS. An HIV screen was positive. Tim was treated with 2.4 million units of benzathine penicillin. One month later his rash had resolved, but his repeat RPR titer was 1:64.

FAST FACTS

Syphilis in Philly

"In Philadelphia, from 1985 through 1989, the number of reported cases of early syphilis (primary, secondary, and early latent stages) increased 551%, from 696 to 4,528 cases per year."

Source: *MMWR* 40(5):77, 1991.

Conclusion

Although syphilis is an "old" infection for which adequate antibiotic therapy exists, it continues to account for significant morbidity and mortality. In addition, syphilis is becoming more difficult to control from an epidemiologic standpoint as the populations affected by it become more difficult to reach. Renewed emphasis is needed in training health-care personnel to recognize and treat syphilis, and increased resources and new strategies are required to screen and educate those populations at risk.

To these ends, work is underway to perfect highly sensitive diagnostic tests for fast and accurate detection of active infection. Moreover, newer antibiotics (eg, azithromycin, ofloxacin, ceftriaxone, and cefotetan)[12-15] currently undergoing clinical trials may prove efficacious in simultaneously eradicating gonorrhea, chlamydia, and syphilis. Finally, the CDC is developing innovative community-based educational programs designed to reach high-risk individuals.[16]

References

1. Centers for Disease Control: Congenital syphilis—New York City, 1986–1988. *MMWR* 38:825, 1989.
2. Rolfs RT, Nakashima AK: Epidemiology of primary and secondary syphilis in the United States, 1981 through 1989. *JAMA* 264:1432, 1990.
3. Handsfield HH, Schwebke JR: Trends in sexually transmitted diseases in homosexually active men in King County Washington, 1980–1990. *Sex Transm Dis* 17:211, 1990.
4. Farley TA, Hadler JL, Gunn RA: The syphilis epidemic in Connecticut: Relationship to drug use and prostitution. *Sex Transm Dis* 17:163, 1990.
5. Andrews JK, Fleming DW, Harger DR, et al: Partner notification: Can it control epidemic syphilis? *Ann Intern Med* 112:539, 1990.
6. Sparling PF: Natural history of syphilis, in Holmes KK, March P, Sparling PF, et al (eds): *Sexually Transmitted Diseases*. New York, McGraw-Hill, 1990, pp 213–219.
7. Hicks CB, Benson PM, Lupton GP, et al: Seronegative secondary syphilis in a patient infected with the human immunodeficiency virus (HIV) with Kaposi sarcoma. *Ann Intern Med* 107:492, 1987.
8. Haas JS, Balan G, Larsen SA, et al: Sensitivity of treponemal tests for detecting prior treated syphilis during human immunodeficiency virus infection. *J Infect Dis* 162:862, 1990.

9. Musher DM, Hamill RJ, Baughn RE: Effect of human immunodeficiency virus (HIV) infection on the course of syphilis and on the response to treatment. *Ann Intern Med* 113:872, 1990.

10. Hart G: Syphilis tests in diagnostic and therapeutic decision making. *Ann Intern Med* 104:368, 1986.

11. Centers for Disease Control: 1989 Sexually transmitted diseases treatment guidelines. *MMWR* 38(S-8):5, 1989.

12. Lukehart SA, Fohn MJ, Baker-Zander SA: Efficacy of azithromycin for therapy of active syphilis in the rabbit model. *J Antimicrob Chemother* 25(suppl A):91, 1990.

13. Lassus A: Comparative studies of azithromycin in skin and soft-tissue infections and sexually transmitted infections by Neisseria and Chlamydia species. *J Antimicrob Chemother* 25(suppl A):115, 1990.

14. Youssef RZ, Murray M, Holmes B, et al: Cefotetan therapy for gonococcal urethritis and cervicitis. *Sex Trans Dis* 17:99, 1990.

15. Covino JM, Cummings M, Smith B: Comparison of ofloxacin and ceftriaxone in the treatment of complicated gonorrhea caused by penicillinase-producing and nonpenicillinase-producing strains. *Antimicrob Agents Chemother* 34:148, 1990.

16. Goldsmith MF: Target: Sexually transmitted diseases. *JAMA* 264:2179, 1990.

A Question of Abuse

MORE AND MORE AMERICANS ARE SUDDENLY RECALLING TRAUMATIC CHILDHOOD EVENTS. BUT HOW MANY OF THEIR MEMORIES ARE REAL?

Nancy Wartik

Nancy Wartik is a Contributing Editor at AMERICAN HEALTH.

Amy Yates's* two-year-old son awoke in the middle of the night screaming from a nightmare. After rocking him back to sleep, Yates, 35, closed her eyes. Suddenly, the room seemed to fill with the odor of kerosene. With the scent came a vivid childhood image of a neighbor's garage, warmed by a kerosene heater. At first Yates didn't know what to make of the memory, but when she described it to her therapist, more pieces of the past emerged: Nearly three decades after the fact, she recalled that the neighbor had given her cocoa and then molested her in that garage, starting when she was six.

"There was a circular gravel driveway and I rode my bike around it," Yates remembers. "Then he'd call me, and I'd have to go into the garage with him. He said if I told, he'd kill me, my mother, my cat." The abuse went on for three years, but she didn't tell anyone or even remember it—until the evening of her son's nightmare: "His screams were like the ones I had then," Yates says, "but mine were silent."

Yates's experience is far from unique. In the past few years, thousands of Americans—among them TV star Roseanne Arnold and former Miss America Marilyn Van Derbur Atler—have reported retrieving long-buried memories of childhood sexual abuse by neighbors, parents, teachers, baby sitters and even clergymen. Such flashbacks may arrive in stages: A person first experiences so-called "body memories" of physical pain or discomfort suffered during the abuse, for example, followed by sights or sounds, and finally, by the return of the emotions involved. Many say that while the memories were submerged, they were afflicted by chronic anxiety, depression, substance abuse or eating disorders. Some have even reported symptoms of multiple personality disorder, a severe mental illness.

*Not her real name.

The so-called recovered-memory phenomenon has divided the psychotherapeutic community. Although some psychiatrists think shocking experiences really *can* vanish from consciousness for decades at a time, only to reemerge with crystal clarity years later, others do not. Skeptical scientists say too many accounts of long-past abuse conflict with findings on how we store and recall experience. Most recovered "memories," they believe, are subconscious efforts to actually resolve old hurts, in many cases spurred by an increasing number of books, support groups and psychotherapists who see sexual abuse at the root of virtually all psychic pain.

Many therapists see sexual abuse at the root of most psychic pain.

In Philadelphia the recently formed False Memory Syndrome Foundation has logged calls from more than 2,800 people claiming they've been falsely accused of incest or abuse. "Generally the person accusing them has cut off all contact," says Pamela Freyd, the executive director. "Families are being torn apart and destroyed. It's a very sad situation." Indeed, some of those who once claimed to have recovered memories of molestation have now changed their minds. Three years ago, after retrieving evermore elaborate recollections of abuse with the help of a therapist, Melody Lee Gavigan of Reno, Nev., consulted a lawyer about filing incest charges against her father. Today she says the memories were "a figment of my imagination encouraged by my therapists and the pop psychology books I was reading." She's struggling to make amends to her family.

But what about Amy Yates's case? When Yates told her mother about what she'd remembered, the older woman recalled that the same neighbor who molested Yates had also tried to kiss her sister. The mother had forbidden the sister to go to the neighbor's house again. "I was horrified," says Yates. "I said, 'But Mom, why did you send me over there for cocoa?' She could have stopped what happened but she didn't."

6. OLD/NEW SEXUAL CONCERNS: Sexual Abuse and Violence

Like Yates, other victims have found evidence that suggests or confirms the reality of their recollections. That leaves psychiatrists concerned about the nature of recovered memory confronting a puzzle whose pieces don't seem to create a cohesive whole.

The study of how we deal with disturbing memories dates back to the turn of the century, when Sigmund Freud theorized that the conscious mind can push away, or repress, anxiety-provoking ideas. When submerged, he said, potent memories can cause both physical and mental symptoms. In fact, on the basis of his early work with women, Freud concluded that many had been abused as children, and that blocked memories of the abuse were causing their ailments and upsets. When colleagues scoffed at the notion, however, Freud diluted his repression theory by saying memories of sexual abuse were fantasies; today critics accuse Freud of betraying women by backing down on his original premise.

Psychologists increasingly acknowledge the alarming incidence of sexual abuse among children, but they still argue over whether the mind really can wall off disturbing material from conscious awareness. "We see evidence of repression every day," insists Dr. Lenore Terr, a professor of psychiatry at the University of California at San Francisco (UCSF) and the author of the forthcoming book *Unchained Memories* (Basic Books, 1994). "A man in a troubled marriage forgets his wife's birthday—don't tell me it's a coincidence. All of us have experienced conflicts or pain that cause us to forget things."

But Dr. David Holmes, a professor of psychology at the University of Kansas who has reviewed decades of studies on the subject, disagrees. "We've been absolutely unable to find solid evidence of repression," he says. "That's not to say that memory isn't selective or that we don't forget things. But memory doesn't follow the pattern Freud suggested. There's not an active process that holds repulsive thoughts back."

Holmes and other researchers assert that recovered recollections of abuse don't conform to current conceptions of how memory works. For instance, people who report recovered memories often recollect them in vivid detail: "It unrolled like a picture, a video," says Barbara Leger*, who was 24 and talking with her therapist when she first retrieved long-buried images of her father molesting her. "A year and a half later, the memories are still coming." But, argues Dr. George Bonanno, a postdoctoral fellow at the program on conscious and unconscious mental processes at UCSF, events which are so elaborately recalled shouldn't have been forgotten in the first place. "When a memory is that strong, it should have been accessible all along," Bonanno notes. "You're unlikely to remember something years later, in a tremendous amount of detail, if you haven't done so already. So for someone to suddenly say, 'Oh, now I remember it all' just doesn't ring true."

Some scientists have similar reservations about the fact that recovered memories often span many years. A recent University of Southern California (USC) study found that among 450 women and men reporting histories of childhood abuse, the period of molestation averaged 10.5 years; 1958 Miss America Van Derbur Atler reports her father abused her from early childhood until she left for college at 18. How, some psychologists wonder, could someone forget something that went on over such a long period of time, and so late into adolescence? "While repression is sometimes possible, I find it a little hard to believe that you can be raped every week of your life for 15 years and repress it all," says Dr. Elizabeth Loftus, a professor of psychology at the University of Washington and a leading memory researcher. "What the studies tend to show is that when kids have a horrible experience, they're plagued by recurrent thoughts of it."

Countering their skeptical colleagues, scientists who study the psychological effects of trauma point out that, in some individuals, a highly stressful experience disrupts the process by which the brain normally stores and retrieves a memory. They draw an analogy between recovered memory and post traumatic stress disorder (PTSD), once known as shell shock. "We've already gone through this whole phenomenon in the PTSD research with combat veterans," says Dr. Frank Putnam, chief of dissociative disorders at the National Institute of Mental Health (NIMH). "Some soldiers had literally no memory of whole days until 15 or 20 years later, when it was safe to remember. I think there's no question that some abuse victims experience the same thing that vets do. It's sexist for us to believe the men soldiers and doubt the women victims."

A recent University of New Hampshire study supports Putnam's theory that traumatic events can be stored in the brain in a way that doesn't conform to the standard model of memory. Researchers located the hospital records of 100 girls under 13 who were treated for sexual abuse between 1973 and 1975. When these same individuals were interviewed nearly 20 years later, the researchers found more than a third had no memory of being molested. The USC study of 450 people who had been sexually abused showed that an even higher number—nearly 60%—had experienced some amnesia regarding the events.

When a person has been repeatedly traumatized, as some abuse victims have been, he is likelier to suffer from amnesia, says UCSF's Terr: "It seems odd and illogical, but if someone has access to hurt you again and again, you can anticipate what's going to happen and muster up very amazing defenses against it. The mind makes massive adjustments to something it expects to recur." Once these practiced defenses become a way of life, Terr adds, people can successfully keep traumatic knowledge from penetrating daily consciousness for long periods.

But what causes the past to surface suddenly? One leading theory is based on research showing that information people process in one state of mind—when agitated or drunk, for instance—can't be easily recalled in another state. To access the material again, one must trigger the mental state in which it was learned.

NIMH's Putnam believes some people who suffer trauma experience this so-called state-dependent learning: To buffer themselves from the horror of it, they enter a mental state that's dissociated, or walled off, from their ordinary consciousness. When they return to a normal state, memories of that trauma are largely inaccessible. The same dissociative process that helps victims endure repeated trauma apparently culminates in some cases in multiple personality disorder, in which two or more disconnected "personalities"—discrete psychological and physiological states—coexist in one person. "The information about what happened to the traumatized person is there," says Putnam, "but it's not retrievable unless he or she taps into the state—or personality—it was acquired in." This explanation for the sudden recollection of trauma helps account for the fact that many people recover memories of abuse after hearing someone describe a similar experience or during periods of unusual stress—both triggers of the traumatic state. Roseanne Arnold, for instance, says she didn't remember her own abuse until her husband, Tom, revealed to her that he'd once been sexually molested by a baby sitter.

Even researchers who accept the idea of recovered memory find that some claims stretch plausibility. First there's the "early age" factor. Not uncommonly, incest survivors say they recall abuse that began when they were six months to a year old. Yet scientists generally agree that people can't recall specific events that occurred before age two. Before that point, the hippocampus, a brain structure fundamental to memory formation, isn't fully developed. The full-fledged faculty doesn't begin to function cohesively until a child is about three or four. "Very young children aren't able to set a memory in the narrative framework that's needed to sustain it" says Dr. Ulric Neisser, a professor of psychology at Emory University in Atlanta. "You have to remember that for every person who makes a vivid assertion of what happened before their first birthday, there are people who make vivid assertions of what happened to them in a prior life, say, at Cleopatra's court."

Hundreds of people have recalled being victims of ritual or satanic abuse.

Recent research suggests that because memory isn't an objective mirror of reality, but a fluid process in which stored information is constantly reinterpreted in the light of the present, it's subject to distortion. In fact, it's easy to inject false memories into the mind. In one study, Loftus and her University of Washington colleagues asked five people to tell younger relatives a false story. The relatives were told that once, as children, they'd gotten lost on a shopping trip. Over the next few days, the researchers asked the grown children to recall more about the experience. In a typical case, one person not only "remembered" being lost and afraid, but also what a man who had returned him to his family had been wearing.

Our memories can also be distorted by popular books, magazine stories, movies and other social influences. In recent years, hundreds of people have recalled being victims of ritual or satanic abuse—sexual torture by covens of adults who also practice murder, mutilation and cannibalism—prompting the FBI to examine the issue in a 1992 report. While acknowledging that those who make the claims may indeed have suffered some sort of trauma or abuse, the report stated that after eight years of investigation by law enforcement officials, there was still "little or no corroborative evidence" for the inflated nature of the allegations. While no one knows exactly what's responsible for the recent raft of satanic memories, a kind of contagion process may be operating: "There's an enormous proliferation of occult themes out there," says Putnam, "appearing on shows from *Geraldo* to Saturday morning cartoons. On top of that you've got therapists going to symposia on ritual abuse and buying into what they hear. The therapists and patients are reinforcing each other in their belief systems."

Many psychologists believe the tidal wave of publicity about incest and sexual abuse sweeping the country today has created an atmosphere that encourages bizarre recollection. Celebrities and regular folk rush to tell their abuse stories in print and on the talk show circuit. In the past five years, a slew of popular self-help books has assured readers that even suspecting abuse is grounds for assuming it happened. "If you think you were abused and your life shows the symptoms, then you were," assert Ellen Bass and Laura Davis in *The Courage to Heal* (Harper Perennial, 1988, $20), which has sold 638,000 copies. (Bass says she'd tone down such emphatic language if she were writing this book today.) Meanwhile, a network of incest-survivor groups has sprung up around the country. One long-time member notes it was common in sessions she attended "for people to come in without clear memories, and to get them during meetings."

Experts tend to lay much of the blame for false memories at the doorstep of well-meaning but incompetent therapists. In their zeal to find a quick solution to psychic angst, these practitioners spur patients to dig for signs that they might have been abused. Some routinely use hypnosis or even one of the so-called truth serum drugs to help clients recover memories. Such techniques, critics note, have also helped people "remember" past lives, the experience of birth or being abducted by aliens.

"A lot of therapists now assume that an incredibly broad range of symptoms—eating disorders, sexual anxiety, alcoholism, depression—can be attributed to repressed sexual memory," says Dr. Janice Haaken, a professor of psychology at Portland State University who has studied the incest-recovery movement. "They ferret out what they believe to be repressed memories of sexual abuse to explain these symptoms. But it's a very naive view of therapy to think you can discover a moment in the past that unlocks everything—the magic key. I've become convinced that some recovered memories are really patients' unconscious gifts for therapists."

As recovered memories move out of the therapist's office and into the legal system, culling false accusations from real ones becomes increasingly urgent. In 1988, in response to lobbying by lawyers, therapists and victims, the state legislature of Washington extended its statute of limitations so that people who retrieve a memory of abuse can sue an alleged perpetrator for up to three years after the memory returns. To date, 21 other states have enacted similar laws. In at least two of these states, judgments or settlements of more than

6. OLD/NEW SEXUAL CONCERNS: Sexual Abuse and Violence

$1 million have been awarded to women bringing abuse claims after old memories resurfaced.

Abuse is not the only crime being tried on the basis of suddenly recovered memories: In a landmark 1991 California trial, Eileen Franklin-Lipsker, 30, put her father behind bars after testifying that she'd suddenly recalled seeing him molest and kill her best friend 20 years earlier. (Franklin-Lipsker also testified that she had been abused by her father as a child.) "I don't have any idea if [Franklin-Lipsker's] father did it or not," says Loftus, an expert witness for the defense in that trial. "In theory it's conceivable she could have had this memory. But I was stunned that all it took to convict someone of murder was one person's word, uncorroborated, about something that happened 20 years ago. It sets a dangerous precedent."

Many people who recover memories have no supporting evidence for events that may have occurred 20 to 30 years earlier. There are as yet few criteria doctors and lawyers can use to evaluate the validity of their claims, nor is there agreement on what such criteria should be. For example, some experts suggest being wary of any recollections that occur in therapy, in a support group, or after a person has read a self-help book on incest. Yet, as NIMH's Putnam points out, these same influences can help people retrieve real memories. Until they have developed the necessary research tools, psychiatrists will be hard pressed to tell genuine from ersatz memories. That also means, notes Loftus, "that right now there are no legitimate figures on how many accounts are based on real events and how many aren't."

True or not, there's little debate that a retrieved memory represents an individual's painful struggle with the past. Psychologists agree that most memories *feel* real to those who harbor them. In addition, "Even if the memory is false, the fact that someone could produce it suggests something wasn't quite right with the family relationship to begin with," says Dr. John Kihlstrom, a professor of psychology at the University of Arizona. "There are lots of ways to take a memory seriously without jumping on it as if it were unquestionably true. Maybe the real problem will go unexamined and untreated by focusing on this memory. Maybe you're a woman whose parents didn't let you take math in school, and so you're locked into a dead-end job, and that's why you are mad at them. It would be a tragedy to believe a problem was caused by incest when it was caused by something else."

On the basis of existing scientific evidence, it's difficult to generalize about the validity of recovered memories. No one, says Terr, should either dismiss out of hand or accept without question a recollection of long-ago trauma. "Some memories are true, some mostly true with distortion, some wholly false," she says. "It's silly that the whole thing has gotten to be so one-way-or-the-other. Really good people are arguing over something that's not a black-or-white situation. Each and every case must be looked at for itself."

The Sexually Abused Boy: Problems in Manhood

Sexual dysfunction, vague somatic complaints, substance abuse, and STDs secondary to extensive sexual activity may signal a hidden history of childhood sexual abuse.

DIANA M. ELLIOTT, PhD, AND JOHN BRIERE, PhD

Diana M. Elliott is with the Sexual Abuse Crisis Center, Harbor–UCLA Medical Center, Los Angeles, and John Briere is Assistant Professor, Department of Psychiatry and the Behavioral Sciences, University of Southern California School of Medicine, Los Angeles.

Sexual abuse is often viewed as a crime by adult males against female children. Over the last 10 years, however, this generalization has been challenged. Recent research suggests that one out of every four children in this country experiences some form of sexual victimization before age 17, and that 15% to 20% of these victims are boys.[1,2] Yet the implications of such abuse in the lives of male survivors have essentially gone unrecognized. This article discusses common clinical and psychological disorders experienced by many men who were sexually abused as children.

How Sexual Abuse In Boys And Girls May Differ

Research suggests that the victimization of boys and girls differs in the following ways:
- Boys are more frequently molested outside the family than are girls.[2]
- The onset of abuse tends to occur at a younger age for boys, and to end at an earlier age.[2–4]
- Physical force is used more frequently against male victims.[3]
- Since the majority of sexual abusers are male, boys are more frequently abused by someone of their own sex, while girls are typically abused by someone of the opposite sex.[5]
- Males tend to minimize their victimization or view it less negatively than do females, despite studies indicating equivalent levels of later psychological damage.[4,5]

> Conservative estimates suggest that 15% to 20% of all victims of sexual abuse are boys.

- One of the most striking gender differences is the male's greater reluctance to disclose the sexual abuse during childhood or adulthood,[6,7] which may make him, as a "hidden victim," susceptible to a number of somatic and emotional disorders years after the abuse.

Immediate Aftereffects

The victim's silence. In our culture, boys are taught at a very young age that masculinity is based, at least in part, on being strong and powerful. When threatened, they are taught to fight back and show no fear. Sexual abuse, on the other hand, is an act in which a smaller, weaker, and younger person is dominated by a bigger, stronger, older one. This inequity may instill a sense of powerlessness in victims, regardless of gender.[1] Unfortunately, while the boy may know that it would be dangerous to fight back, internalized stereotypes of masculinity can lead to self-perceptions of weakness based on his "failure" to successfully resist the perpetrator. Ultimately, the male abuse survivor may be less likely to disclose sexual abuse because he believes he will be confessing to weakness and diminished masculinity.[5]

Early psychological problems. As is true of female victims, sexually abused boys often come to the attention of caretakers because of their unusual or problematic behavior. Many young victims openly masturbate, for example, or become aggressive, excessively demanding, or withdrawn in social settings. Regressive behavior, such as enuresis, encopresis, or "baby talk," may occur, and the boy may complain of nightmares, sleep disturbances, and/or excessive fears. If the child is in school, difficulties in peer relationships, a decline in school performance, and "acting-out" behaviors are common.

Although sex differences in these areas are not always apparent, clinicians note that molested boys are more prone to externalize sexual and aggressive behavior than are abused girls. Similarly, as teenagers, male victims are more likely to report school problems, trouble with the police, substance abuse, fantasies of aggression, sexual preoccupation, excessive masturbation, and sexual aggression.[6,8]

Long-Term Sequelae

The damaging aftereffects of sexual abuse on adults have been well-documented.[1,4,7,9] In general, it appears that psychological distress levels are approximately equivalent in adult male and female survivors of childhood abuse, although there may be behavioral differences.[4,9-11]

Self-perception. Individuals who were sexually molested as children are, as a group, more likely to have negative self-perceptions than their nonabused peers. Irrespective of gender, abuse survivors tend to exhibit denial, confusion, low self-esteem, guilt, and shame regarding their abuse. Survivors may view themselves as, in a sense, defective. Eating disorders, negative body image, and self-consciousness regarding appearance are seen in both male and female victims.

As with female survivors, male sexual abuse survivors have a greater tendency toward self-destructive behaviors such as substance abuse, involvement in high-risk activities, self-mutilation, and suicidal behavior than do individuals with no history of abuse.[4,9-11]

Post-traumatic stress symptoms. Adult survivors of sexual abuse may display post-traumatic stress symptoms such as hypervigilance, nightmares, flashbacks, dissociation, and sleep disturbances. Mood disturbances are frequently reported and include symptoms of depression and anxiety.

It is not uncommon for sexual abuse survivors to be amnestic regarding the circumstances of their abuse. In fact, they may have no recollection of large periods of their childhood.

Physical complaints. Several studies have documented higher rates of diffuse physical complaints among male abuse survivors, including anxiety disorders, sleep and eating disturbances, gastrointestinal problems, and fatigue.

Sexual problems. Sexual dysfunction is very common in male sexual abuse survivors, and clinical experience indicates that concern about sexual matters is often the problem that brings these men into the physician's office. Compared with their nonabused peers, abuse survivors report much greater difficulty establishing sexual relationships, significantly more sexual problems, and less satisfaction in their current sexual relationships. For men, sexual dysfunction may take the form of compulsive masturbation, decreased sexual desire, erectile dysfunction, premature ejaculation, and/or anorgasmia.

Confusion about sexuality is very common among male sexual abuse survivors. Our society encourages men to be aggressive and dominant partners. However, as noted earlier, sexual victimization of males can carry with it a hidden implication of reduced manhood. Heterosexual men who were victimized by another man commonly fear that sexual molestation has made (or will make) them homosexual. The incidence of homosexuality does, in fact, appear to be higher among sexually abused men than among nonabused men, although the reason for this relationship is unclear.[12] Male victims who are homosexual may fear that their sexual preference caused the abuse or that the abuse caused their sexual preference. Because our society does not fully accept homosexuality as an alternative lifestyle, both homosexual and heterosexual survivors who fear homosexuality may experience confusion and shame in connection with their sexuality.[5]

Many men and women report that the abuse they experienced as children included sexual stimulation that was, to some extent, pleasurable. While not surprising from a biologic perspective, this physiologic response often creates feelings of confusion and guilt for victims. Men are more likely to view the overall experience as neutral or even pleasurable; women, in contrast, usually view the experience as negative.[5,7,10] In a society that does not allow men to be victims, the best alternative is to turn an abusive event into a pleasurable one.[5] Yet despite this more positive perception of the abuse, in adulthood male victims frequently exhibit the same problems associated with sexual abuse that are found among female survivors.[4,9]

It is a common clinical impression that male victims are more prone to "act out" their sexual abuse trauma in terms of violence toward others, whereas female victims may be more likely to internalize their trauma, making them more vulnerable to revictimization.[13] Indeed, a number of studies indicate a substantial overrepresentation of sexual abuse victims among male sex offenders.[14] Additionally, a recent study of sexual abuse survivors in treatment revealed that abused males were far more likely to report having sexually

> Sexual dysfunction may take the form of decreased sexual desire, premature ejaculation, or anorgasmia.

abused a child than were either female abuse survivors in the same study or males in the general population.[9] It should be noted, however, that the statistical relationship between molestation as a child and subsequent molestation of others is small: Most men who were sexually abused as children do not go on to sexually abuse others.

Difficulty in relationships. Partly because sexual molestation typically occurs in the context of a relationship, and partly because the survivor may view himself as "damaged," many victims of abuse report an inability to develop intimate, sustained, and meaningful adult relationships. Poor social skills and an inability to trust others perpetuate their sense of isolation. As a result, they are more likely to remain single and, if married, are more likely to divorce or separate from their partners than are nonabused individuals.[4,5,7,11,13]

Manifestations of interpersonal difficulties vary. Some male survivors express a greater need to be in control of interpersonal contacts, perhaps as a means of dealing with the extreme vulnerability they felt as children. They may fear the motives of others and remain hypervigilant in relationships. Some survivors live very isolated lives, preferring little intimate contact. Others, however, are more afraid of being alone. This can result in either multiple short-term relationships or longer relationships that recapitulate the abusive relationship with the childhood perpetrator. In some cases, this recapitulation may result in physically or sexually abusive relationships with spouse, partner, and/or children. Many abuse survivors, while aware of their interpersonal difficulties, are unable to overcome these patterns without professional intervention.[4,5,9,11]

Implications for the Physician

Given the high prevalence of sexual abuse in the general population, clinicians should be prepared to deal with abuse-related difficulties in their patients. Awareness of the potential impact of such abuse may affect both assessment and treatment of the survivor.

Assessment. During routine history-taking, include specific questions about childhood sexual experiences such as, "Before age 17, did anyone coax, bribe, or force you to have sexual contact?" or "Did you ever have sexual contact with someone five or more years your senior?"

Be alert to a history of diffuse somatic complaints with no associated physical findings, complaints of sexual dysfunction, wounds seemingly due to self-mutilation, suicidal behavior, substance addictions, and sexually transmitted diseases secondary to extensive sexual activity. All can point to a history of childhood abuse.

Address specific characteristics of the abuse. Current research suggests that adult symptoms are more severe when the sexual abuse (1) occurred at an especially early age; (2) involved the use of force or violence; (3) was committed by a parental figure; (4) continued for a long time; (5) included anal, oral, or vaginal penetration; or (6) involved concomitant physical abuse or emotional neglect.[1,5,7]

Treatment. Many male survivors who would not consider seeking formal psychotherapeutic assistance may respond to a clinician whose ostensible role is to provide medical care. The physician can help such men by pointing out to them the possible connections between their childhood trauma and their current problems, and encouraging them to seek professional assistance. If the male abuse survivor is willing to seek formal psychotherapy, the physician should refer him to a mental health professional specializing in sexual abuse (see box, "How to Locate a Specialist").

Conclusion

The long-term impact of early sexual abuse in men often goes unrecognized. By identifying abuse-related problems that may bring patients to their office, primary care physicians can play an important role in providing the education, support, referrals, and treatment that will help these men overcome their early trauma.

How to Locate a Specialist

Information regarding clinical specialists in the area of childhood sexual abuse can be obtained from:

American Professional Society on the Abuse of Children
332 South Michigan Avenue
Suite 1600
Chicago, IL 60604

Telephone: 312-554-0166

References

1. Finkelhor D: *A Sourcebook on Child Sexual Abuse.* Beverly Hills, Calif, Sage, 1986.
2. Finkelhor D: *Child Sexual Abuse: New Research and Theory.* New York, Free Press, 1984.
3. Pierce R, Pierce L: The sexually abused child: A comparison of male and female victims. *Child Abuse & Neglect* 9:191, 1985.
4. Briere J, Evans D, Runtz M, et al: Symptomatology in men who were molested as children: A comparison study. *American Journal of Orthopsychiatry* 58:457, 1987.
5. Lew M: *Victims No Longer: Men Recovering From Incest and Other Sexual Child Abuse.* New York, HarperCollins, 1990.
6. Vander Mey BJ: The sexual victimization of male children: A review of previous research. *Child Abuse & Neglect* 12:61, 1988.
7. Elliott DM: *The Effects of Childhood Sexual Experiences on Adult Functioning in a National Sample of Professional Women.* La Mirada, Calif, Rosemead School of Psychology, 1990, doctoral dissertation.
8. Pyle EA, Goodman GS: Sex differences in the initial impact of child sexual abuse. Read before the Annual Convention of the American Psychological Association, New York City, August, 1987.
9. Conte J, Briere J, Sexton D: Sex differences in the long-term effects of sexual abuse: The results of a national clinical survey. Read before the National Conference on Child Abuse and Neglect, Salt Lake City, October, 1989.
10. Fritz GS, Stoll K, Wagner NN: A comparison of males and females who were sexually molested as children. *Journal of Sex and Marital Therapy* 7:54, 1981.
11. Urquiza AJ, Crowley C: Sex differences in the survivors of childhood sexual abuse. Read before the National Conference on the Sexual Victimization of Children, New Orleans, May, 1986.
12. Simari CG, Baskin D: Incest experiences within homosexual populations: A preliminary study. *Arch Sex Behav* 11:329, 1982.
13. Briere J: *Therapy for Adults Molested as Children: Beyond Survival.* New York, Springer Pub Co, 1989.
14. Gebhard P, Gagnon J, Pomeroy W, et al: *Sex Offenders.* New York, HarperCollins, 1965.

Who's to Blame for SEXUAL VIOLENCE?

"Society shows greater tolerance for violence against women than for other crimes, especially when perpetrated by acquaintances, dates, lovers, or husbands. . . ."

Diana Scully

Dr. Scully, associate professor of sociology and coordinator of women's studies, Virginia Commonwealth University, Richmond, is the author of Understanding Sexual Violence: A Study of Convicted Rapists.

SHORTLY after his election to the presidency, George Bush gave a speech to an assembly of the American Association of University Women during which he strongly condemned violence against women. At a news conference the next day, referring to the AAUW speech, a female reporter asked the President if he intended to give such a message to gatherings of men also. Women, she correctly observed, already know about violence against them; we don't need your message, Mr. President, talk to the men.

This incident reflects a critical public policy issue—that rape and other forms of sexual violence widely are regarded as women's problem. Despite the fact that men are the perpetrators of the overwhelming majority of sexual violence, it is women who continue to be perceived as responsible, either by causing rape to happen or by failing to avoid it, and it is women who are supposed to solve "their" problem. Instead, the central issue should be why men don't define sexual violence as *their* problem.

Part of the answer has to do with common beliefs about the origins of sexual violence. The disease model of rape, in particular, has been a powerful force in maintaining the *status quo* by promulgating the view that sexual violence is a psychopathologically isolated, idiosyncratic act limited to a few "sick" men. From this perspective, the source of sexual violence is an aggregation of individual problems. When sexually violent behavior is presumed to be confined to a few sick men, drugs, surgery, shock therapy, psychotherapy, or prison are the "cures" prescribed to solve this complex social problem. The consequence of defining responsibility this way is that men collectively never have to confront sexual violence as *their* problem.

This type of thinking was evident in the reporting of the tragic 1989 massacre at the engineering school of the University of Montreal. A young man, armed with a semi-automatic assault rifle, murdered 14 women while shouting, "You are all a bunch of feminists. I hate feminists." Articles in the mass media attributed his actions to a disturbed psyche and the beatings his mother, sister, and he had received from his father. The politics of these murders—that this event, though extreme, is part of the general pattern of daily sexual violence and the liability that women collectively face—was ignored for the most part.

Despite its popularity, there is little empirical support for the disease model of rape. Twenty years of psychological research have failed to find a consistent pattern of personality type or character disorder that reliably discriminates rapists from other groups of men. Various research has found that fewer than five percent of convicted rapists were psychotic at the time of their crime, thus leaving 95% of known rapists unexplained, not to mention sexually violent men who successfully have avoided conviction and confinement.

Other evidence indicates that rape is not a behavior confined to a few sick men. Instead, current research confirms that sexual aggression/violence instead should be thought of as a continuum of behaviors ranging from verbal abuse in the street and harassment in the workplace to wife battering, rape, and murder. Similarly, men can be thought of as varying along a continuum of sexual aggression with some more likely than others to commit sexually aggressive acts against women. Striking evidence for this proposition is found in a growing body of research that indicates that many men in this society are capable of varying degrees of sexual aggression and violence.

Studies on college men have found that as many as 30-40% admit to having engaged in sexually aggressive behavior, some of it legally rape, in dating situations. A similar number indicate some likelihood that they themselves would rape if they could be assured of not being caught. These figures are consistent with research on college women and their reports of experiencing sexual aggression in dating situations. If this level of sexual aggression exists in the college population, it is unlikely the same age, non-college population is less sexually violent.

Some anthropologists observe that, in pre-industrial societies, there appears to be substantial variation in the frequency of abusive treatment of women from one culture to another. Rape and other forms of sexual violence toward women are found as regular features of violent patriarchal societies organized around the social, political, economic, and sexual subordination and devaluation of women.

Among contemporary societies, the frequency of rape also varies, and the U.S. is the most rape-prone of all. In 1988, for example, the rate of reported rape in the U.S. was 20 times higher than in England and Wales. According to figures presented by Sen. Joseph R. Biden, Jr. (D.-Del.), chairman of the Senate Judiciary Committee investigating violence against women, the rate of assaults against young women has jumped 48% since 1974 while declining 12% for young men. Over all, rapes have increased four times faster than the total crime rate. How can this amount of crime against women be the result of "a few sick men"?

Thus, the problem with the disease model of rape is that, in addition to being empirically true of only a very small number of sexually violent men, it attempts to understand a complex social problem by focusing on individual offenders, ignores the role of culture and social structure as predisposing factors, and shifts the responsibility for rape from men to women.

Why do men commit sexual violence?

My research on convicted rapists was grounded in a feminist socio-cultural model and the assumption that all behavior, including sexual aggression, is learned. Instead of examining the case histories of sexually violent men for evidence of pathology or individual motives, I used convicted rapists collectively as experts on a sexually violent culture. Rather than assuming that rape is a disease, I asked instead what goals men have learned to achieve through sexually violent means.

During a two-year period, my research associate, Joseph Marolla, and I conducted individual, private, face-to-face interviews with a volunteer sample of 114 convicted rapists and a contrast group of 75 other felons in seven Virginia prisons. All of the former were serving sentences for the rape or attempted rape of an adult woman and most had been convicted of more than one crime. Twelve percent had been convicted of more than one rape or attempted rape; 39% also had convictions for burglary or robbery, 29% for abduction, 25% for sodomy, and 11% for first- or second-degree murder. Their sentences for rape and accompanying crimes ranged from 10 years to multiple life sentences. The vast majority were young, between the ages of 18 and 35 years, when they were interviewed; 46% were white and 54% were black.

The interviews consisted of a complete background history, including childhood, family, religious, marital, educational, employment, sexual, psychiatric, and criminal; a series of scales measuring attitudes toward women, masculinity, interpersonal violence, and rape; and, for the rapists, 40 pages of open-ended questions about the rape and the victim. The project involved approximately 700 hours of interviews and produced some 15,000 pages of data.

There is much to be learned about what men gain from sexual violence by talking to those who rape. The answers rapists gave to questions about their acts revealed a number of ways they had used sexual violence to achieve goals. A number employed rape as a means of revenge and punishment. Implicit in revenge-rapes was the collective liability of women. In some cases, victims were substitutes for significant women on whom the men desired to take revenge. In others, victims represented all women, and rape was used to punish, humiliate, and "put them in their place." In either case, victims were seen as objects, not individuals.

For some men, rape was an afterthought or a bonus they added to a burglary or robbery. From their perspective, the rape was "no big deal," just another part of the routine. Others used sexual violence to gain access to unavailable or unwilling women—a tactic when an acquaintance, date, lover, or wife says no. Some raped in groups and enjoyed the act as a male bonding activity. For them, it was a form of recreation, just something else to do. For some of the men, rape was a fantasy come true, a particularly exciting form of impersonal sex which they enjoyed because it enabled them to dominate and control women by exercising a singularly male form of power. These men also talked of the pleasures of raping—how it is a challenge, an adventure, a dangerous and "ultimate" experience. Rape made them feel good and, in some cases, elevated their self-image. One stated, "After rape, I always felt like I had just conquered something. . . . "

Thus, convicted rapists tell us that men rape because, in this society, sexual violence is useful and rewarding behavior. Significantly, almost no one thought he would go to prison for it. In fact, they perceived rape as a rewarding, low-risk act. As one man said, "I knew what I was doing. I just said, the hell with the consequences. I told myself what I was going to do was rape . . . but I didn't think I would go to prison. I thought I had gotten away with it."

In addition to asking what men gain from rape, it is necessary to consider how it is made possible in sexually violent cultures. An analysis of the way men who rape construct reality reveals the culturally derived excuses and justifications that sustain, if not promote, sexual violence.

Two types of sexually violent men can be distinguished. One type, admitters, confessed to having had sexual contact with the victim and defined what they did as rape. They expressed the belief that rape is reprehensible and they understood women's fear of sexual violence. They explained themselves and their actions by appealing to forces beyond their control, forces they said that reduced their capacity to act rationally and thus compelled them to rape. Two types of excuses predominated—minor emotional problems and drunkenness or temporary loss of inhibition. Admitters used these excuses to view their sexual violence as idiosyncratic, rather than typical, behavior. This allowed them to conceptualize themselves as nice guys who had made a mistake, but were not rapists.

When they were raping, however, admitters were far from nice guys. Aware of the generalized image women have of rapists as violent and subhuman creatures, they used this to their advantage, terrifying the victim and rendering her subdued and compliant. Admitters also were aware of the emotional impact of rape on women and they took satisfaction in the belief that their victim felt powerless, humiliated, and degraded—the way they wanted her to feel. These are men who know what they are doing when they rape, use their perceptiveness to enhance the satisfaction they experience from sexual violence, and later use excuses to remove the blame.

In contrast, deniers, the other type of sexually violent men, rape because their value system provides no compelling reason not to. These are men who admitted to sexual contact with the victim, in some cases admitted to the use of a weapon and the infliction of serious physical injury in addition to the rape, but did not define what they did as rape. Instead, deniers used justifications to argue that their behavior, even if not quite right, was situationally appropriate. Their denials were drawn from common cultural rape stereotypes and took two forms, both of which ultimately denied the existence of a victim.

The first form of denial was buttressed by the cultural view of men as sexually masterful and women as coy, but seductive. Injury was denied by portraying the victim as willing, even enthusiastic, or as politely resistant at first, but eventually yielding to "relax and enjoy it." Force was made to appear as merely a technique of seduction. Rape was disclaimed because, rather than harm a woman, they had made her dreams come true.

In the second form of denial, the victim was portrayed as the type of woman who "got what she deserved." Through attacks on the woman's sexual reputation and emotional state, these men argued that, since the victim wasn't "a nice girl," they were not rapists. Consistent with both forms of denial was the use of alcohol and drugs as an explanation. In contrast to admitters, who accentuated such use as an excuse for their behavior, deniers emphasized the victim's consumption in an effort both to discredit her and make her appear more responsible while making their own action appear justified.

Deniers are characterized by a relative absence of awareness. They didn't know or didn't care how their victim perceived them, and they were either unaware of their victim's feelings or, consistent with cultural stereotypes, they assumed that, once the rape began, she relaxed and enjoyed it. Deniers, then, represent the type of man who is so imperceptive he is incapable of understanding the meaning of sexual violence to women.

Admitters and deniers present a contrast in the type of man who rapes. Yet, despite the differences, they were the same in one most essential way. The majority did not

6. OLD/NEW SEXUAL CONCERNS: Sexual Abuse and Violence

experience guilt or shame as a result of raping nor did they report feeling any emotions for their victims during or following the rape. A typical comment was, "I felt like I had got what I wanted and had to get on with my business. She was of no more concern." Another stated, "It just blew past. I played some basketball and then went to my girl's house and had sex with her. I wasn't worried or sorry."

Instead of the kind of social control emotions that might have constrained their sexually violent behavior, these men indicate that rape caused them to feel nothing or to feel good. To understand the absence of emotions, it is necessary to examine the perspective of these men toward women in general and toward their victim in particular.

Rapists were characterized by an intensely rigid double standard of moral and sexual conduct—a standard which both denies women the rights that are accorded to men and requires them to have the protection of a man. These "pedestal" values, far from reflecting positive feelings for women, as myth would have us believe, speak of rigid intolerance and were associated with other very hostile and violent attitudes towards women. Rapists identified with traditional images of masculinity and male gender role privilege, believed very strongly in rape stereotypes, and, for them, being male carried the right to discipline and punish women.

These men were able to rape because their victims had no real or symbolic value outside of the role they forced them to perform. The satisfaction they derived from sexual violence reveals the extent to which they have learned to objectify women. Women are jokes, objects, targets, sexual commodities, and pieces of property to be used or conquered, not human beings with rights or feelings. Hierarchial gender relations and the corresponding mores that devalue women and diminish them to exploitable objects or property are the factors that empower men to rape.

If sexual violence in all its manifestations is the consequence of patriarchal social structure, it is clear that the imprisonment and/or treatment of individual offenders is not the solution to eliminating the vast amount of violence directed at women which, despite these measures, continues to show an unprecedented increase. Only profound change at both the individual and collective levels of society is capable of eroding the rape supportive elements of our culture. To accomplish this, public policy must confront the central problem, not just treat the symptoms. To date, no legislation, including that recently proposed by Sen. Biden—although a promising first step—has done this.

Clearly, the conditions that support sexual violence must be changed, and rape and other forms of woman abuse have to be made unacceptable and less rewarding for men. To do so, the factors that abridge women's rights as full human beings and the barriers that prevent women from acting in accordance with and achieving full parity with men must be removed. Any policy that improves the collective situation of women by generating greater equality and independence, such as the Equal Rights Amendment and/or comparable worth, civil rights, national child care, or abortion rights legislation, should have a long-term effect on diminishing sexual violence.

Changing attitudes

Structural change must be accompanied by efforts to alter dangerous and destructive attitudes that support sexual violence and the abuse of women. The attitudinal factors that research, including mine, has shown to encourage sexual violence include: traditional attitudes towards women such as the belief that they shouldn't have the same rights and privileges as men; belief in rape myths—i.e., the idea that women enjoy and are responsible for rape; and acceptance of interpersonal violence as a method of disciplining women and handling conflict. Changing these attitudes requires programs that educate children, starting at an early age, and resocialize adults.

Successful education would help to eliminate the excuses and justifications used to trivialize and neutralize sexual violence toward all women, including family members, and that allow men to avoid responsibility. The origins of violence in the family are located in the structural subordination of women. Thus, this violence is the result, not the cause, of the larger social problem. Society shows greater tolerance for sexual violence against women than for other crimes, especially when perpetrated by acquaintances, dates, lovers, or husbands because of lingering attitudes about women as men's property and because these crimes are accepted as different and idiosyncratic, rather than a regular feature of a male dominant society. In many jurisdictions, for example, assault, even murder, of a wife is treated as a less serious crime than similar acts directed towards a stranger.

All crimes against women, including assaults by husbands, lovers, friends, bosses, and acquaintances, must be treated with the same degree of severity as those against men and against strangers. When they are taken seriously, women will feel safer in reporting all those crimes against them that now go unnoticed. Alcohol and drugs are not the cause of sexual violence. They are an excuse to divert attention away from the real issue—the way women are perceived and treated in this society. Some of the money currently going into drug and alcohol programs could be used to provide resources for women to extricate themselves and their children from situations of dependence on abusive men.

Finally, men must take responsibility for sexual violence and work collectively to change those aspects of male culture that support the abuse of women. Legislation to ban firearms would be a welcome preliminary step and also would result in a reduction of most types of crime, including domestic violence. Even though it is threatening, taking responsibility means accepting that the majority of sexually violent men are otherwise normal and that, for some men, rape is sex; in fact, sex is rape.

A recent popular talk show featured as its theme "men who have been harmed by rape." All of the guests were fathers, husbands, or brothers of women who had been raped, and they talked passionately about the anguish it had caused them, how it had changed their lives. Predictably, they also expressed anger and the desire for revenge, and they stated the need to protect their daughters, wives, and sisters.

I was struck by the similarity between the reaction of these men, sincere as they were, and the response of convicted rapists when I asked them how they would feel and what they would do if their significant woman was raped. The overwhelming reaction among rapists was an expression of anger and violence. Additionally, the majority of these men said they would find a way to get personal revenge. Many of them simply answered, "I would find the guy and kill him."

This type of male reaction to rape appears common and is part of the problem. Since men who rape also abuse, rape, and are violent toward their own significant women, this reaction hardly expresses care or the desire to protect "their" woman for her own sake. It is a reflection of traditional attitudes and values that condone violence and mandate revenge when a man's property is violated by another man because he, not the woman, is the offended party. In the final analysis, women's interests are not served by individual efforts to protect some of them, a tactic which also increases female dependence, or by individual acts of revenge, but only by efforts to eliminate sexual violence toward all women.

Women have not been passive recipients of sexual violence. We have organized, protested, fought, counseled, cried together, marched, debunked myths, tried to "take back the night," challenged authority, learned self defense, operated hot lines, changed laws, raised money, built shelters, written voluminously—everything but commit acts of violence. Yet, I believe that no fundamental change will occur until men are forced to admit that sexual violence is *their* problem.

Sex Differences in the Brain

Cognitive variations between the sexes reflect differing hormonal influences on brain development. Understanding these differences and their causes can yield insights into brain organization

Doreen Kimura

Doreen Kimura studies the neural and hormonal basis of human intellectual function. She is professor of psychology and honorary lecturer in the department of clinical neurological sciences at the University of Western Ontario in London. Kimura, a fellow of the Royal Society of Canada, received the 1992 John Dewan Award for outstanding research from the Ontario Mental Health Foundation. She recently finished a book on neuromotor mechanisms in communication.

Women and men differ not only in physical attributes and reproductive function but also in the way in which they solve intellectual problems. It has been fashionable to insist that these differences are minimal, the consequence of variations in experience during development. The bulk of the evidence suggests, however, that the effects of sex hormones on brain organization occur so early in life that from the start the environment is acting on differently wired brains in girls and boys. Such differences make it almost impossible to evaluate the effects of experience independent of physiological predisposition.

Behavioral, neurological and endocrinologic studies have elucidated the processes giving rise to sex differences in the brain. As a result, aspects of the physiological basis for these variations have in recent years become clearer. In addition, studies of the effects of hormones on brain function throughout life suggest that the evolutionary pressures directing differences nevertheless allow for a degree of flexibility in cognitive ability between the sexes.

Major sex differences in intellectual function seem to lie in patterns of ability rather than in overall level of intelligence (IQ). We are all aware that people have different intellectual strengths. Some are especially good with words, others at using objects—for instance, at constructing or fixing things. In the same fashion, two individuals may have the same overall intelligence but have varying patterns of ability.

Men, on average, perform better than women on certain spatial tasks. In particular, men have an advantage in tests that require the subject to imagine rotating an object or manipulating it in some other way. They outperform women in mathematical reasoning tests and in navigating their way through a route. Further, men are more accurate in tests of target-directed motor skills—that is, in guiding or intercepting projectiles.

Women tend to be better than men at rapidly identifying matching items, a skill called perceptual speed. They have greater verbal fluency, including the ability to find words that begin with a specific letter or fulfill some other constraint. Women also outperform men in arithmetic calculation and in recalling landmarks from a route. Moreover, women are faster at certain precision manual tasks, such as placing pegs in designated holes on a board.

Although some investigators have reported that sex differences in problem solving do not appear until after puberty, Diane Lunn, working in my laboratory at the University of Western Ontario, and I have found three-year-old boys to be better at targeting than girls of the same age. Moreover, Neil V. Watson, when in my laboratory, showed that the extent of experience playing sports does not account for the sex difference in targeting found in young adults. Kimberly A. Kerns, working with Sheri A. Berenbaum of the University of Chicago, has found that sex differences in spatial rotation performance are present before puberty.

Differences in route learning have been systematically studied in adults in laboratory situations. For instance, Liisa Galea in my department studied undergraduates who followed a route on a tabletop map. Men learned the route in fewer trials and made fewer errors than did women. But once learning was complete, women remembered more of the landmarks than did men. These results, and those of other researchers, raise the possibility that women tend to use landmarks as a strategy to orient themselves in everyday life. The prevailing strategies used by males have not yet been clearly established, although they must relate to spatial ability.

Marion Eals and Irwin Silverman of York University studied another function that may be related to landmark memory. The researchers tested the ability of individuals to recall objects and their locations within a confined space—such as in a room or on a tabletop. Women were better able to remember whether an item had been displaced or not. In addition, in my laboratory, we measured the accuracy of object location: subjects were shown an array of objects and were later asked to replace them in their exact positions. Women did so more accurately than did men.

Problem-Solving Tasks Favoring Women

Women tend to perform better than men on tests of perceptual speed, in which subjects must rapidly identify matching items—for example, pairing the house on the far left with its twin:

In addition, women remember whether an object, or a series of objects, has been displaced:

On some tests of ideational fluency, for example, those in which subjects must list objects that are the same color, and on tests of verbal fluency, in which participants must list words that begin with the same letter, women also outperform men:

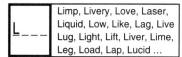

Women do better on precision manual tasks—that is, those involving fine-motor coordination—such as placing the pegs in holes on a board:

And women do better than men on mathematical calculation tests:

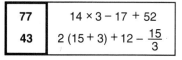

It is important to place the differences described above in context: some are slight, some are quite large. Because men and women overlap enormously on many cognitive tests that show average sex differences, researchers use variations within each group as a tool to gauge the differences between groups. Imagine, for instance, that on one test the average score is 105 for women and 100 for men. If the scores for women ranged from 100 to 110 and for men from 95 to 105, the difference would be more impressive than if the women's scores ranged from 50 to 150 and the men's from 45 to 145. In the latter case, the overlap in scores would be much greater.

One measure of the variation of scores within a group is the standard deviation. To compare the magnitude of a sex difference across several distinct tasks, the difference between groups is divided by the standard deviation. The resulting number is called the effect size. Effect sizes below 0.5 are generally considered small. Based on my data, for instance, there are typically no differences between the sexes on tests of vocabulary (effect size 0.02), nonverbal reasoning (0.03) and verbal reasoning (0.17).

On tests in which subjects match pictures, find words that begin with similar letters or show ideational fluency—such as naming objects that are white or red—the effect sizes are somewhat larger: 0.25, 0.22 and 0.38, respectively. As discussed above, women tend to outperform men on these tasks. Researchers have reported the largest effect sizes for certain tests measuring spatial rotation (effect size 0.7) and targeting accuracy (0.75). The large effect size in these tests means there are many more men at the high end of the score distribution.

Since, with the exception of the sex chromosomes, men and women share genetic material, how do such differences come about? Differing patterns of ability between men and women most probably reflect different hormonal influences on their developing brains. Early in life the action of estrogens and androgens (male hormones chief of which is testosterone) establishes sexual differentiation. In mammals, including humans, the organism has the potential to be male or female. If a Y chromosome is present, testes or male gonads form. This development is the critical first step toward becoming a male. If the gonads do not produce male hormones or if for some reason the hormones cannot act on the tissue, the default form of the organism is female.

Once testes are formed, they produce two substances that bring about the development of a male. Testosterone causes masculinization by promoting the male, or Wolffian, set of ducts and, indirectly through conversion to dihydrotestosterone, the external appearance of scrotum and penis. The Müllerian regression factor causes the female, or Müllerian, set of ducts to regress. If anything goes wrong at any stage of the process, the individual may be incompletely masculinized.

Not only do sex hormones achieve the transformation of the genitals into male organs, but they also organize corresponding male behaviors early in life. Since we cannot manipulate the hormonal environment in humans, we owe much of what we know about the details of behavioral determination to studies in other animals. Again, the intrinsic tendency, according to studies by Robert W. Goy of the University of Wisconsin, is to develop the female pattern that occurs in the absence of masculinizing hormonal influence.

If a rodent with functional male genitals is deprived of androgens immediately after birth (either by castration or by the administration of a compound that blocks androgens), male sexual behavior, such as mounting, will be reduced. Instead female sexual behavior, such as lordosis (arching of the back), will be enhanced in adulthood. Similarly, if androgens are administered to a female directly after birth, she displays more male sexual behavior and less female behavior in adulthood.

Bruce S. McEwen and his co-workers at the Rockefeller University have shown that, in the rat, the two processes of defeminization and masculinization require somewhat different biochemical changes. These events also occur at somewhat different times. Testosterone can be converted to either estrogen (usually considered a female hormone) or dihydrotestosterone. Defeminization takes place primarily after birth in rats and is mediated by estrogen, whereas masculinization involves both dihydrotestosterone and estrogen and occurs for the most part before birth rather than after, according to studies by McEwen. A substance called alpha-fetoprotein may protect female brains from the masculinizing effects of their estrogen.

The area in the brain that organizes female and male reproductive behavior is the hypothalamus. This tiny structure at the base of the brain connects to the pituitary, the master endocrine gland. Roger A. Gorski and his colleagues at the University of California at Los Angeles have shown that a region of the pre-optic area of the hypothalamus is visibly larger in male rats than in females. The size increment in males is promoted by the presence of androgens in the immediate postnatal, and to some extent prenatal, period. Laura S. Allen in Gorski's laboratory has found a similar sex difference in the human brain.

Other preliminary but intriguing studies suggest that sexual behavior may reflect further anatomic differences. In 1991 Simon LeVay of the Salk Institute for Biological Studies in San Diego reported that one of the brain regions that is usually larger in human males than in females—an interstitial nucleus of the anterior hypothalamus—is smaller in homosexual than in heterosexual men. LeVay points out that this finding supports suggestions that sexual preference has a biological substrate.

Homosexual and heterosexual men may also perform differently on cognitive tests. Brian A. Gladue of North Dakota State University and Geoff D. Sanders of City of London Polytechnic report that homosexual men perform less well on several spatial tasks than do heterosexual men. In a recent study in my laboratory, Jeff Hall found that homosexual men had lower scores on targeting tasks than did heterosexual men; however, they were superior in ideational fluency—listing things that were a particular color.

This exciting field of research is just starting, and it is crucial that investigators consider the degree to which differences in life-style contribute to group differences. One should also keep in mind that results concerning group differences constitute a general statistical statement; they establish a mean from which any individual may differ. Such studies are potentially a rich source of information on the physiological basis for cognitive patterns.

The lifelong effects of early exposure to sex hormones are characterized as organizational, because they appear to alter brain function permanently during a critical period. Administering the same hormones at later stages has no such effect. The hormonal effects are not limited to sexual or reproductive behaviors: they appear to extend to all known behaviors in which males and females differ. They seem to govern problem solving, aggression and the tendency to engage in rough-and-tumble play—the boisterous body contact that young males of some mammalian species display. For example, Michael J. Meaney of McGill University finds that dihydrotestosterone, working through a structure called the amygdala rather than through the hypothalamus, gives rise to the play-fighting behavior of juvenile male rodents.

Male and female rats have also been found to solve problems differently. Christina L. Williams of Barnard College has shown that female rats have a greater tendency to use landmarks in spatial learning tasks—as it appears women do. In Williams's experiment, female rats used landmark cues, such as pictures on the wall, in preference to geometric cues, such as angles and the shape of the room. If no landmarks were available, however, females used geometric cues. In contrast, males did not use landmarks at all, preferring geometric cues almost exclusively.

Interestingly, hormonal manipulation during the critical period can alter these behaviors. Depriving newborn males of testosterone by castrating them or administering estrogen to newborn females results in a complete reversal of sex-typed behaviors in the adult animals. (As mentioned above, estrogen can have a masculinizing effect during brain development.) Treated females behave like males, and treated males behave like females.

Natural selection for reproductive advantage could account for the evolution of such navigational differences. Steven J. C. Gaulin and Randall W. FitzGerald of the University of Pittsburgh have suggested that in species of voles in which a male mates with several females rather than with just one, the range he must traverse is greater. Therefore, navigational ability seems critical to reproductive success. Indeed, Gaulin and FitzGerald found sex differences in laboratory maze learning only in voles that were polygynous, such as the meadow vole, not in monogamous species, such as the prairie vole.

Again, behavioral differences may parallel structural ones. Lucia F. Jacobs in Gaulin's laboratory has discovered that the hippocampus—a region thought to be involved in spatial learning in both birds and mammals—is larger in male polygynous voles than in females. At present, there are no data on possible sex differences in hippocampal size in human subjects.

Evidence of the influence of sex hormones on adult behavior is less direct in humans than in other animals. Researchers are instead guided by what may be parallels in other species and by spontaneously occurring exceptions to the norm in humans.

One of the most compelling areas of evidence comes from studies of girls exposed to excess androgens in the prenatal or neonatal stage. The production of abnormally large quantities of adrenal androgens can occur because of a genetic defect called congenital adrenal hyperplasia (CAH). Before the 1970s, a similar condition also unexpectedly appeared when pregnant women took various synthetic steroids. Although the consequent masculinization of the geni-

Problem-Solving Tasks Favoring Men

Men tend to perform better than women on certain spatial tasks. They do well on tests that involve mentally rotating an object or manipulating it in some fashion, such as imagining turning this three-dimensional object

or determining where the holes punched in a folded piece of paper will fall when the paper is unfolded:

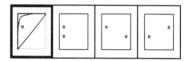

Men also are more accurate than women in target-directed motor skills, such as guiding or intercepting projectiles:

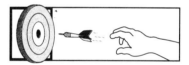

They do better on disembedding tests, in which they have to find a simple shape, such as the one on the left, once it is hidden within a more complex figure:

And men tend to do better than women on tests of mathematical reasoning:

| 1,100 | If only 60 percent of seedlings will survive, how many must be planted to obtain 660 trees? |

tals can be corrected early in life and drug therapy can stop the overproduction of androgens, effects of prenatal exposure on the brain cannot be reversed.

Studies by researchers such as Anke A. Ehrhardt of Columbia University and June M. Reinisch of the Kinsey Institute have found that girls with excess expo-

6. OLD/NEW SEXUAL CONCERNS: Males/Females—War and Peace

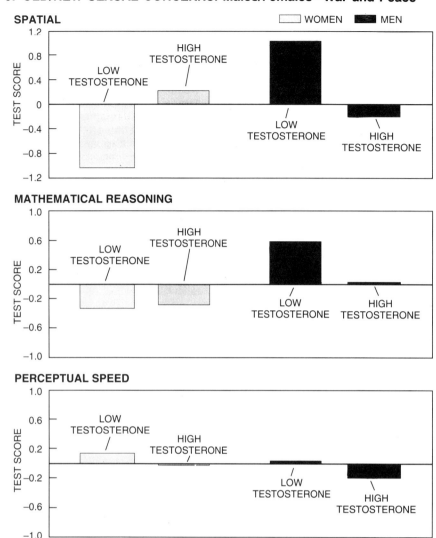

TESTOSTERONE LEVELS can affect performance on some tests (see boxes on pages 214 and 215 for examples of tests). Women with high levels of testosterone perform better on a spatial task (*top*) than do women with low levels; men with low levels outperform men with high levels. On a mathematical reasoning test (*middle*), low testosterone corresponds to better performance in men; in women there is no such relation. On a test in which women usually excel (*bottom*), no relation is found between testosterone and performance.

sure to androgens grow up to be more tomboyish and aggressive than their unaffected sisters. This conclusion was based sometimes on interviews with subjects and mothers, on teachers' ratings and on questionnaires administered to the girls themselves. When ratings are used in such studies, it can be difficult to rule out the influence of expectation either on the part of an adult who knows the girls' history or on the part of the girls themselves.

Therefore, the objective observations of Berenbaum are important and convincing. She and Melissa Hines of the University of California at Los Angeles observed the play behavior of CAH-affected girls and compared it with that of their male and female siblings. Given a choice of transportation and construction toys, dolls and kitchen supplies or books and board games, the CAH girls preferred the more typically masculine toys—for example, they played with cars for the same amount of time that normal boys did. Both the CAH girls and the boys differed from unaffected girls in their patterns of choice. Because there is every reason to think that parents would be at least as likely to encourage feminine preferences in their CAH daughters as in their unaffected daughters, these findings suggest that the toy preferences were actually altered in some way by the early hormonal environment.

Spatial abilities that are typically better in males are also enhanced in CAH girls. Susan M. Resnick, now at the National Institute on Aging, and Berenbaum and their colleagues reported that affected girls were superior to their unaffected sisters in a spatial manipulation test, two spatial rotation tests and a disembedding test—that is, the discovery of a simple figure hidden within a more complex one. All these tasks are usually done better by males. No differences existed between the two groups on other perceptual or verbal tasks or on a reasoning task.

Studies such as these suggest that the higher the androgen levels, the better the spatial performance. But this does not seem to be the case. In 1983 Valerie J. Shute, when at the University of California at Santa Barbara, suggested that the relation between levels of androgens and some spatial capabilities might be nonlinear. In other words, spatial ability might not increase as the amount of androgen increases. Shute measured androgens in blood taken from male and female students and divided each into high- and low-androgen groups. All fell within the normal range for each sex (androgens are present in females but in very low levels). She found that in women, the high-androgen subjects were better at the spatial tests. In men the reverse was true: low-androgen men performed better.

Catherine Gouchie and I recently conducted a study along similar lines by measuring testosterone in saliva. We added tests for two other kinds of abilities: mathematical reasoning and perceptual speed. Our results on the spatial tests were very similar to Shute's: low-testosterone men were superior to high-testosterone men, but high-testosterone women surpassed low-testosterone women. Such findings suggest some optimum level of androgen for maximal spatial ability. This level may fall in the low male range.

No correlation was found between testosterone levels and performance on perceptual speed tests. On mathematical reasoning, however, the results were similar to those of spatial ability tests for men: low-androgen men tested higher, but there was no obvious relation in women.

Such findings are consistent with the suggestion by Camilla P. Benbow of Iowa State University that high mathematical ability has a significant biological determinant. Benbow and her colleagues have reported consistent sex differences in mathematical reasoning ability favoring males. These differences are especially sharp at the upper end of the distribution, where males outnumber females 13 to one. Benbow argues that these differences are not readily explained by socialization.

It is important to keep in mind that the relation between natural hormonal levels and problem solving is based on correlational data. Some form of connection between the two measures exists, but how this association is determined or what its causal basis may be is unknown. Little is currently understood about the relation between adult levels of hormones and those in early life, when abilities appear to be organized in the nervous system. We have a lot to learn about the precise mechanisms underlying cognitive patterns in people.

Another approach to probing differences between male and female brains is to examine and compare the functions of particular brain systems. One noninvasive way to accomplish this goal is to study people who have experienced damage to a specific brain region. Such studies indicate that the left half of the brain in most people is critical for speech, the right for certain perceptual and spatial functions.

It is widely assumed by many researchers studying sex differences that the two hemispheres are more asymmetrically organized for speech and spatial functions in men than in women. This idea comes from several sources. Parts of the corpus callosum, a major neural system connecting the two hemispheres, may be more extensive in women; perceptual techniques that probe brain asymmetry in normal-functioning people sometimes show smaller asymmetries in women than in men, and damage to one brain hemisphere sometimes has a lesser effect in women than the comparable injury has in men.

In 1982 Marie-Christine de Lacoste, now at the Yale University School of Medicine, and Ralph L. Holloway of Columbia University reported that the back part of the corpus callosum, an area called the splenium, was larger in women than in men. This finding has subsequently been both refuted and confirmed. Variations in the shape of the corpus callosum that may occur as an individual ages as well as different methods of measurement may produce some of the disagreements. Most recently, Allen and Gorski found the same sex-related size difference in the splenium.

The interest in the corpus callosum arises from the assumption that its size may indicate the number of fibers connecting the two hemispheres. If more connecting fibers existed in one sex, the implication would be that in that sex the hemispheres communicate more fully. Although sex hormones can alter callosal size in rats, as Victor H. Denenberg and his associates at the University of Connecticut have demonstrated, it is unclear whether the actual number of fibers differs between the sexes. Moreover, sex differences in cognitive function have yet to be related to a difference in callosal size. New ways of imaging the brain in living humans will undoubtedly increase knowledge in this respect.

The view that a male brain is functionally more asymmetric than a female brain is long-standing. Albert M. Galaburda of Beth Israel Hospital in Boston and the late Norman Geschwind of Har-

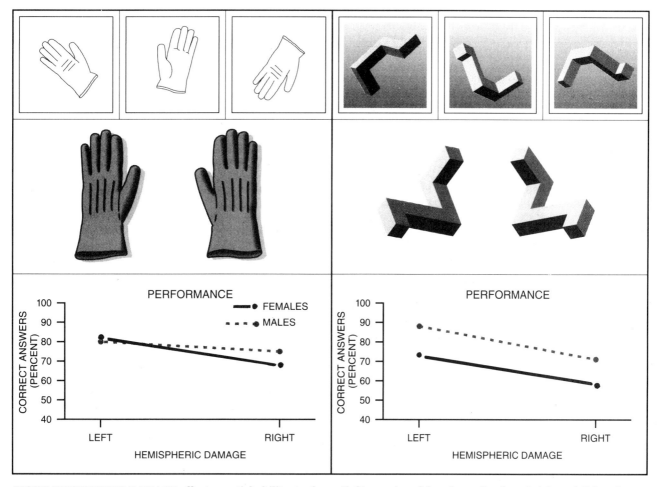

RIGHT HEMISPHERIC DAMAGE affects spatial ability to the same degree in both sexes (*graphs at bottom*), suggesting that women and men rely equally on that hemisphere for certain spatial tasks. In one test of spatial rotation performance (*left*), a series of drawings of a gloved right or left hand must be matched to a right- or left-handed glove. In a second test (*right*), photographs of a three-dimensional object must be matched to one of two mirror images of the same object.

6. OLD/NEW SEXUAL CONCERNS: Males/Females—War and Peace

vard Medical School proposed that androgens increased the functional potency of the right hemisphere. In 1981 Marian C. Diamond of the University of California at Berkeley found that the right cortex is thicker than the left in male rats but not in females. Jane Stewart of Concordia University in Montreal, working with Bryan E. Kolb of the University of Lethbridge in Alberta, recently pinpointed early hormonal influences on this asymmetry: androgens appear to suppress left cortex growth.

Last year de Lacoste and her colleagues reported a similar pattern in human fetuses. They found the right cortex was thicker than the left in males. Thus, there appear to be some anatomic reasons for believing that the two hemispheres might not be equally asymmetric in men and women.

Despite this expectation, the evidence in favor of it is meager and conflicting, which suggests that the most striking sex differences in brain organization may not be related to asymmetry. For example, if overall differences between men and women in spatial ability were related to differing right hemispheric dependence for such functions, then damage to the right hemisphere would perhaps have a more devastating effect on spatial performance in men.

My laboratory has recently studied the ability of patients with damage to one hemisphere of the brain to rotate certain objects mentally. In one test, a series of line drawings of either a left or a right gloved hand is presented in various orientations. The patient indicates the hand being depicted by simply pointing to one of two stuffed gloves that are constantly present.

The second test uses two three-dimensional blocklike figures that are mirror images of one another. Both figures are present throughout the test. The patient is given a series of photographs of these objects in various orientations, and he or she must place each picture in front of the object it depicts. (These nonverbal procedures are employed so that patients with speech disorders can be tested.)

As expected, damage to the right hemisphere resulted in lower scores for both sexes on these tests than did damage to the left hemisphere. Also as anticipated, women did less well than men on the block spatial rotation test. Surprisingly, however, damage to the right hemisphere had no greater effect in men than in women. Women were at least as affected as men by damage to the right hemisphere. This result suggests that the normal differences between men and women on such rotational tests are

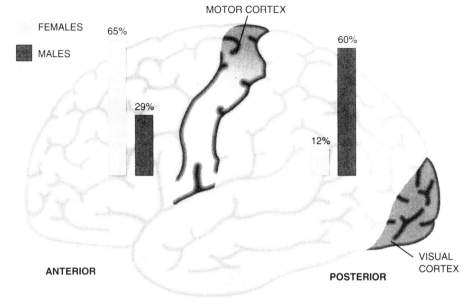

not the result of differential dependence on the right hemisphere. Some other brain systems must be mediating the higher performance by men.

Parallel suggestions of greater asymmetry in men regarding speech have rested on the fact that the incidence of aphasias, or speech disorders, are higher in men than in women after damage to the left hemisphere. Therefore, some researchers have found it reasonable to conclude that speech must be more bilaterally organized in women. There is, however, a problem with this conclusion. During my 20 years of experience with patients, aphasia has not been disproportionately present in women with right hemispheric damage.

In searching for an explanation, I discovered another striking difference between men and women in brain organization for speech and related motor function. Women are more likely than men to suffer aphasia when the front part of the brain is damaged. Because restricted damage within a hemisphere more frequently affects the posterior than the anterior area in both men and women, this differential dependence may explain why women incur aphasia less often than do men. Speech functions are thus less likely to be affected in women not because speech is more bilaterally organized in women but because the critical area is less often affected.

A similar pattern emerges in studies of the control of hand movements, which are programmed by the left hemisphere. Apraxia, or difficulty in selecting appropriate hand movements, is very

APHASIAS, or speech disorders, occur most often in women when damage is to the front of the brain. In men, they occur more frequently when damage is in the posterior region. The data presented above derive from one set of patients.

common after left hemispheric damage. It is also strongly associated with difficulty in organizing speech. In fact, the critical functions that depend on the left hemisphere may relate not to language per se but to organization of the complex oral and manual movements on which human communication systems depend. Studies of patients with left hemispheric damage have revealed that such motor selection relies on anterior systems in women but on posterior systems in men.

The synaptic proximity of women's anterior motor selection system (or "praxis system") to the motor cortex directly behind it may enhance fine-motor skills. In contrast, men's motor skills appear to emphasize targeting or directing movements toward external space—some distance away from the self. There may be advantages to such motor skills when they are closely meshed with visual input to the brain, which lies in the posterior region.

Women's dependence on the anterior region is detectable even when tests involve using visual guidance—for instance, when subjects must build patterns with blocks by following a visual model. In studying such a complex task, it is possible to compare the effects of damage to the anterior and posterior regions of both hemispheres because per-

45. Sex Differences in the Brain

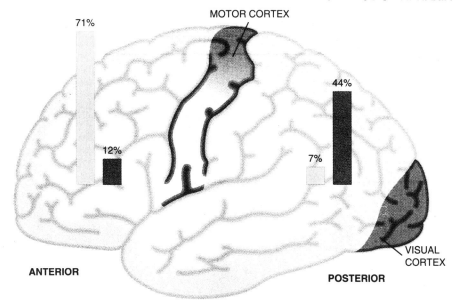

INCIDENCE OF APRAXIA

APRAXIA, or difficulty in selecting hand movements, is associated with frontal damage to the left hemisphere in women and with posterior damage in men. It is also associated with difficulties in organizing speech.

formance is affected by damage to either hemisphere. Again, women prove more affected by damage to the anterior region of the right hemisphere than by posterior damage. Men tend to display the reverse pattern.

Although I have not found evidence of sex differences in functional brain asymmetry with regard to basic speech, motor selection or spatial rotation ability, I have found slight differences in more abstract verbal tasks. Scores on a vocabulary test, for instance, were affected by damage to either hemisphere in women, but such scores were affected only by left-sided injury in men. This finding suggests that in reviewing the meanings of words, women use the hemispheres more equally than do men.

In contrast, the incidence of non-right-handedness, which is presumably related to lesser left hemispheric dependence, is higher in men than in women. Even among right-handers, Marion Annett, now at the University of Leicester in the U.K., has reported that women are more right-handed than men—that is, they favor their right hand even more than do right-handed men. It may well be, then, that sex differences in asymmetry vary with the particular function being studied and that it is not always the same sex that is more asymmetric.

Taken altogether, the evidence suggests that men's and women's brains are organized along different lines from very early in life. During development, sex hormones direct such differentiation. Similar mechanisms probably operate to produce variation within sexes, since there is a relation between levels of certain hormones and cognitive makeup in adulthood.

One of the most intriguing findings is that cognitive patterns may remain sensitive to hormonal fluctuations throughout life. Elizabeth Hampson of the University of Western Ontario showed that the performance of women on certain tasks changed throughout the menstrual cycle as levels of estrogen went up or down. High levels of the hormone were associated not only with relatively depressed spatial ability but also with enhanced articulatory and motor capability.

In addition, I have observed seasonal fluctuations in spatial ability in men. Their performance is improved in the spring when testosterone levels are lower. Whether these intellectual fluctuations are of any adaptive significance or merely represent ripples on a stable baseline remains to be determined.

To understand human intellectual functions, including how groups may differ in such functions, we need to look beyond the demands of modern life. We did not undergo natural selection for reading or for operating computers. It seems clear that the sex differences in cognitive patterns arose because they proved evolutionarily advantageous. And their adaptive significance probably rests in the distant past. The organization of the human brain was determined over many generations by natural selection. As studies of fossil skulls have shown, our brains are essentially like those of our ancestors of 50,000 or more years ago.

For the thousands of years during which our brain characteristics evolved, humans lived in relatively small groups of hunter-gatherers. The division of labor between the sexes in such a society probably was quite marked, as it is in existing hunter-gatherer societies. Men were responsible for hunting large game, which often required long-distance travel. They were also responsible for defending the group against predators and enemies and for the shaping and use of weapons. Women most probably gathered food near the camp, tended the home, prepared food and clothing and cared for children.

Such specializations would put different selection pressures on men and women. Men would require long-distance route-finding ability so they could recognize a geographic array from varying orientations. They would also need targeting skills. Women would require short-range navigation, perhaps using landmarks, fine-motor capabilities carried on within a circumscribed space, and perceptual discrimination sensitive to small changes in the environment or in children's appearance or behavior.

The finding of consistent and, in some cases, quite substantial sex differences suggests that men and women may have different occupational interests and capabilities, independent of societal influences. I would not expect, for example, that men and women would necessarily be equally represented in activities or professions that emphasize spatial or math skills, such as engineering or physics. But I might expect more women in medical diagnostic fields where perceptual skills are important. So that even though any one individual might have the capacity to be in a "nontypical" field, the sex proportions as a whole may vary.

FURTHER READING

SEX DIFFERENCES IN THE BRAIN: THE RELATION BETWEEN STRUCTURE AND FUNCTION. Edited by G. J. DeVries, J.P.C. DeBruin, H.B.M. Uylings and M. A. Corner in *Progress in Brain Research*, Vol. 61. Elsevier, 1984.

MASCULINITY/FEMININITY. Edited by J. M. Reinisch, L. A. Rosenblum and S. A. Sanders. Oxford University Press, 1987.

BEHAVIORAL ENDOCRINOLOGY. Edited by Jill B. Becker, S. Marc Breedlove and David Crews. The MIT Press/Bradford Books, 1992.

Women & Men

Can we get along? Should we even try?

Lawrence Wright

Texas Monthly

On top of the intense racial and economic tensions that plague American life today, it seems that animosity between men and women has hit a boiling point once again. Less obvious than in the early '70s, when the emerging women's movement challenged and forever changed relationships between the sexes, the gender war is now being played out more subtly against a series of public events that have inspired heated debates in the workplace, on the park bench, and in the bedroom. The tensions started a year or so ago, when it was possible to watch both the Clarence Thomas hearings and the William Kennedy Smith trial on television during the day, then go see Thelma and Louise *after dinner, and curl up with either Robert Bly's* Iron John *or Susan Faludi's* Backlash *before bed. Add to this backdrop more recent events like the Navy's Tailhook incident and Dan Quayle's attack on Murphy Brown, and it should come as no surprise that men and women are having a hard time getting along at all.*

In his early studies on the origins of neurosis, Sigmund Freud came to a damning conclusion about men. So many of his patients had revealed stories about sexual experiences in infancy or childhood that Freud decided the "seduction" of children must be the root of all neurotic behavior. When his own sister began to exhibit signs of neurosis, Freud declared: "In every case the father, not excluding my own, had to be blamed as a pervert."

I consider this statement as I stroke my daughter's hair. Caroline is 10 years old. Her eyes are closed, and her head is in my lap. This should be a tender, innocent scene, but we no longer live in a time when anyone believes in innocence. Blame and suspicion color the atmosphere. As a man and a father, I feel besieged and accused. I am appallingly aware of the trust I hold, in the form of my daughter's sleeping body. The line between affection and abuse is in the front of my mind. I feel like a German coming to grips with Nazi guilt. Yes, some men are perverts—but all men? Am I?

Freud later rejected his early hypothesis after his own father died. He suspected that many of the stories his patients had related were fantasized. But now there are those who say in effect that Freud was closer to the truth the first time.

"Men are pigs and they like it that way," an angry writer stated in the op-ed section of the *New York Times*. At a 1991 women's political symposium, Texas governor Ann Richards' ethics adviser, Barbara Jordan, decreed: "I believe that women have a capacity for understanding and compassion which a man structurally does not have, does not have it because he cannot have it. He's just incapable of it." At the same meeting, Houston mayor Kathy Whitmire said that men are less intelligent than women. If these female chauvinists had been speaking of any constituency other than men, they would be run out of public life. But men feel too guilty to defend themselves.

Contempt for men pervades the most obscure strata of our society. A magazine called *House Rabbit Journal* devoted a recent issue to the failings of men as nurturers. "We assume that women perform the primary care-giving role with the house rabbit (as with the kids), and they form the strongest bonds with the bunny," wrote one author. The magazine advised wom-

I'm mad at men too, but I'm also mad at being the object of slanders.

en rabbit owners who want their men to share in their rabbit pleasures to avoid talking about the warm, fuzzy, cuddly aspects of the animal and instead emphasize its traits of integrity, fortitude, and spirit. "I have found, in my relationship with my husband, that having large numbers of animals living with us has put a strain on our relationship," admitted the writer. "With each piece of furniture that has been destroyed, each time we had to avoid the urine puddles in our bed at night, each time we've spent 300 dollars at the vet for a rabbit I picked up at the pound, there has been some initial resentment on the part of my husband. But ultimately he, too, has learned the value of caring."

There is plenty of evidence of the damage men do. Look at the battered women in the shelters. Every year about 20,000 women in Texas alone seek refuge in the shelters from physical abuse in their homes, but the shelters are able to accommodate *fewer* than half of them. In 1990 more than 100,000 women were reported raped in America, the highest total in history and an increase of 12.6 percent over the number of reported rapes per capita in 1980. A prosecutor I know works in the Family Justice Division of the Travis County, Texas, district attorney's office. In the '70s, that office prosecuted only a handful of child-abuse cases a year. Now, Frank Bryan says, he has more than 200 indictments on his desk and a backlog of cases he doesn't want to discuss. "Generally, my impression of men has plummeted," he told me. "I tell all my friends with children never to hire a male baby-sitter. The things these guys do . . ."

But my 15-year-old son is a baby-sitter. That's how Gordon earns his pocket money. It saddens me that he would be shunned because he's a male and therefore a candidate for perversion. On the other hand, I might not hire a male to watch Caroline. Her safety and self-esteem are too important to place in jeopardy.

"Come over and sit in my lap," a grandfatherly preacher friend of mine said out of a lifetime of habit to a little girl he knew. He was at a gathering of friends and family. Suddenly, the room went dead quiet and every woman turned to stare daggers at him. In that moment, the preacher realized that he would never ask a little girl to sit in his lap again. His presumption of innocence had been revoked—not because of his past behavior, which had been exemplary, but simply because he is a man. He has suffered a loss, and so has the little girl. She is being held apart from the love and comfort he has to offer. And at some level she must have understood the subliminal message that hung in the air: Don't trust men.

Is it possible that of the two genders nature created, one is nearly perfect and the other is badly flawed? Well, yes, say the psychobiologists. Unlike women, who carry two X chromosomes, men have an X and a Y. The latter has relatively little genetic information except for the gene that makes us men. A woman who has a recessive gene on one X chromosome might have a countering dominant gene on the other. That's not true for men, who are therefore more vulnerable to biological and environmental insults, as well as more prone to certain behavioral tendencies that may be genetically predetermined. Although male hormones (called androgens) don't cause violent criminal or sexual behavior, they apparently create an inclination in that direction. A low level of arousability—that is to say, a lack of responsiveness to external stimuli—is more common in men than in women. It is reflected in the greater number of male children who die of sudden infant death syndrome and the much larger proportion of boys who are hyperactive and require far more excitement than most children to keep from becoming bored. In adults, this biological need for extra stimulation seems to be connected to higher rates of criminality. Androgens are associated with a number of other male traits (in humans as well as animals), including assertive sexual behavior, status-related aggression, spatial reasoning, territoriality, pain tolerance, tenacity, transient bonding, sensation seeking, and predatory behavior. Obviously, this list posts many of the most common female complaints about men, and yet androgens make a man a man; one can't separate maleness from characteristic male traits.

"Why have any men at all?" wrote Sally Miller Gearhart in a 1982 manifesto titled "The Future—If There Is One—Is Female." Gearhart is an advocate of ovular merging, a process that involves the mating of two eggs, which has been successfully accomplished with mice. Only female offspring are produced. I've always worried that one day women would figure out how to get along without us and they would be able to reproduce unilaterally, like sponges. It's not genocide, exactly. It's more like job attrition, the way employers cut back positions without actually firing anyone. "A 75 percent female to 25 percent male ratio could be achieved in one generation if one half of a population reproduced heterosexually and one half by ovular merging," according to Gearhart. "Such a prospect is attractive to women who feel that if they bear sons, no amount of love and care and non-sexist training will save those sons from a culture where male violence is institutionalized and revered. These women are saying, 'No more sons. We will not spend 20 years of our lives raising a potential rapist, a potential batterer, a potential Big Man.'"

Every man is a "potential" rapist; the only way to eliminate the potential is to get rid of potency. During the Clarence Thomas hearings, Tom Brokaw asked a female legal expert about the lack of a pattern of sexual harassment in Thomas' behavior. "He's not dead yet," she snapped.

I'm mad at men too. I am disgusted by the rise in child-abuse cases and reported rapes. I deplore sexual

Contempt for men pervades even the most obscure strata of our society.

harassment. I'm grateful for the ascendancy of women in business and politics, which may yet advance the humanity of those callings. I have to issue these disclaimers because I'm a man writing on the subject. But I'm also mad at being the object of slanders such as that men are incapable of compassion. Anyone looking at men today should be able to see that they are confused and full of despair. It's not just our place in society or the family that we are struggling for; we're fighting against our own natures. We didn't create the instincts that make us aggressive, that make us value action over

6. OLD/NEW SEXUAL CONCERNS: Males/Females—War and Peace

consensus, that make us more inclined toward strength than sympathy. Nature and human history have rewarded those qualities and in turn have created the kind of people men are. Moreover, these competitive qualities have been necessary for the survival of the species, and despite the debate over masculinity, they are still valued today. Some trial lawyers now include their levels of testosterone, the most abundant of the androgens, on their résumés.

A couple of weeks ago I went to pick Caroline up at her after-school day-care center at the neighborhood Presbyterian church. She had a new teacher, a man, in fact. I made a point of going over to introduce myself and making him feel welcome. The new man was out on the playground with a walkie-talkie. "I'm Caroline's dad," I said, but before I could get around to my welcoming speech, he said, "I'm sorry, but I'm going to have to ask you for a picture ID." As Caroline's father, I appreciated the security, but as a man, I took offense at having to prove that I was not a pervert. True, women are also asked to show identification, but I suspect that it is done in the spirit of fairness. Everybody assumes it is men who are the problem.

Advocacy groups have been using hugely inflated statistics to bludgeon the public into believing that men are waging a war against women and children. "One out of four men is a rapist" is an anecdotal statistic I've heard on several occasions that is not tied to any real survey that I can find.

The real statistics are bad enough. According to Department of Justice victimization studies, the actual chance that a woman will experience a rape or an attempted rape during her lifetime is 8 percent (1 out of 12). Yet a recent study by Neil Gilbert, a professor of social welfare at the University of California at Berkeley, showed that the incidence of date rape, though still far too high, has actually declined substantially since 1980.

Figures about child abuse and domestic violence have been similarly inflated and biased against men. I've read that one out of four females, and one in six males, will be molested or raped by the time they are 18. Most of these scary figures are conjectures based on reports of abuse received by police and child-abuse hotlines. More than a million such reports are filed every year. About 60 percent of them turn out to be unfounded. More than half of the cases that are categorized as neglect or abuse are actually "deprivation of necessities," such as poor medical care, inadequate clothing or shelter, malnutrition—problems of poverty, in other words. Only about 5 percent involve serious physical battering of the sort that we think of as child abuse, and about the same amount turns out to be actual sexual abuse.

Many women dismiss female violence in the home either as innocuous and different in nature from male violence or as self-defense, but Suzanne Steinmetz, a sociology professor at Indiana University, found that some men become targets of abuse when they attempt to protect their children from the mother's violence—the reverse of the stereotype. Other studies have concluded that women typically are just as assaultive as their husbands. Although the men characteristically cause more injuries, wives strike the first blow in 48 percent of the cases, according to one study. The effect of using inflated and, in some cases, falsified statistics to make rape, child abuse, and wife beating seem more prevalent than they actually are—to make them seem, in some dreadful manner, the norm—is to slander the character of men, who are presumed to be the perpetrators of domestic violence, which is not an exclusive feature of male character.

Women also have power that they sometimes discredit. Most men I know feel overwhelmed by women—and by their own need for women. Therefore the rage women feel at men can be terrifying and sexually daunting. Lately I hear women complaining about wimps, about men being uninterested and emotionally withdrawn and sexually unavailable. The ancient stereotype of the frigid woman is being replaced by that of the impotent male. It's not just a fear of intimacy that causes men to founder sexually, nor the dread of AIDS. Men are discovering what women have always known: Sex is a dangerous theater. When women felt powerless, they were sexually passive. Increasingly, now it's the men who are passive and for the same reasons women were in the past. They're afraid. They're afraid of being punished, of being engulfed by women's anger. They feel paralyzed by changes in the social fabric that leave them confused about how to behave around women or even how to talk to them. They sense that the relations between the sexes have become politicized and legalized as never before. Men are going to have to learn how to come to terms with powerful women, how to get used to women with muscles and anger and sexual demands. At the same time, women are going to have to find a way of celebrating manliness without putting it down.

My wife has 15 children in her kindergarten class, and it's rare that she has more than two with a father at home. Sometimes when I visit Roberta's class, the children stare at me as if I were another species. Until a male art teacher arrived last year, I was the only man many of these children would see all day. "You look like Superman in those glasses," a 5-year-old boy told me. He meant Clark Kent. I'm an average-sized man, but to children who rarely see men except from a distance or on television, all men look alike—huge and forbidding and hiding explosive, supernatural strength. This is just one of the harmful effects of the absence of men in children's lives: We've become mythologized.

Here in public school you can see the appalling truth that the traditional family is dead. The men have gone; in many cases, they were never there in the first place. Christine Williams and Debra Umberson, sociologists at the University of Texas, undertook a study of why men disengaged from their children. "We wanted to interview 50 divorced men who did not have custody of their children," says Umberson. "It took us a year to find 43. The reason is it was too painful for them to talk about. When you finally reach them, you hear a lot of complaints about the system, how they are treated as 'just a pocket,' a source of money. They're not invited to be a part of the family, and they don't feel the system

---

The majority of children sexually abused by men are girls; the majority of children battered and killed by women are boys.

Women earn about 30 percent less than men; 20 men are killed on the job for each woman who is fatally injured at work.

Most of these statistics come from U.S. Statistical Abstracts. For specific citations, see the reference section of the book Knights Without Armor *(Jeremy P. Tarcher, Inc., 1991) by Aaron R. Kipnis and* Backlash *(Doubleday, 1991) by Susan Faludi.*

appreciates the effort they do make to take care of their kids."

"Was there anything about this study that surprised you?" I asked Umberson.

"Well, yes," she said. "It was that some of these men were so involved with their children—their kids were really incredibly important to them. That surprised me, because when you read the literature, you get the picture that men just don't care."

It's true that traditional male roles have been compromised or usurped. "The feeling men had that

Figures about child abuse and domestic violence have been inflated and biased against men.

their home is their castle can't be sustained any longer when more than 50 percent of married women work outside the home," said Williams. "Men don't get their authority handed to them on a platter anymore. Women demand to be listened to now. It's no wonder that men feel under siege and that the sense of gratification in being a man is being taken away from them. And who better to blame than women?"

But blame is not the point, for men or women. The point is that families without men are more likely to be poor, and children without fathers are more likely to be deprived—not just of the material comforts but of the sense of the mutuality of the sexes.

Somehow men have got to find a place for themselves again in the family. We're only beginning to see some of the consequences of fatherlessness, especially where boys are concerned. My personal fear is that fatherlessness will have unanticipated political and spiritual consequences, such as a longing for authoritarianism and a further lack of attachment between the sexes. The rise in gangs seems to be connected to the absence of male role models. There is a well-established connection between children of broken homes (a term that seems quaint these days) and the likelihood of committing serious criminal offenses. In any case, children who grow up not knowing who men are pay a price as well. I'm not saying that single mothers—or single fathers—can't do a good job of raising children. But a society of children who don't understand men produces men who don't understand themselves.

I lift up Caroline and take her to bed. Nothing in the world means more to me than our love for each other. I love the difference between us, her femaleness and my maleness. It is a powerful and curious experience to see parts of myself manifested in little-girl form; she is a sort of mirror for me, across time and gender.

I'm afraid of what life has to offer her. I'm worried that the family idea is finished and that the sexes have pulled so far apart that some radical and soulless bureaucratic arrangement is in the process of replacing it. I want Caroline to find love and to experience the joy that I have in being her parent. I want her to find a man who will love her as deeply as I do, who will take care of her and nurture her and stay with her the rest of her life. But I think the chances of that happening are small.

I know that her relationships with men will depend, in large measure, on what she gets from me. That is the most important thing I can give her, a sense of being with a man, trace memories of having me tickle her and toss her in the air, of my taking her temperature when she's sick and rubbing her face with a cool cloth, of her dancing on my shoes. She will remember these things in some almost unrememberable way: They will be a part of her character; she will be the kind of person these things happened to. Therefore she will probably be more trusting of men. That may be a mistake. Who knows what kind of men she is going to meet?

But perhaps her generation will come to a different conclusion. They may decide that the sexes have something special to offer each other, and they'll be able to look at the very things that separate men and women and appreciate them, even savor them. In that case, the language they will learn to speak to each other will be that of love, not blame.

It's a jungle out there, so get used to it!

Women need to realize that men are testosterone-driven animals

Camille Paglia

Rape is an outrage that cannot be tolerated in civilized society. Yet feminism, which has waged a crusade for rape to be taken more seriously, has put young women in danger by hiding the truth about sex from them.

In dramatizing the pervasiveness of rape, feminists have told young women that before they have sex with a man, they must give consent as explicit as a legal contract's. In this way, young women have been convinced that they have been the victims of rape. On elite campuses in the Northeast and on the West Coast, they have held consciousness-raising sessions, petitioned administrations, demanded inquests. At Brown University, outraged, panicky "victims" have scrawled the names of alleged attackers on the walls of women's rest rooms. What marital rape was to the '70s, "date rape" is to the '90s.

The incidence and seriousness of rape do not require this kind of exaggeration. Real acquaintance rape is nothing new. It has been a horrible problem for women for all of recorded history. Once fathers and brothers protected women from rape. Once the penalty for rape was death. I come from a fierce Italian tradition where, not so long ago in the motherland, a rapist would end up knifed, castrated, and hung out to dry.

But the old clans and small rural communities have broken down. In our cities, on our campuses far from home, young women are vulnerable and defenseless. Feminism has not prepared them for this. Feminism keeps saying the sexes are the same. It keeps telling women they can do anything, go anywhere, say anything, wear anything. No, they can't. Women will always be in sexual danger.

One of my male students recently slept overnight with a friend in a passageway of the Great Pyramid in Egypt. He described the moon and sand, the ancient silence and eerie echoes. I will never experience that. I am a woman. I am not stupid enough to believe I could ever be safe there. There is a world of solitary adventure I will never have. Women have always known these

Women will always be in sexual danger.

somber truths. But feminism, with its pie-in-the-sky fantasies about the perfect world, keeps young women from seeing life as it is.

We must remedy social injustice whenever we can. But there are some things we cannot change. There are sexual differences that are based in biology. Academic feminism is lost in a fog of social constructionism. It believes we are totally the product of our environment. This idea was invented by Rousseau. He was wrong. Emboldened by dumb French language theory, academic feminists repeat the same hollow slogans over and over to each other. Their view of sex is naive and prudish. Leaving sex to the feminists is like letting your dog vacation at the taxidermist's.

The sexes are at war. Men must struggle for identity against the overwhelming power of their mothers. Women have menstruation to tell them they are women. Men must do or risk something to be men. Men become masculine only when other men say they are. Having sex with a woman is one way a boy becomes a man.

College men are at their hormonal peak. They have just left their mothers and are questing for their male identity. In groups, they are dangerous. A woman going to a fraternity party is walking into Testosterone Flats, full of prickly cacti and blazing guns. If she goes, she should be armed with resolute alertness. She should arrive with girlfriends and leave with them. A girl who lets herself get dead drunk at a fraternity party is a fool.

47. It's a Jungle Out There

A girl who goes upstairs alone with a brother at a fraternity party is an idiot. Feminists call this "blaming the victim." I call it common sense.

For a decade, feminists have drilled their disciples to say, "Rape is a crime of violence but not of sex." This sugar-coated Shirley Temple nonsense has exposed young women to disaster. Misled by feminism, they do not expect rape from the nice boys from good homes who sit next to them in class.

Aggression and eroticism are deeply intertwined. Hunt, pursuit, and capture are biologically programmed into male sexuality. Generation after generation, men must be educated, refined, and ethically persuaded away from their tendency toward anarchy and brutishness. Society is not the enemy, as feminism ignorantly claims. Society is woman's protection against rape. Feminism, with its solemn Carry Nation repressiveness, does not see what is for men the eroticism or fun element in rape, especially the wild, infectious delirium of gang rape. Women who do not understand rape cannot defend themselves against it.

The date-rape controversy shows feminism hitting the wall of its own broken promises. The women of my '60s generation were the first respectable girls in history to swear like sailors, get drunk, stay out all night—in short, to act like men. We sought total sexual freedom and equality. But as time passed, we woke up to cold reality. The old double standard protected women. When anything goes, it's women who lose.

Today's young women don't know what they want. They see that feminism has not brought sexual happiness. The theatrics of public rage over date rape are their way of restoring the old sexual rules that were shattered by my generation. Because nothing about the sexes has really changed. The comic film *Where the Boys Are* (1960), the ultimate expression of '50s man-chasing, still speaks directly to our time. It shows smart, lively women skillfully anticipating and fending off the dozens of strategies with which horny men try to get them into bed. The agonizing date-rape subplot and climax are brilliantly done. The victim, Yvette Mimieux, makes mistake after mistake, obvious to the other girls. She allows herself to be lured away from her girlfriends and into isolation with boys whose character and intentions she misreads. *Where the Boys Are* tells the truth. It shows courtship as a dangerous game in which the signals are not verbal but subliminal.

Neither militant feminism, which is obsessed with politically correct language, nor academic feminism, which believes that knowledge and experience are "constituted by" language, can understand pre-verbal or non-verbal communication. Feminism, focusing on sexual politics, cannot see that sex exists in and through the body. Sexual desire and arousal cannot be fully translated into verbal terms. This is why men and women misunderstand each other.

Trying to remake the future, feminism cut itself off from sexual history. It discarded and suppressed the sexual myths of literature, art, and religion. Those myths show us the turbulence, the mysteries and passions of sex. In mythology we see men's sexual anxiety, their fear of women's dominance. Much sexual violence is rooted in men's sense of psychological weakness toward women. It takes many men to deal with one woman. Woman's voracity is a persistent motif. Clara Bow, it was rumored, took on the USC football team on weekends. Marilyn Monroe, singing "Diamonds Are a Girl's Best Friend," rules a conga line of men in tuxes. Half-clad Cher, in the video for "If I Could Turn Back Time," deranges a battleship of screaming sailors and straddles a pink-lit cannon. Feminism, coveting social power, is blind to woman's cosmic sexual power.

To understand rape, you must study the past. There never was and never will be sexual harmony. Every woman must take personal responsibility for her sexuality, which is nature's red flame. She must be prudent and cautious about where she goes and with whom. When she makes a mistake, she must accept the consequences and, through self-criticism, resolve never to make that mistake again. Running to Mommy and Daddy on the campus grievance committee is unworthy of strong women. Posting lists of guilty men in the toilet is cowardly, infantile stuff.

The Italian philosophy of life espouses high-energy confrontation. A male student makes a vulgar remark about your breasts? Don't slink off to whimper and simper with the campus shrinking violets. Deal with it. On the spot. Say, "Shut up, you jerk! And crawl back to the barnyard where you belong!" In general, women who project this take-charge attitude toward life get harassed less often. I see too many dopey, immature, self-pitying women walking around like melting sticks of butter. It's the Yvette Mimieux syndrome: Make me happy. And listen to me weep when I'm not.

The date-rape debate is already smothering in propaganda churned out by the expensive Northeastern colleges and universities, with their overconcentration of boring, uptight academic feminists and spoiled, affluent students. Beware of the deep manipulativeness of rich students who were neglected by their parents. They love to turn the campus into hysterical psychodramas of sexual transgression, followed by assertions of parental authority and concern. And don't look for sexual enlightenment from academe, which spews out mountains of books but never looks at life directly.

As a fan of football and rock music, I see in the simple, swaggering masculinity of the jock and in the noisy posturing of the heavy-metal guitarist certain fundamental, unchanging truths about sex. Masculinity is aggressive, unstable, combustible. It is also the most creative cultural force in history. Women must reorient themselves toward the elemental powers of sex, which can strengthen or destroy.

The only solution to date rape is female self-awareness and self-control. A woman's number one line of defense is herself. When a real rape occurs, she should report it to the police. Complaining to college committees because the courts "take too long" is ridiculous. College administrations are not a branch of the judiciary. They are not equipped or trained for legal inquiry. Colleges must alert incoming students to the problems and dangers of adulthood. Then colleges must stand back and get out of the sex game.

> Girls have lower self-esteem and attempt suicide more frequently than boys; five times more boys than girls successfully commit suicide.

> Breast cancer kills more than 40,000 women annually; prostate cancer kills more than 30,000 men.

The blame game

The cause of equality will not be served by pointing fingers

SAM KEEN

Any man who hasn't spent the past 25 years watching Rambo movies or lobbying for the NRA has noticed by now that feminism has profoundly changed our cultural climate. Feminism isn't a passing storm; it is a permanent shift in the weather patterns. A generation of men has been deluged by feminist analysis, rhetoric, demands, and political programs. The women we most admire, fear, and struggle with are feminists. Many men were intimidated by feminism and chose to ignore its challenge. Others passively acquiesced to it, or uncritically accepted it, surviving on a spoon-fed diet of blame and guilt. What the majority of men has not done is confront the feminist analysis and worldview and sort out the healing treasures from the toxic trash. It is time to do it.

"Feminism" is a label describing a kaleidoscope, the many-faceted responses of a multitude of women wrestling with the question of self-definition and seeking social changes that will give greater justice, power, and dignity to women. But we need to make a rough and ready distinction between the best and worst of feminism, between feminism as a prophetic protest and feminism as an ideology.

Prophetic feminism is a model for the changes men are beginning to experience.

Ideological feminism is a continuation of a pattern of general enmity and scapegoating that men have traditionally practiced against women.

As a prophetic movement, feminism has been a cry of the agony of women, a vision of what women may become, and a celebration of the feminine. For more than 20 years, a powerful community of feminist activists has worked to secure economic and legal justice for women. And feminist theorists have revisioned history, philosophy, language, and the arts in an effort to recover women's contribution to the intellectual life of Western culture. Because it was born out of a painful awareness of the indignities and political disenfranchisement of women, prophetic feminism has remained aware of the wounding nature of our social and economic system itself—wounding to both men and women.

Ideological feminism, by contrast, is animated by a spirit of resentment, the tactic of blame, and the desire for vindictive triumph over men that comes out of the dogmatic assumption that women are the innocent victims of a male conspiracy. Perhaps the best rule of thumb to use in detecting ideological feminism is to pay close attention to the ideas, moral sentiments, arguments, and mythic history that cluster around the notion of "patriarchy." All of the great agonies of our time are attributed to the great Satan of patriarchy. The rule of men is solely responsible for poverty, injustice, violence, warfare, technomania, pollution, and the exploitation of the Third World.

This type of demonic theory of history renders men responsible for all of the ills of society, and women innocent. If there is warfare it is because men are naturally hostile and warlike, not because when tribes and peoples come into conflict it is the males who have historically been conditioned, trained, and expected to fulfill the role of warrior. If there is environmental pollution it is not because it is the inevitable result of the urban-technological-industrial life-style that modern people have chosen, but because of patriarchal technology. And if there is injustice in society it is not because some men and some women are insensitive to the

Just friends

Can men and women do it—without doing it?

WHY IS IT OFTEN SO DIFFICULT FOR MEN AND women who like each other to keep sex from muddying the waters of friendship?

Friendship itself is ambiguous, and the ambiguities of male/female friendships are even stronger. There's no set standard and precious little discussion about the terms for friendships across gender lines. If you've slept together, does that rule out friendship? If you're attracted to one another but don't consummate the relationship, are you "friends" or merely flirts?

Part of the problem may be a lack of role models depicting men and women as friends. With the exception of Huck Finn and Becky Thatcher, what other characters from books or films or the tube are strictly platonic friends? Harry and Sally ended up in bed; Sam and Diane thought things were through but, in fact, they weren't; *Moonlighting* went off the air as soon as Maddy and David started fooling around; and we all know what happened to Annie Hall.

The instinctive (and typically female) tendency toward self-blame can also be a factor, causing a woman to feel guilty and confused if a man starts blurring the line between friendship and romance. The question of who's sent what signals to whom can get awfully complicated when two hormonally equipped parties are involved. In all fairness, I know plenty of women who, having been attracted to men who just wanted to "be friends" (whatever their definition may be), were disappointed when their more romantic hopes weren't reciprocated. Still, I have never met a woman who ended a friendship because the man refused to hop into bed.

"There are many different ways to express intimacy," says Carla Golden, associate professor of psychology at Ithaca College. "There's sexual expressiveness and then there's just plain talking. I think women are far more versed in expressing themselves intimately without being sexual."

So what do women get from friendships with men? One very basic thing is safety. Traveling with a man is a completely different experience from traveling alone or with another woman. Even walking down the street takes on a different meaning when I'm with a man—there are no whistles, no catcalls, no horns honking.

But that's not all there is to it. My relationships with men are also important to me because they provide a respite from introspection. As Golden puts it, "Sometimes women want a sense of difference, an edge. While many women enjoy talking about personal things [with other women], sometimes it gets to be too much."

There's also the much bigger—and murkier—issue of self-acceptance and self-worth. Since men possess the bulk of the world's power and status, it's no wonder women value their attention so much. I hate to admit it, but I feel a lot more connected to the rest of the world when I'm out with a man, even if he is just a pal.

And then there's just a basic fascination with an often alien world. In grade school I envied the boys because they always seemed to have more fun than we girls. Theirs was a world filled with vigorous activities like sports, while the female world was filled with Barbie dolls, cooking sets, and talk. By the same token, I resented those boys because they really didn't seem to care if we girls were there. They were perfectly content to toss a football around without us.

Later, in high school and college, the guys did invite us to participate in their games, but by that time touch football had a completely different meaning. Of course, that was part of the fun—tackling someone was perfectly legitimate, a safe way to release pent-up sexual aggression.

It's that same sexual energy that's so alluring—and frustrating—about cross-gender friendships. The inevitable flirtations and sexual tensions add spice to the relationship, an excitement that often can't be found in same-sex friendships. Because, if the truth be told, there's usually some sort of attraction between any two people who become friends, and if you're straight, it's that much more intense with people of the opposite sex. Whether you choose to act on the attraction is something else, but the issue is bound to pop up at some point. (My gay male friends, by the way, have said that they encounter the same problems establishing "just friends" relationships with other gay males.)

In the best of friendships, both parties address these issues and try to overcome them, either by acknowledging the sexual tension but refusing to act on it or by acting on it and getting it out of the way. Pointedly ignoring the sexual vibes rarely works, for ultimately it's too hard to pretend nothing's going on when you know something is.

—Abby Ellin
Boston Phoenix

Excerpted with permission from Alternet News Service. First published in Boston Phoenix *(June 19, 1992). Subscriptions: $41.50/yr. (52 issues) from 126 Brookline Av., Boston, MA 02215. Back issues available from the same address.*

6. OLD/NEW SEXUAL CONCERNS: Males/Females—War and Peace

sufferings of others, but because white male oppressors dominate all women and people of color. Should a Margaret Thatcher, Indira Gandhi, or Imelda Marcos exhibit signs of the pathology of power, an ideological feminist will hasten to explain that she has been colonized by patriarchal attitudes.

Early on, many feminists borrowed the categories oppressed minorities and colonial peoples used in their fight against racism and imperialism to press their case against "the patriarchy." All men, by definition, became guilty of sexism and all women, by definition, were victims. Including all women in the same oppressed "class" seated Marie Antoinette and Rosa Parks side by side in the back of the same bus.

The notion that women are a class or a repressed minority like migrant workers, blacks, Indians in America, or Jews in Germany trivializes the pain involved in class structure and the systematic abuse suffered by ethnic minorities. The injustices that go with class and race are too severe to be confused with the gender problem. All upper classes are composed of equal numbers of men and women. The fruits of exploitation are enjoyed equally by men, women, and children of the upper classes. The outrages of exploitation are borne equally by men and women and children of the lower classes. Both class and ethnic minorities suffer real oppression. It is an insult to the oppressed of the world to have rich and powerful women included within the congregation of the downtrodden merely because they are female.

Ever since Adam, men have been blaming their problems on women. Women have been systematically accused of being temptresses, seducers, powerful contaminants. And if they were viewed as the weaker sex, they were nevertheless blamed whenever a scapegoat was needed.

But the cause of reclaiming female dignity and achieving greater economic justice for women will not be well served by a switch in the dialectics of blame from women to men. This only keeps the old game alive and ensures that the battle between the sexes will continue. We need to find new ways of thinking about men and women, and about the painful and marvelous ways in which we have related and may relate to each other.

Feminists of all sorts are right to be outraged by the dehumanization, destruction, and desecration caused by the modern corporate-industrial-warfare system. They are also right to indict men for their role in creating and maintaining this system, and right to insist on masculine guilt. They, however, are disastrously wrong in excusing women from responsibility for the destructive aspects of a cultural system that can only be created and perpetuated by consensual interaction of men and women (especially the men and women of the elite, powerful, privileged, and ruling classes).

Men are angry at women because they resent being blamed for everything that has gone wrong since Adam ate the apple. Yes, we felt guilty because we went to useless wars (but what right do women who were in no danger of draft or combat have to criticize?). Yes, we feel guilty because we created technologies that proved to be destructive (but didn't women, the poor, and the underdeveloped nations crave all those new cars and labor-saving devices?). Yes, we feel guilty because we were born white, middle-class, and on the fast track to power and prestige (but haven't the women we married encouraged us to succeed and provide?).

It's easy enough to scoff at men's new awareness of wounds and feelings of victimhood, to suspect them of crocodile tears. But I suggest something far more interesting than hypocrisy is happening when men begin to feel the pain and poverty of their positions. When the powerful begin to feel their impotence, when the masters begin to feel their captivity, we have reached a point where we are finally becoming conscious that the social system we have all conspired to create is victimizing us all. At the moment, sensitive women and men are both somewhat depressed by the overwhelming complexity and seeming intractability of the techno-economic-gender system that is oppressing our psyches and destroying our ecosphere. But our depression can turn into a sense of empowerment when we begin to look carefully at the way men and women interact in a codependent way to maintain the system.

The world is dangerous, threatened, and wounding. There can be no question but that the historical humiliation of women is a fact. But men's suffering from gender roles and the social system is also a fact. The healing of the relationship between the sexes will not begin until men and women cease to use their suffering as a justification for their hostility. It serves no useful purpose to argue about who suffers most. Before we can begin again together, we must repent separately. In the beginning we need simply to listen to each other's stories, the histories of wounds.

Then we must examine the social-economic-political system that has turned the mystery of man and woman into the alienation between the genders. And, finally, we must grieve together. Only repentance, mourning, and forgiveness will open our hearts to each other and give us the power to begin again.

Ending the battle between the sexes

First separate, then communicate

Aaron R. Kipnis & Elizabeth Herron

SPECIAL TO *UTNE READER*

Aaron R. Kipnis, Ph.D., *is a consultant to numerous organizations concerned with gender issues and lectures for many clinical training institutes. He is author of the critically acclaimed book* Knights Without Armor: A Practical Guide for Men in Quest of Masculine Soul, *(Jeremy Tarcher, 1991, Putnam, 1992).*

Elizabeth Herron, M.A., *specializes in women's empowerment and gender reconciliation. She is co-director of the Santa Barbara Institute for Gender Studies and is co-author of* Gender War/Gender Peace; The Quest for Love and Justice Between Women and Men *(Feb. 1994, Morrow).*

Have you noticed that American men and women seem angrier at one another than ever? Belligerent superpowers have buried the hatchet, but the war between the sexes continues unabated. On every television talk show, women and men trade increasingly bitter accusations. We feel the tension in our homes, in our workplaces, and in our universities.

The Clarence Thomas-Anita Hill controversy and the incidents at the Navy's Tailhook convention brought the question of sexual harassment into the foreground of national awareness, but it now appears that these flaps have merely fueled male-female resentment instead of sparking a productive dialogue that might enhance understanding between the sexes.

Relations between women and men are rapidly changing. Often, however, these changes are seen to benefit one sex at the expense of the other, and the mistrust that results creates resentment. Most men and women seem unable to entertain the idea that the two sexes' differing perspectives on many issues can be equally valid. So polarization grows instead of reconciliation, as many women and men fire ever bigger and better-aimed missiles across the gender gap. On both sides there's a dearth of compassion about the predicaments of the other sex.

For example:
- Women feel sexually harassed; men feel their courting behavior is often misunderstood.
- Women fear men's power to wound them physically; men fear women's power to wound them emotionally.
- Women say men aren't sensitive enough; men say women are too emotional.
- Women feel men don't do their fair share of housework and child care; men feel that women don't feel as much pressure to provide the family's income and do home maintenance.
- Many women feel morally superior to men; many men feel that they are more logical and just than women.
- Women say men have destroyed the environment; men say the women's movement has destroyed the traditional family.
- Men are often afraid to speak about the times that they feel victimized and powerless; women frequently deny their real power.
- Women feel that men don't listen; men feel that women talk too much.
- Women resent being paid less than men; men are concerned about the occupational hazards and stress that lead to their significantly shorter life spans.
- Men are concerned about unfairness in custody and visitation rights; women are concerned about fathers who shirk their child support payments.

It is very difficult to accept the idea that so many conflicting perspectives could all have intrinsic value. Many of us fear that listening to the story of another will somehow weaken our own voice, our own initiative, even our own identity. The fear keeps us locked in adversarial thinking and patterns of blame and alienation. In this frightened absence of empathy, devaluation of the other sex grows.

In an attempt to address some of the discord between the sexes, we have been conducting gender workshops around the country. We invite men and women to spend some time in all-male and all-female groups, talking about the opposite sex. Then we bring the two groups into an encounter with one another. In one of our mixed groups this spring, Susan, a 35-year-old advertising executive, told the men, "Most men these days are insensitive jerks. When are men going to get it that we are coming to work to make a living, not

6. OLD/NEW SEXUAL CONCERNS: Males/Females—War and Peace

to get laid? Anita Hill was obviously telling the truth. Most of the women I work with have been harassed as well."

Michael, her co-worker, replied, "Then why didn't she tell him ten years ago that what he was doing was offensive? How are we supposed to know where your boundaries are if you laugh at our jokes, smile when you're angry, and never confront us in the direct way a man would? How am I supposed to learn what's not OK with you, if the first time I hear about it is at a grievance hearing?"

We've heard many permutations of this same conversation:

> "I just can't listen to women's issues anymore while passively watching so many men go down the tubes."

Gina, a 32-year-old school teacher in Washington, D.C., asks, "Why don't men ever take *no* for an answer?"

Arthur, a 40-year-old construction foreman, re-

His-and-hers politics

A woman explains her unease about the men's movement

I'm in the habit of noticing the symbols on rest room doors because I believe they are cultural icons. Not long ago in a restaurant, I noted that the "Men's" symbol was the routine international stick figure: one head, two arms, two legs, the sort of thing that could be beamed into deep space to show alien life forms what we look like here on Earth. The "Women's," on the other hand, showed a silhouette of someone wearing a beehive hairdo and applying lipstick. I observed: These things are not equivalent.

My daughter often brings home work sheets from preschool that remind me how we all acquire this habit of looking for matching pairs. Draw a circle around the things that go together. Salt and pepper. Cup and saucer. Left and right. Romeo and Juliet. His and Hers. When I try to understand the collection of ideas and goals that has come to be called the men's movement, what disturbs me is that it generally stands as an "other half" to the women's movement, and in my mind it doesn't belong there. It is not an equivalent. Women are fighting for their lives, and men are looking for some peace of mind.

I do believe that men face some cultural problems that come to them solely on the basis of gender: They are so strictly trained to be providers that many other areas of their lives are neither cultivated nor validated. They usually have to grow up without the benefit of close bonding with a same-sex parent. They struggle with guilt and doubts associated with a history of privilege.

Women struggle with the fact that they are statistically likely to be impoverished, worked to the bone, and raped.

If there is kindness in us, we will not belittle another's pain, regardless of its size. When a friend calls me to moan that she's just gotten a terrible haircut, I'll give her some sympathy. But I will give her a lot more if she calls to say she's gotten ovarian cancer. Let's keep some perspective. The men's movement and the women's movement aren't salt and pepper; they are hangnail and hand grenade.

I met a friend for lunch today, in a restaurant I'm fond of for its rest room iconography and other reasons. The co-owners, a gay man and woman, are the parents of a child whose family also includes their gay and lesbian partners. Once when my own daughter asked me if every child needed to have a mom and a dad, I pointed to this family to widen her range. I told her that in a world where people didn't hurt each other for reasons of color or gender, families could look all kinds of different ways, and they could be happy. We're still waiting for that world, obviously, but in the meantime I like the restaurant: The service is friendly and the vegetables are pesticide-free. The bathroom has nothing at all on the door, because there is only one. It serves one customer at a time, fairly and well, regardless of gender.

I had come to the restaurant to meet a friend, and while I waited my mind ran from rest rooms to Iron John, so that when she arrived and sat down I asked abruptly, "What do you think of the men's movement?"

My friend blinked a couple of times and said, "I think it's a case of people thinking that feminism is only for women, and if there's a 'Hers' there has to be a 'His,' too."

That's it, exactly. The tragedy is that the formation of a men's movement to "respond" to feminism creates antipathy in place of cooperation. The women's movement is called by that name because women are its heartiest proponents—of necessity, because of lives on the line—but what it asks is simply for all humans to be treated fairly and well, regardless of gender. If its goals could be met, those of the men's movement would be moot points: When women and men are partners in the workplace and the home, sons will be nurtured by fathers; the burden of breadwinning will be shared; the burdens of privilege, if there are any, will surely be erased when power comes up as evenly as grass.

To reach this place, we don't need a "His" and "Hers." What we need is for both sides, the beehive hairdos and the creatures of planet Earth, to claim the goal of equal rights as "Ours."

—*Barbara Kingsolver*

Excerpted from Women Respond to the Men's Movement: A Feminist Collection *by Kay Leigh Hagan. Copyright © 1992 by Kay Leigh Hagan. "Cabbages and Kings" © 1992 by Barbara Kingsolver. Reprinted by arrangement with HarperSanFrancisco, a division of HarperCollins Publishers, Inc.*

49. Ending the Battle

plies that in his experience, "some women *do* in fact say no when they mean yes. Women seem to believe that men should do all the pursuing in the mating dance. But then if we don't read her silent signals right, we're the bad guys. If we get it right, though, then we're heroes."

Many men agree that they are in a double bind. They are labeled aggressive jerks if they come on strong, but are rejected as wimps if they don't. Women feel a similar double bind. They are accused of being teases if they make themselves attractive but reject the advances of men. Paradoxically, however, as Donna, a fortyish divorcée, reports, "When I am up front about my desires, men often head for the hills."

As Deborah Tannen, author of the best-seller about male-female language styles *You Just Don't Understand*, has observed, men and women often have entirely different styles of communication. How many of us have jokingly speculated that men and women actually come from different planets? But miscommunication alone is not the source of all our sorrow.

Men have an ancient history of enmity toward women. For centuries, many believed women to be the cause of our legendary fall from God's grace. "How can he be clean that is born of woman?" asks the Bible. Martin Luther wrote that "God created Adam Lord of all living things, but Eve spoiled it all." The "enlightened" '60s brought us Abbie Hoffman, who said: "The only alliance I would make with the women's liberation movement is in bed." And from the religious right, Jerry Falwell still characterizes feminism as a "satanic attack" on the American family.

In turn, many feel the women's movement devalues the role of men. Marilyn French, author of *The Women's Room*, said, "All men are rapists and that's all they are." In response to the emerging men's movement, Betty Friedan commented, "Oh God, sick . . . I'd hoped by now men were strong enough to accept their vulnerability and to be authentic without aping Neanderthal cavemen."

This hostility to the men's movement is somewhat paradoxical. Those who are intimately involved with the movement say that it is primarily dedicated to ending war and racism, increasing environmental awareness, healing men's lives and reducing violence, promoting responsible fatherhood, and creating equal partnerships with women—all things with which feminism is ideologically aligned. Yet leaders of the men's movement often evoke indignant responses from women. A prominent woman attorney tells us, "I've been waiting 20 years for men to hear our message. Now instead of joining us at last, they're starting their *own* movement. And now they want us to hear that they're wounded too. It makes me sick."

On the other hand, a leader of the men's movement says, "I was a feminist for 15 years. Recently, I realized that all the men I know are struggling just as much as women. Also, I'm tired of all the male-bashing. I just can't listen to women's issues anymore while passively watching so many men go down the tubes."

Some of our gender conflict is an inevitable byproduct of the positive growth that has occurred in our society over the last generation. The traditional gender roles of previous generations imprisoned many women and men in soul-killing routines. Women felt dependent and disenfranchised; men felt distanced from feelings, family, and their capacity for self-care.

Even in our own culture, women and men have traditionally had places to meet apart from members of the other sex.

With almost 70 percent of women now in the work force, calls from Barbara Bush and Marilyn Quayle for women to return to the home full time seem ludicrous, not to mention financially impossible. In addition, these calls for the traditional nuclear family ignore the fact that increasing numbers of men now want to downshift from full-time work in order to spend more time at home. So if we can't go back to the old heroic model of masculinity and the old domestic ideal of femininity, how then do we weave a new social fabric out of the broken strands of worn-out sexual stereotypes?

Numerous participants in the well-established women's movement, as well as numbers of men in the smaller but growing men's movement, have been discovering the strength, healing, power, and sense of security that come from being involved with a same-sex group. Women and men have different social, psychological, and biological realities and receive different behavioral training from infancy through adulthood.

In most pre-technological societies, women and men both participate in same-sex social and ceremonial groups. The process of becoming a woman or a man usually begins with some form of ritual initiation. At the onset of puberty, young men and women are brought into the men's and women's lodges, where they gain a deep sense of gender identity.

Even in our own culture, women and men have traditionally had places to meet apart from members of the other sex. For generations, women have gathered over coffee or quilts; men have bonded at work and in taverns. But in our modern society, most heterosexuals believe that a member of the opposite sex is supposed to fulfill all their emotional and social needs. Most young people today are not taught to respect and honor the differences of the other gender, and they arrive at adulthood both mystified and distrustful, worried about the other sex's power to affect them. In fact, most cross-gender conflict is essentially *conflict between different cultures.* Looking at the gender war from this perspective may help us develop solutions to our dilemmas.

6. OLD/NEW SEXUAL CONCERNS: Males/Females—War and Peace

In recent decades, cultural anthropologists have come to believe that people are more productive members of society when they can retain their own cultural identity within the framework of the larger culture. As a consequence, the old American "melting pot" theory of cultural assimilation has evolved into a new theory of diversity, whose model might be the "tossed salad." In this ideal, each subculture retains its essential identity, while coexisting within the same social container.

Applying this idea to men and women, we can see the problems with the trend of the past several decades toward a sex-role melting pot. In our quest for gender equality through sameness, we are losing both the beauty of our diversity and our tolerance for differences. Just as a monoculture is not as environmentally stable or rich as a diverse natural ecosystem, androgyny denies the fact that sexual differences are healthy.

> **Both men and women feel they are in a double bind.**

In the past, perceived differences between men and women have been used to promote discrimination, devaluation, and subjugation. As a result, many "we're all the same" proponents—New Agers and humanistic social theorists, for example—are justifiably suspicious of discussions that seek to restore awareness of our differences. But pretending that differences do not exist is not the way to end discrimination toward either sex.

Our present challenge is to acknowledge the value of our differing experiences as men and women, and to find ways to reap this harvest in the spirit of true equality. Carol Tavris, in her book *The Mismeasure of Women*, suggests that instead of "regarding cultural and reproductive differences as problems to be eliminated, we should aim to eliminate *the unequal consequences that follow from them.*"

Some habits are hard to change, even with an egalitarian awareness. Who can draw the line between what is socially conditioned and what is natural? It may not be possible, or even desirable, to do so. What seems more important is that women and men start understanding each other's different cultures and granting one another greater freedom to experiment with whatever roles or lifestyles attract them.

Lisa, a 29-year-old social worker from New York participating in one of our gender workshops, told us, "Both Joel [her husband] and I work full time. But it always seems to be me who ends up having to change my schedule when Gabe, our son, has a doctor's appoint-

---

Who's on top?

A revisionist history of men and women

ROBERT GRAVES, THE POET AND HISTORIAN, SAYS, "The most important history of all for me is the changing relationship between men and women down the centuries."

For thousands of years there has been a tragic situation—the domination by men and the degradation of women. We are so used to it we do not notice it.

This was not always so. Now there is an underlying feeling that true equality is impossible because men and women are so different. We can never be like each other. But I disagree. We once were and we must again become noble equals.

Go back three thousand years to Asia Minor, the first civilization that was somewhat stable. In those happy and far-off days women were deeply respected and loved by men and had a kind of wise command over things. This was evidenced by the greatest queen of all time, perhaps, Semiramis of Assyria, a great wise and beneficent ruler. And she had another quality of women then—bravery, for she was also a great soldier. In fact that was what especially charmed her husband. She reigned 42 years.

There were not startling physical differences between men and women then. The statue of the Winged Victory of Samothrace had not knock-knees, poor musculature, nor enormously exaggerated breasts. There is a beautiful Greek statue of Orestes and Electra who were brother and sister, their arms over each other's shoulders. They are the same height, built identically alike with the same limber prowess and athletic beauty.

In Ancient Egypt, Diodorus Siculus tells us, the women ruled their husbands. There is no ambiguity about it; the wives were absolutely supreme. Herodotus said: "With them the women go to market, the men stay home and weave. The women discharged all kinds of public affairs. The men dealt with domestic affairs." In Sparta women were the dominant sex. They alone could own property. This was the case among the Iroquois, the Kamchadale people of Siberia, and countless others. "When women ruled in Kamchatka, the men not only did the cooking but all the housework, docilely doing everything assigned to them," according

49. Ending the Battle

ment or a teacher conference, is sick at home or has to be picked up after school. It's simply taken for granted that in most cases my time is less important than his. I know Joel tries really hard to be an engaged father. But the truth is that I feel I'm always on the front line when it comes to the responsibilities of parenting and keeping the home together. It's just not fair."

Joel responds by acknowledging that Lisa's complaint is justified; but he says, "I handle all the home maintenance, fix the cars, do all the banking and bookkeeping and all the yard work as well. These things aren't hobbies. I also work more overtime than Lisa. Where am I supposed to find the time to equally co-parent too? Is Lisa going to start mowing the lawn or help me build the new bathroom? Not likely."

In many cases of male-female conflict, as with Lisa and Joel, there are two differing but *equally valid* points of view. Yet in books, the media, and in women's and men's groups, we only hear about most issues from a woman's point of view or from a man's. This is at the root of the escalating war between the sexes.

For us, the starting point in the quest for gender peace is for men and women to spend more time with members of the same sex. We have found that many men form intimate friendships in same-sex groups. In addition to supporting their well-being, these connections can take some of the pressure off their relationships with women. Men in close friendships no longer expect women to satisfy *all* their emotional needs. And when women meet in groups they support one another's need for connection and also for empowerment in the world. Women then no longer expect men to provide their sense of self-worth. So these same-sex groups can enhance not only the participants' individual lives, but their relationships with members of the other sex as well.

If men and women *remain* separated, however, we risk losing perspective and continuing the domination or scapegoating of the other sex. In women's groups, male-bashing has been running rampant for years. At a recent lecture we gave at a major university, a young male psychology student said, "This is the first time in three years on campus that I have heard anyone say a single positive thing about men or masculinity."

Many women voice the same complaint about their experiences in male-dominated workplaces. Gail, a middle management executive, says, "When I make proposals to the all-male board of directors, I catch the little condescending smirks and glances the men give one another. They don't pull that shit when my male colleagues speak. If they're that rude in front of me, I can only imagine how degrading their comments are when they meet in private."

There are few arenas today in which women and men can safely come together on common ground to frankly discuss our rapidly changing ideas about gender justice. Instead of more sniping from the sidelines, what is needed is for groups of women and men to communicate directly with one another. When we take this

---

to the historian C. Meiners. "Men are so domesticated that they greatly dislike being away from home for more than one day. Should a longer absence than this become necessary, they try to persuade their wives to accompany them, for they cannot get on without the women folk."

In Abyssinia, in Lapland, men did what seems to us women's work. Tacitus, describing the early Teutons, tells how women did all the work, the hunting, tilling the soil, while men idled and looked after the house, equivalent now to playing bridge and taking naps. The heirlooms in the family, a harnessed horse, a strong spear, a sword and shield passed on to the women. They were the fighters.

And so they were in Libya, in the Congo. In India under the Queens of Nepal only women soldiers were known. In Dahomey, the king had a bodyguard of warrior women and these were braver than any of his men warriors. And physiologically, things were reversed: The women, more active and strenuous, became taller, stronger, tougher than the sedentary home-body men.

Now about women's current

> **One of the best ways to be a great man would be to be a true friend of women.**

physical inferiority. To feel superior, men chose wives with low-grade physical prowess, unable to walk or run decently, with feeble feet, ruined knees, and, as at present, enormously exaggerated breasts. Their offspring, of course, dwindle and become inferior. And men chose such women, as Bertrand Russell said, "because it makes them feel so big and strong without incurring any real danger."

It seems to me one of the best ways to be a great man would be to be a true friend of women. How? Neither pamper nor exploit them. Love in women their greatness which is the same as it is in men. Insist on bravery, honor, grandeur, generosity in women.

I say this because I think there is a state of great unhappiness between us. If we can be true equals, we will be better friends, better lovers, better wives and husbands.

—*Brenda Ueland*

From the book Strength to Your Sword Arm: Selected Writings *by Brenda Ueland. Copyright © 1992 by The Estate of Brenda Ueland. Reprinted by permission of Holy Cow! Press. The book is available for $16.70 postpaid from Holy Cow! Press, Box 3170, Mt. Royal Station, Duluth, MN 55803; 218/724-1653.*

next step and make a commitment to spend time apart and then meet with each other, then we can begin to build a true social, political, and spiritual equality. This process also instills a greater appreciation for the unique gifts each sex has to contribute.

Husband-and-wife team James Sniechowski and Judith Sherven conduct gender reconciliation meetings—similar to the meetings we've been holding around the country—each month in Southern California. In a recent group of 25 people (11 women, 14 men), participants were invited to explore questions like: What did you learn about being a man/woman from your mother? From your father? Sniechowski reports that, "even though, for the most part, the men and women revealed their confusions, mistrust, heartbreaks, and bewilderments, the room quickly filled with a poignant beauty." As one woman said of the meeting, "When I listen to the burdens we suffer, it helps me soften my heart toward them." On another occasion a man said, "My image of women shifts as I realize they've been through some of the same stuff I have."

Discussions such as these give us an opportunity to really hear one another and, perhaps, discover that many of our disagreements come from equally valid, if different, points of view. What many women regard as intimacy feels suffocating and invasive to men. What many men regard as masculine strength feels isolating and distant to women. Through blame and condemnation, women and men shame one another. Through compassionate communication, however, we can help one another. This mutual empowerment is in the best interests of both sexes, because when one sex suffers, the other does too.

Toward the end of our meetings, men and women inevitably become more accountable for the ways in which they contribute to the problem. Gina said, "I've never really heard the men's point of view on all this before. I must admit that I rarely give men clear signals when they say or do something that offends me."

Arthur then said, "All my life I've been trained that my job as a man is to keep pursuing until 'no' is changed to 'yes, yes, yes.' But I hear it that when a woman says no, they want me to respect it. I get it now that what I thought was just a normal part of the dance is experienced as harassment by some women. But you know, it seems that if we're ever going to get together now, more women are going to have to start making the first moves."

After getting support from their same-sex groups and then listening to feedback from the whole group, Joel and Lisa realize that if they are both going to work full time they need to get outside help with family tasks, rather than continuing to blame and shame one another for not doing more.

Gender partnership based on strong, interactive, separate but equal gender identities can support the needs of both sexes. Becoming more affirming or supportive of our same sex doesn't have to lead to hostility toward the other sex. In fact, the acknowledgment that gender diversity is healthy may help all of us to become more tolerant toward other kinds of differences in our society.

Through gender reconciliation—both formal workshops and informal discussions—the sexes can support each other, instead of blaming one sex for not meeting the other's expectations. Men and women clearly have the capacity to move away from the sex-war rhetoric that is dividing us as well as the courage necessary to create forums for communication that can unite and heal us.

Boys and girls need regular opportunities in school to openly discuss their differing views on dating, sex, and gender roles. In universities, established women's studies courses could be complemented with men's studies, and classes in the two fields could be brought together from time to time to deepen students' understanding of both sexes. The informal discussion group is another useful format in which men and women everywhere can directly communicate with each other (see *Utne Reader* issue no. 44 [March/April 1991]). In the workplace the struggle for gender understanding needs to go beyond the simple setting up of guidelines about harassment; it is essential that women and men regularly discuss their differing views on gender issues. Outside help is often needed in structuring such discussions and getting them under way. Our organization, the Santa Barbara Institute for Gender Studies, trains and provides "reconciliation facilitators" for that purpose.

These forums must be fair. Discussions of women's wage equity must also include men's job safety. Discussions about reproductive rights, custody rights, or parental leave must consider the rights of both mothers and fathers—and the needs of the children. Affirmative action to balance the male-dominated political and economic leadership must also bring balance to the female-dominated primary-education and social-welfare systems.

We call for both sexes to come to the negotiating table from a new position of increased strength and self-esteem. Men and women do not need to become more like one another, merely more deeply themselves. But gender understanding is only a step on the long road that must ultimately lead to fundamental institutional change. We would hope, for example, that in the near future men and women will stop arguing about whether women should go into combat and concentrate instead on how to end war. The skills and basic attitudes that will lead to gender peace are the very ones we need in order to meet the other needs of our time—social, political, and environmental—with committed action.

Glossary

— A —

abnormal: anything considered not normal, i.e., not conforming to the subjective standards a social group has established as the norm

abortifacients: substances that cause termination of pregnancy

acquaintance (date) rape: when a sexual encounter is forced by someone who is known to the victim

acquired immunodeficiency syndrome: fatal disease caused by a virus that is transmitted through the exchange of bodily fluids primarily in sexual activity, and intravenous drug use

activating effect: the direct influence some hormones can have on activating or deactivating sexual behavior

actual use failure rate: a measure of how often a birth control method can be expected to fail when human error and technical failure are considered

adolescence: period of emotional, social, and physical transition from childhood to adulthood

adultery toleration: marriage partners extend the freedom to each other to have sex with others

affectional: relating to feelings or emotions, such as romantic attachments

afterbirth: the tissues expelled after childbirth including the placenta, the remains of the umbilical cord and fetal membranes

agenesis (absence) of the penis (ae-JEN-a-ses): a congenital condition in which the penis is undersized and nonfunctional

AIDS: acquired immunodeficiency syndrome

ambisexual: alternate term for bisexual

amniocentesis: a process whereby medical problems with a fetus can be determined while it is still in the womb; a needle is inserted into the amniotic sac, amniotic fluid is withdrawn, and its cells examined

amnion (AM-nee-on): a thin membrane that forms a closed sac to enclose the embryo; the sac is filled with amniotic fluid that protects and cushions the embryo

anal intercourse: insertion of the penis into the rectum of a partner

androgen: a male hormone, such as testosterone, that affects physical development, sexual desire, and behavior. It is produced by both male and female sex glands and influences each sex in varying degrees

androgyny (an-DROJ-a-nee): combination of traditional feminine and masculine traits in a single individual

anejaculation: lack of ejaculation at the time of orgasm

anorchism (a-NOR-kiz-um): rare birth defect in which both testes are lacking

aphrodisiacs (af-ro-DEE-zee-aks): foods or chemicals purported to foster sexual arousal; they are believed to be more myth than fact

apotemnophilia: a rare condition characterized by the desire to function sexually after having a leg amputated

areola (a-REE-a-la): darkened, circular area of skin surrounding the nipple

artificial embryonation: a process in which the developing embryo is flushed from the uterus of the donor woman 5 days after fertilization and placed in another woman's uterus

artificial insemination: injecting the sperm cells of a male into a woman's vagina, with the intention of conceiving a child

asceticism (a-SET-a-siz-um): usually characterized by celibacy, this philosophy emphasizes spiritual purity through self-denial and self-discipline

asexuality: characterized by a low interest in sex

autoerotic asphyxiation: accidental death from pressure placed around the neck during masturbatory behavior

autofellatio (fe-LAY-she-o): a male providing oral stimulation to his own penis, an act most males do not have the physical agility to perform

— B —

Bartholin's glands (BAR-tha-lenz): small glands located in the opening through the minor lips that produce some secretion during sexual arousal

behavior therapy: used of techniques to learn new patterns of behavior, often employed in sex therapy

berdache (bare-DAHSH): anthropological term for cross-dressing in other cultures

bestiality (beest-ee-AL-i-tee): a human being having sexual contact with an animal

birth canal: term applied to the vagina during the birth process

birthing rooms: special areas in the hospital, decorated and furnished in a nonhospital way, set aside for giving birth; the woman remains here to give birth rather than being taken to a separate delivery room

bisexual: refers to some degree of sexual attraction to or activities with members of both sexes

blastocyst: the morula, after five days of cell division—has developed a fluid-filled cavity in its interior and entered the uterine cavity

bond: the emotional link between parent and child created by cuddling, cooing, physical and eye contact early in the newborn's life

bondage: tying, restraining, or applying pressure to body parts for sexual arousal

brachioproctic activity (brake-ee-o-PRAHK-tik): known in slang as "fisting"; a hand is inserted into the rectum of a partner

brothels: houses of prostitution

bulbourethral glands: also called Cowper's glands

— C —

call boys: highly paid male prostitutes

call girls: more highly paid prostitutes who work by appointment with a more exclusive clientele

cantharides (kan-THAR-a-deez): a chemical extracted from a beetle that, when taken internally, creates irritation of blood vessels in the genital region; it can cause physical harm

case studies: an in-depth look at a particular individual and how he or she might have been helped to solve a sexual or other problem. They may offer new and useful ideas for counselors to use with other patients

catharsis theory: suggests that viewing pornography provides a release for sexual tension, thus preventing antisocial behavior

celibacy (SELL-a-ba-see): choosing not to share sexual activity with others

cervical cap: a device that is shaped like a large thimble and fits over the cervix; not a particularly effective contraceptive because it can dislodge easily during intercourse

cervical intraepithelial neoplasia (CIN) (ep-a-THEE-lee-al nee-a-PLAY-zhee-a): abnormal, precancerous cells sometimes identified in a Pap smear

cervix (SERV-ix): lower "neck" of the uterus that extends into the back part of the vagina

cesarian section: a surgical method of childbirth in which delivery occurs through an incision in the abdominal wall and uterus

chancroid (SHAN-kroyd): a venereal disease caused by the bacterium *Hemophilus ducreyi*: and characterized by sores on the genitals which, if left untreated, could result in pain and rupture of the sores

child molesting: sexual abuse of a child by an adult

chlamydia (kluh-MID-ee-uh): now known to be a common STD, this organism is a major cause of urethritis in males; in females it often presents no symptoms

chorion (KOR-ee-on): the outermost extra-embryonic membrane essential in the formation of the placenta

chorionic villi sampling (CVS): a technique for diagnosing medical problems in the fetus as early as the 8th week of pregnancy; a sample of the chorionic membrane is removed through the cervix and studied

cilia: microscopic hair-like projections that help move the ovum through the fallopian tube

circumcision: in the male, surgical removal of the foreskin from the penis; in the female, surgical procedure that cuts the prepuce, exposing the clitoral shaft

climacteric: mid-life period experienced by both men and women when there is greater emotional stress than usual and sometimes physical symptoms

climax: another term for orgasm

clinical research: the study of the cause, treatment or prevention of a disease or condition by testing large numbers of people

clitoridectomy: surgical removal of the clitoris; practiced routinely in some cultures

clitoris (KLIT-a-rus): sexually sensitive organ found in the female vulva; it becomes engorged with blood during arousal

clone: the genetic duplicate organism produced by the cloning process

cloning: a process involving the transfer of a full complement of chromosomes from a body cell of an organism into an ovum from which the chromosomal material has been removed; if allowed to develop into a new organism, it is an exact genetic duplicate of the one from which the original body cell was taken; the process is not yet used for humans, but has been performed in lower animal species

cohabitation: living together and sharing sex without marrying

235

coitus (KO-at-us *or* ko-EET-us): heterosexual, penis-in-vagina intercourse

coitus interruptus (ko-EET-us *or* KO-ut-us): a method of birth control in which the penis is withdrawn from the vagina prior to ejaculation

comarital sex: also called mate-swapping, a couple swaps sexual partners with another couple

combining of chromosomes: occurs when a sperm unites with an egg, normally joining 23 pairs of chromosomes to establish the genetic "blueprint" for a new individual. The sex chromosomes establish its sex: XX for female and XY for male

coming out: to acknowledge to oneself and to others that one is sexually attracted to others of the same sex

Comstock Laws: enacted in the 1870s, this federal legislation prohibited mailing information about contraception

condom: a sheath worn over the penis during intercourse to collect semen and prevent disease transmission

consensual adultery: permission given to at least one partner within the marital relationship to participate in extramarital sexual activity

controlled experiment: research in which the investigator examines what is happening to one variable while all other variables are kept constant

conventional adultery: extramarital sex without the knowledge of the spouse

coprophilia: sexual arousal connected with feces

core gender-identity/role: a child's early sense and expression of its maleness, femaleness, or ambivalence, prior to puberty

corona: the ridge around the penile glans

corpus luteum: follicle cell cluster that remains after the ovum is released, secreting hormones that help regulate the menstrual cycle

Cowper's glands: two small glands in the male which secrete alkaline fluid into the urethra during sexual arousal

cross-genderists: transgenderists

cryptorchidism (krip-TOR-ka-diz-um): condition in which the testes have not descended into the scrotum prior to birth

cunnilingus (kun-a-LEAN-gus): oral stimulation of the clitoris, vaginal opening, or other parts of the vulva

cystitis (sis-TITE-us): a nonsexually transmitted infection of the urinary bladder

— D —

decriminalization: reducing the legal sanctions for particular acts while maintaining the possibility of legally regulating behavior through testing, licensing, and reporting of financial gain

deoxyribonucleic acid (DNA) (dee-AK-see-rye-bow-new-KLEE-ik): the chemical in each cell that carries the genetic code

deprivation homosexuality: can occur when members of the opposite sex are unavailable

desire phase: Kaplan's term for the psychological interest in sex that precedes a physiological, sexual arousal

deviation: term applied to behavior or orientations that do not conform to a society's accepted norms; it often has negative connotations

diaphragm (DY-a-fram): a latex rubber cup, filled with spermicide, that is fitted to the cervix by a clinician; the woman must learn to insert it properly for full contraceptive effectiveness

diethylstilbestrol (DES) (dye-eth-a-stil-BES-trole): synthetic estrogen compound given to mothers whose pregnancies are at high risk of miscarrying

dilation and curettage (D & C): a method of induced abortion in the second trimester of pregnancy that involves a scraping of the uterine wall

dilation and evacuation (D & E): a method of induced abortion in the second trimester of pregnancy; it combines suction with a scraping of the inner wall of the uterus

discrimination: the process by which an individual extinguishes a response to one stimulus while preserving it for other stimuli

dysfunction: when the body does not function as expected or desired during sex

dysmenorrhea (dis-men-a-REE-a): painful menstruation

— E —

E. coli: bacteria naturally living in the human colon, often causes urinary tract infection

ectopic pregnancy (ek-TOP-ik): the implantation of a blastocyst somewhere other than in the uterus, usually in the fallopian tube

ejaculation: muscular expulsion of semen from the penis

ejaculatory inevitability: the sensation in the male that ejaculation is imminent

ELISA: the primary test used to determine the presence of AIDS in humans

embryo (EM-bree-o): the term applied to the developing cells, when about a week after fertilization, the blastocyst implants itself in the uterine wall

endometrial hyperplasia (hy-per-PLAY-zhee-a): excessive growth of the inner lining of the uterus (endometrium)

endometriosis (en-doe-mee-tree-O-sus): growth of the endometrium out of the uterus into surrounding organs

endometrium: interior lining of the uterus, innermost of three layers

epidemiology (e-pe-dee-mee-A-la-jee): the branch of medical science that deals with the incidence, distribution, and control of disease in a population

epididymis (ep-a-DID-a-mus): tubular structure on each testis in which sperm cells mature

epididymitis (ep-a-did-a-MITE-us): inflammation of the epididymis of the testis

episiotomy (ee-piz-ee-OTT-a-mee): a surgical incision in the vaginal opening made by the clinician or obstetrician if it appears that the baby will tear the opening in the process of being born

epispadias (ep-a-SPADE-ee-as): birth defect in which the urinary bladder empties through an abdominal opening, and the urethera is malformed

erectile dysfunction: difficulty achieving or maintaining penile erection (impotence)

erection: enlargement and stiffening of the penis as blood engorges the columns of spongy tissue, and internal muscles contract

erogenous zone (a-RAJ-a-nus): any area of the body that is sensitive to sexual arousal

erotica: artistic representations of nudity or sexual activity

estrogen (ES-tro-jen): hormone produced abundantly by the ovaries; it plays an important role in the menstrual cycle

estrogen replacement therapy (ERT): controversial treatment of the physical changes of menopause by administering dosages of the hormone estrogen

ethnocentricity: the tendency of the members of one culture to assume that their values and norms of behavior are the "right" ones in comparison to other cultures

excitement: the arousal phase of Masters and Johnson's 4-phase model of the sexual response cycle

exhibitionism: exposing the genitals to others for sexual pleasure

external values: the belief systems available from one's society and culture

extramarital sex: married person having sexual intercourse with someone other than her or his spouse; adultery

— F —

fallopian tubes: structures that are connected to the uterus and lead the ovum from an ovary to the inner cavity of the uterus

fellatio: oral stimulation of the penis

fetal alcohol syndrome (FAS): a condition in a fetus characterized by abnormal growth, neurological damage, and facial distortion caused by the mother's heavy alcohol consumption

fetal surgery: a surgical procedure performed on the fetus while it is still in the uterus

fetishism (FET-a-shizm): sexual arousal triggered by objects or materials not usually considered to be sexual

fetus: the term given to the embryo after two months of development in the womb

fibrous hymen: unnaturally thick, tough tissue composing the hymen

follicles: capsule of cells in which an ovum matures

follicle-stimulating hormone (FSH): pituitary hormone that stimulates the ovaries or testes

foreplay: sexual activities shared in early stages of sexual arousal, with the term implying that they are leading to a more intense, orgasm-oriented form of activity such as intercourse

foreskin: fold of skin covering the penile glans; also called prepuce

fraternal: a twin formed from two separate ova which were fertilized by two separate sperm

frenulum (FREN-yu-lum): thin, tightly-drawn fold of skin on the underside of the penile glans; it is highly sensitive

frottage (fro-TAZH): gaining sexual gratification from anonymously pressing or rubbing against others, usually in crowded settings

frotteur: one who practices frottage

— G —

gamete intra-fallopian transfer (GIFT): direct placement of ovum and concentrated sperm cells into the woman's fallopian tube, increasing the chances of fertilization

gay: slang term referring to homosexual persons and behaviors

gender dysphoria (dis-FOR-ee-a): term to describe gender-identity/role that does not conform to the norm considered appropriate for one's physical sex

gender transposition: gender dysphoria

gender-identity/role (G-I/R): a person's inner experience and outward expression of maleness, femaleness, or some ambivalent position between the two

general sexual dysfunction: difficulty for a woman in achieving sexual arousal

generalization: application of specific learned responses to other, similar situations or experiences

genetic engineering: the modification of the gene structure of cells to change cellular functioning

genital herpes (HER-peez): viral STD characterized by painful sores on the sex organs

genital warts: small lesions on genital skin caused by papilloma virus, this STD increases later risks of certain malignancies

glans: in the male, the sensitive head of the penis; in the female, sensitive head of the clitoris, visible between the upper folds of the minor lips

gonadotropin releasing hormone (GnRH) (go-nad-a-TRO-pen): hormone from the hypothalamus that stimulates the release of FSH and LH by the pituitary

gonorrhea (gon-uh-REE-uh): bacterial STD causing urethral pain and discharge in males; often no symptoms in females

granuloma inguinale (gran-ya-LOW-ma in-gwa-NAL-ee or -NALE): venereal disease characterized by ulcerations and granulations beginning in the groin and spreading to the buttocks and genitals

group marriage: three or more people in a committed relationship who share sex with one another

G spot: a vaginal area that some researchers feel is particularly sensitive to sexual stimulation

— H —

hard-core pornography: pornography that makes use of highly explicit depictions of sexual activity or shows lengthy scenes of genitals

hedonists: believers that pleasure is the highest good

hemophiliac (hee-mo-FIL-ee-ak): someone with the hereditary sex-linked blood defect hemophilia, affecting males primarily and characterized by difficulty in clotting

heterosexual: attractions or activities between members of opposite sexes

HIV: human immunodeficiency virus

homophobia (ho-mo-PHO-bee-a): strongly held negative attitudes and irrational fears relating to homosexuals

homosexual: term applied to romantic and sexual attractions and activities between members of the same sex

homosexualities: a term that reminds us there is not a single pattern of homosexuality, but a wide range of same-sex orientations

hookers: street name for female prostitutes

hormone implants: contraceptive method in which hormone-releasing plastic containers are surgically inserted under the skin

hot flash: a flushed, sweaty feeling in the skin caused by dilated blood vessels, often associated with menopause

human chorionic gonadotropin (HCG): a hormone detectable in the urine of a pregnant woman

human immunodeficiency virus: the virus that initially attacks the human immune system, eventually causing AIDS

hustlers: male street prostitutes

H-Y antigen: a biochemical produced in an embryo when the Y chromosome is present; it causes fetal gonads to develop into testes

hymen: membranous tissue that can cover part of the vaginal opening

hypersexuality: exaggeratedly high level of interest in and drive for sex

hyposexuality: an especially low level of sexual interest and drive

hypospadias (hye-pa-SPADE-ee-as): birth defect caused by incomplete closure of the urethra during fetal development

— I —

identical: a twin formed by a single ovum which was fertilized by a single sperm before the cell divided in two

imperforate hymen: lack of any openings in the hymen

impotence (IM-pa-tens): difficulty achieving or maintaining erection of the penis

in loco parentis: a Latin phrase meaning in the place of the parent

in vitro fertilization (IVF): a process whereby the union of the sperm and egg occurs outside the mother's body

incest (IN-sest): sexual activity between closely related family members

incest taboo: cultural prohibitions against incest, typical of most societies

induced abortion: a termination of pregnancy by artificial means

infertility: the inability to produce offspring

infibulation: surgical procedure, performed in some cultures, that seals the opening of the vagina

informed consent: complete information about the purpose of a study and how they will be asked to perform given to prospective human research subjects

inhibited sexual desire (ISD): loss of interest and pleasure in formerly arousing sexual stimuli

internal values: the individualized beliefs and attitudes that a person develops by sorting through external values and personal needs

interstitial cells: cells between the seminiferous tubules that secrete testosterone and other male hormones

interstitial-cell-stimulating hormone (ICSH): pituitary hormone that stimulates the testes to secrete testosterone; known as luteinizing hormone (LH) in females

intrauterine devices (IUDs): birth control method involving insertion of a small plastic device into the uterus

introitus (in-TROID-us): outer opening of the vagina

invasive cancer of the cervix (ICC): advanced and dangerous malignancy requiring prompt treatment

— K —

Kaposi's sarcoma: a rare form of cancer of the blood vessels, characterized by small, purple skin lesions

kiddie porn: term used to describe the distribution and sale of photographs and films of children or younger teenagers engaging in some form of sexual activity

kleptomania: extreme form of fetishism, in which sexual arousal is generated by stealing

— L —

labor: uterine contractions in a pregnant woman; an indication that the birth process is beginning

lactation: production of milk by the milk glands of the breasts

Lamaze method (la-MAHZ): a birthing process based on relaxation techniques practiced by the expectant mother; her partner coaches her throughout the birth

laminaria (lam-a-NER-ee-a): a dried seaweed sometimes used in dilating the cervical opening prior to vacuum curettage

laparoscopy: simpler procedure for tubal ligation, involving the insertion of a small scope into the abdomen, through which the surgeon can see the fallopian tubes and close them off

laparotomy: operation to perform a tubal ligation, or female sterilization, involving an abdominal incision

latency period: Freudian concept that during middle childhood, sexual energies are dormant; recent research tends to suggest that latency does not exist

lesbian (LEZ-bee-un): refers to female homosexuals

libido (la-BEED-o or LIB-a-do): a term first used by Freud to define human sexual longing, or sex drive

lumpectomy: surgical removal of a breast lump, along with a small amount of surrounding tissue

luteinizing hormone (LH): pituitary hormone that triggers ovulation in the ovaries and stimulates sperm production in the testes

lymphogranuloma venereum (LGV) (lim-foe-gran-yu-LOW-ma-va-NEAR-ee-um): contagious venereal disease caused by several strains of *Chlamydia* and marked by swelling and ulceration of lymph nodes in the groin

— M —

major lips: two outer folds of skin covering the minor lips, clitoris, urethral opening, and vaginal opening

mammography: sensitive X-ray technique used to discover small breast tumors

marital rape: a woman being forced to have sex by her husband

masochist: the individual in a sadomasochistic sexual relationship who takes the submissive role

massage parlors: places where women can be hired to perform sexual acts

mastectomy: surgical removal of all or part of a breast

menage à trois (may-NAZH-ah-TRWAH): troilism

menarche (MEN-are-kee): onset of menstruation at puberty

menopause (MEN-a-poz): time in midlife when menstruation ceases

menstrual cycle: the hormonal interactions that prepare a woman's body for possible pregnancy at roughly monthly intervals

menstruation (men-stru-AY-shun): phase of menstrual cycle in which the inner uterine lining breaks down and sloughs off; the tissue, along with some blood, flows out through the vagina; also called the period

midwives: medical professionals, both women and men, trained to assist with the birthing process

minor lips: two inner folds of skin that join above the clitoris and extend along the sides of the vaginal and urethral openings

miscarriage: a natural termination of pregnancy

modeling theory: suggests that people will copy behavior they view in pornography

molluscum contagiosum (ma-LUS-kum kan-taje-ee-O-sum): a skin disease transmitted by direct bodily contact, not necessarily sexual, that is characterized by eruptions on the skin that appear similar to whiteheads with a hard seed-like core

monogamous: sharing sexual relations with only one person

monorchidism (ma-NOR-ka-dizm): presence of only one testis in the scrotum

mons: cushion of fatty tissue located over the female's pubic bone

moral values: beliefs associated with ethical issues, or rights and wrongs; they are often a part of sexual decision making

morula (MOR-yul-a): a spherical, solid mass of cells formed by 3 days of embryonic cell division

Müllerian ducts (myul-EAR-ee-an): embryonic structures that develop into female sexual and reproductive organs unless inhibited by male hormones

Müllerian inhibiting substance: hormone produced by fetal testes that prevents further development of female structures from the Müllerian ducts

myometrium: middle, muscular layer of the uterine wall

— N —

National Birth Control League: an organization founded in 1914 by Margaret Sanger to promote use of contraceptives

natural childbirth: a birthing process that encourages the mother to take control thus minimizing medical intervention

necrophilia (nek-ro-FILL-ee-a): having sexual activity with a dead body

nongonococcal urethritis (NGU) (non-gon-uh-KOK-ul yur-i-THRYT-us): urethral infection or irritation in the male urethra caused by bacteria or local irritants

normal: a subjective term used to describe sexual behaviors and orientations. Standards of normalcy are determined by social, cultural, and historical standards

normal asexuality: an absence or low level of sexual desire considered normal for a particular person

normalization: integration of mentally retarded persons into the social mainstream as much as possible

nymphomania (nim-fa-MANE-ee-a): compulsive need for sex in women; apparently quite rare

— O —

obscenity: depiction of sexual activity in a repulsive or disgusting manner

onanism (O-na-niz-um): a term sometimes used to describe masturbation, it comes from the biblical story of Onan who practiced coitus interruptus and "spilled his seed on the ground"

oocytes (OH-a-sites): cells that mature to become ova

open-ended marriage: each partner in the primary relationship grants the other freedom to have emotional and sexual relationships with others

opportunistic infection: a disease resulting from lowered resistance of a weakened immune system

organic disorder: physical disorder caused by the organs and organ systems of the human body

organizing effect: manner in which hormones control patterns of early development in the body

orgasm: (OR-gaz-em) pleasurable sensations and series of contractions that release sexual tension, usually accompanied by ejaculation in men

orgasmic release: reversal of the vasocongestion and muscular tension of sexual arousal, triggered by orgasm

orgy (OR-jee): group sex

os: opening in the cervix that leads into the hollow interior of the uterus

osteoporosis (ah-stee-o-po-ROW-sus): disease caused by loss of calcium from the bones in post-menopausal women, leading to brittle bone structure and stooped posture

ova: egg cells produced in the ovary; in reproduction, it is fertilized by a sperm cell; one cell is an ovum

ovaries: pair of female gonads, located in the abdominal cavity, that produce ova and female hormones

ovulation: release of a mature ovum through the wall of an ovary

ovum transfer: use of an egg from another woman for conception, with the fertilized ovum being implanted in the uterus of the woman wanting to become pregnant

oxytocin: pituitary hormone that plays a role in lactation and in uterine contractions

— P —

pansexual: lacking highly specific sexual orientations or preferences; open to a range of sexual activities

PAP smear: medical test that examines a smear of cervical cells, to detect any cellular abnormalities

paraphilia (pair-a-FIL-ee-a): a newer term used to describe sexual orientations and behaviors that vary from the norm; it means "a love beside"

paraplegic: a person paralyzed in the legs, and sometimes pelvic areas, as the result of injury to the spinal cord

partial zone dissection (PZD): a technique used to increase the chances of fertilization by making a microscopic incision in the zona pellucida of an ovum. This creates a passageway through which sperm may enter the egg more easily

pedophilia (peed-a-FIL-ee-a): another term for child sexual abuse

pelvic inflammatory disease (PID): a chronic internal infection associated with certain types of IUDs

penis: male sexual organ that can become erect when stimulated; it leads urine and sperm to the outside of the body

perimetrium: outer covering of the uterus

perinatally: a term used to describe things related to pregnancy, birth, or the period immediately following the birth

perineal areas (pair-a-NEE-al): the sensitive skin between the genitals and the anus

Peyronie's disease (pay-ra-NEEZ): development of fibrous tissue in spongy erectile columns within the penis

phimosis (fy-MOS-us): a condition in which the penile foreskin is too tight to retract easily

pimps: men who have female prostitutes working for them

placenta (pla-SENT-a): the organ that unites the fetus to the mother by bringing their blood vessels closer together; it provides nourishment and removes waste for the developing baby

plateau phase: the stable, leveled-off phase of Masters and Johnson's 4-phase model of the sexual response cycle

polygamy: practice, in some cultures, of being married to more than one spouse

pornography: photographs, films, or literature intended to be sexually arousing through explicit depictions of sexual activity

potentiation: establishment of stimuli early in life that form ranges of response for later in life

pregnancy-induced hypertension: a disorder that can occur in the latter half of pregnancy marked by a swelling in the ankles and other parts of the body, high blood pressure, and protein in the urine; can progress to coma and death if not treated

premature birth: a birth that takes place prior to the 36th week of pregnancy

premature ejaculation: difficulty that some men experience in controlling the ejaculatory reflex, resulting in rapid ejaculation

premenstrual syndrome (PMS): symptoms of physical discomfort, moodiness, and emotional tensions that occur in some women for a few days prior to menstruation

preorgasmic: a term often applied to women who have not yet been able to reach orgasm during sexual response

prepuce (PREE-peus): in the female, tissue of the upper vulva that covers the clitoral shaft

priapism (pry-AE-pizm): continual, undesired, and painful erection of the penis

primary dysfunction: a difficulty with sexual functioning that has always existed for a particular person

progesterone (pro-JES-ter-one): ovarian hormone that causes uterine lining to thicken

progestin injection: use of injected hormone that can prevent pregnancy for several months; not yet approved for use in the United States

prolactin: pituitary hormone that stimulates the process of lactation

prolapse of the uterus: weakening of the supportive ligaments of the uterus, causing it to protrude into the vagina

promiscuity (prah-mis-KIU-i-tee): sharing casual sexual activity with many different partners

prostaglandin: hormone-like chemical whose concentrations increase in a woman's body just prior to menstruation

prostaglandin or saline-induced abortion: used in the 16-24th weeks of pregnancy, prostaglandins, salt solutions, or urea is injected into the amniotic sac, administered intravenously, or inserted into the vagina in suppository form, to induce contractions and fetal delivery

prostate: gland located beneath the urinary bladder in the male; it produces some of the secretions in semen

prostatitis (pras-tuh-TITE-us): inflammation of the prostate gland

pseudonecrophilia: a fantasy about having sex with the dead

psychosexual development: complex interaction of factors that form a person's sexual feelings, orientations, and patterns of behavior

psychosocial development: the cultural and social influences that help shape human sexual identity

puberty: time of life when reproductive capacity develops and secondary sex characteristics appear

pubic lice: small insects that can infect skin in the pubic area, causing a rash and severe itching

pubococcygeus (PC) muscle (pyub-o-kox-a-JEE-us): part of the supporting musculature of the vagina that is involved in orgasmic response and over which a woman can exert some control

pyromania: sexual arousal generated by setting fires

— Q —

quadriplegic: a person paralyzed in the upper body, including the arms, and lower body as the result of spinal cord injury

— R —

random sample: a representative group of the larger population that is the focus of a scientific poll or study

rape trauma syndrome: the predictable sequence of reactions that a victim experiences following a rape

recreational marriage: extramarital sex with a low level of emotional commitment performed for fun and variety

refractory period: time following orgasm during which a man cannot be restimulated to orgasm

reinforcement: in conditioning theory, any influence that helps shape future behavior as a punishment or reward stimulus

resolution phase: the term for the return of a body to its unexcited state following orgasm

retarded ejaculation: a male who has never been able to reach an orgasm

retrograde ejaculation: abnormal passage of semen into the urinary bladder at the time of ejaculation

retrovirus (RE-tro-vi-rus): a class of viruses that reproduces with the aid of the enzyme reverse transcriptase, which allows the virus to integrate its genetic code into that of the host cell, thus establishing permanent infection

Rh incompatibility: condition in which a blood protein of the infant is not the same as the mother's; antibodies formed in the mother can destroy red blood cells in the fetus

Rho GAM: medication administered to a baby soon after delivery to prevent formation of antibodies when the baby is Rh positive and its mother Rh negative

rhythm method: a natural method of birth control that depends on an awareness of the woman's menstrual-fertility cycle

RU-486: a progesterone antagonist used as a postcoital contraceptive

rubber dam: small square sheet of latex used to cover the vulva, vagina, or anus to help prevent transmission of HIV during sexual activity

— S —

sadist: the individual in a sadomasochistic sexual relationship who takes the dominant role

sadomasochism (sade-o-MASS-o-kiz-um): refers to sexual themes or activities involving bondage, pain, domination, or humiliation of one partner by the other

sample: a small representative group of a population that is the focus of a scientific poll or study

satyriasis (sate-a-RYE-a-sus): compulsive need for sex in men; apparently rare

scabies (SKAY-beez): a skin disease caused by a mite that burrows under the skin to lay its eggs causing redness and itching; transmitted by bodily contact that may or may not be sexual

scrotum (SKROTE-um): pouch of skin in which the testes are contained

secondary dysfunction: develops after some period of normal sexual function

selective reduction: use of abortion techniques to reduce the number of fetuses when there are more than three in a pregnancy, thus increasing the chances of survival for the remaining fetuses

self-gratification: giving oneself pleasure, as in masturbation; a term typically used today instead of more negative descriptors

self-pleasuring: self-gratification; masturbation

semen: (SEE-men): mixture of fluids and sperm cells ejaculated through the penis

seminal vesicle (SEM-un-al): gland at the end of each vas deferens that secretes a chemical that helps sperm to become mobile

seminiferous tubules (sem-a-NIF-a-rus): tightly coiled tubules in the testes in which sperm cells are formed

sensate focus: early phase of sex therapy treatment, in which the partners pleasure each other without involving direct stimulation of sex organs

sex therapist: professional trained in the treatment of sexual dysfunctions

sexual addiction: inability to regulate sexual behavior

sexual dysfunctions: difficulties people have in achieving sexual arousal

sexual harassment: unwanted sexual advances or coercion that can occur in the workplace or academic settings

sexual individuality: the unique set of sexual needs, orientations, fantasies, feelings, and activities that develops in each human being

sexual phobias and aversions: exaggerated fears of forms of sexual expression

sexual revolution: the changes in thinking about sexuality and sexual behavior in society that occurred in the 1960s and 1970s

sexual surrogates: paid partners used during sex therapy with clients lacking their own partners; only rarely used today

shaft: in the female, the longer body of the clitoris, containing erectile tissue; in the male, cylindrical base of penis that contains 3 columns of spongy tissue: 2 corpora cavernosa and a corpus spongiosum

shunga: ancient scrolls used in Japan to instruct couples in sexual practices through the use of paintings

situational homosexuality: deprivation homosexuality

Skene's glands: secretory cells located inside the female urethra

smegma: thick, oily substance that may accumulate under the prepuce of the clitoris or penis

social learning theory: suggests that human learning is influenced by observation of and identification with other people

social scripts: a complex set of learned responses to a particular situation that is formed by social influences

sodomy laws: prohibit a variety of sexual behaviors in some states, that have been considered abnormal or antisocial by legislatures. These laws are often enforced discriminatorily against particular groups, such as homosexuals

sonograms: ultrasonic rays used to project a picture of internal structures such as the fetus; often used in conjunction with amniocentesis or fetal surgery

spectatoring: term used by Masters and Johnson to describe self-consciousness and self-observation during sex

sperm: reproductive cells produced in the testes; in fertilization, one sperm unites with an ovum

spermatocytes (sper-MAT-o-sites): cells lining the seminiferous tubules from which sperm cells are produced

spermicidal jelly (cream): sperm-killing chemical in a gel base or cream, used with other contraceptives such as diaphragms

spermicides: chemicals that kill sperm; available as foams, creams, jellies, or implants in sponges or suppositories

sponge: a thick polyurethane disc that holds a spermicide and fits over the cervix to prevent conception

spontaneous abortion: another term for miscarriage

Staphylococcus aureus (staf-a-low-KAK-us): the bacteria that can cause toxic shock syndrome

statutory rape: a legal term used to indicate sexual activity when one partner is under the age of consent; in most states that age is 18

sterilization: rendering a person incapable of conceiving, usually by interrupting passage of the egg or sperm

straight: slang term for heterosexual

streetwalkers: female prostitutes who work on the streets

suppositories: contraceptive devices designed to distribute their spermicide by melting or foaming in the vagina

syndrome (SIN-drome): a group of signs or symptoms that occur together and characterize a given condition

syphilis (SIF-uh-lus): sexually transmitted disease (STD) characterized by four stages, beginning with the appearance of a chancre

systematic desensitization: step-by-step approaches to unlearning tension-producing behaviors and developing new behavior patterns

— T —

testes (TEST-ees): pair of male gonads that produce sperm and male hormones

testicular cancer: malignancy on the testis that may be detected by testicular self examination

testicular failure: lack of sperm and/or hormone production by the testes

testosterone (tes-TAS-ter-one): major male hormone produced by the testes; it helps to produce male secondary sex characteristics

testosterone replacement therapy: administering testosterone injections to increase sexual interest or potency in older men; not considered safe for routine use

theoretical failure rate: a measure of how often a birth control method can be expected to fail, when used without error or technical problems

thrush: a disease caused by a fungus and characterized by white patches in the oral cavity

toxic shock syndrome (TSS): an acute disease characterized by fever and sore throat, and caused by normal bacteria in the vagina which are activated if tampons or some contraceptive devices such as diaphragms or sponges are left in for long periods of time

transgenderists: people who live in clothing and roles considered appropriate for the opposite sex for sustained periods of time

transsexuals: feel as though they should have the body of the opposite sex

transvestism: dressing in clothes appropriate to the opposite sex, usually for sexual gratification

transvestite: an individual who dresses in clothing considered appropriate for the opposite sex, and adopts similar mannerisms, often for sexual pleasure

trichomoniasis (trik-uh-ma-NEE-uh-sis): a vaginal infection caused by the *Trichomonas* organism

troilism (TROY-i-lizm): sexual activity shared by three people

tubal ligation: a surgical separation of the fallopian tubes to induce permanent female sterilization

— U —

umbilical cord: tubelike tissues and blood vessels arising from the embryo's navel connecting it to the placenta

urethra (yu-REE-thrah): tube that passes from the urinary bladder to the outside of the body

urethral opening: opening through which urine passes to the outside of the body

urophilia: sexual arousal connected with urine or urination

uterus (YUTE-a-rus): muscular organ of the female reproductive system; a fertilized egg implants itself within the uterus

— V —

vacuum curettage: (kyur-a-TAZH): a method of induced abortion performed with a suction pump

vagina (vu-JI-na): muscular canal in the female that is responsive to sexual arousal; it receives semen during heterosexual intercourse for reproduction

vaginal atresia (a-TREE-zha): birth defect in which the vagina is absent or closed

vaginal atrophy: shrinking and deterioration of vaginal lining, usually the result of low estrogen levels during aging

vaginal fistulae (FISH-cha-lee *or* -lie): abnormal channels that can develop between the vagina and other internal organs

vaginismus (vaj-uh-NIZ-mus): involuntary spasm of the outer vaginal musculature, making penetration of the vagina difficult or impossible

vaginitis (vaj-uh-NITE-us): general term for inflammation of the vagina

values: system of beliefs with which people view life and make decisions, including their sexual decisions

variable: an aspect of a scientific study that is subject to change

variation: a less pejorative term to describe nonconformity to accepted norms

varicose veins: overexpanded blood vessels; can occur in veins surrounding the vagina

vas deferens: tube that leads sperm upward from each testis to the seminal vesicles

vasectomy (va-SEK-ta-mee *or* vay-ZEK-ta-mee): a surgical division of the vas deferens to induce permanent male sterilization

villi: the fingerlike projections of the chorion that form a major part of the placenta

viral hepatitis: inflammation of the liver caused by a virus

voyeurism (VOI-yur-izm): gaining sexual gratification from seeing others nude or involved in sexual acts

vulva: external sex organs of the female, including the mons, major and minor lips, clitoris, and opening of the vagina

— W —

Western blot: test used to verify positive AIDS virus detected first by the ELISA

Wolffian ducts (WOOL-fee-an): embryonic structures that develop into male sexual and reproductive organs if male hormones are present

— Y —

yeast infection: a type of vaginitis caused by an overgrowth of a fungus normally found in an inactive state in the vagina

— Z —

zona pellucida (ZO-nah pe-LOO-sa-da): transparent, outer membrane of an ovum

zoophilia (zoo-a-FILL-ee-a): bestiality

Index

abortion: attitudes of African Americans toward, 125; life before and possibly after *Roe v. Wade* and, 119–122; global politics of, 123–127; sexuality education in various countries and, 14, 15, 17; teenagers and, 119–122, 150–151, 158; transforming pro-choice movement and, 25–30
abortion pill, 116–118
abstinence, 107, 110
abuse. *See* sexual abuse
accidental infidelity, 169
acquaintance rape, feminism and, 224–225
Adolescent Family Life Act (AFLA), 52
adolescents. *See* teenagers
adultery. *See* infidelity
African Americans: access of female, to abortion, 28; attitudes of, toward abortion, 125; female condom and, 112
Agape lover, 93, 94
aging, sexuality and, 176–178, 179–183
AIDS, 51, 147; college campuses and, 189–193; medical progress in, 186; prospects for vaccines for, 187; social behavior and, 188; treatment of, 186–187
Allen, Laura, 214, 217
Alperstein, Linda Perlin, 145, 147
alpha-fetoprotein, 214
altruistic lover, 93, 94
American Psychiatric Association, on homosexuality, 68, 76
amygdala, 215
Ancient Near East, homosexuality in, 58
androgens, 221
anxiety, love and, 92
aphasia, 218
apraxia, 218, 219
Arabs, homosexuality among, 59
Aron, Arthur, 92
artha, 7
arthritis, sex and, 181
anthropology, study of courtship and marriage by, 89
attachment style, intimacy and, 86–87
"attachment theory," influence of mothers on love style and, 95–96
attractiveness, physical: effect of aging on sexuality and, 177; male mating strategies and, 82, 83; sexuality of "thirtynothings" and, 163–164

basal body temperature method, 110
Baulieu, Etienne-Emile, 116, 117
Belgium, sexuality education in, 14–15, 18, 19
Berenbaum, Sheri A., 213, 216
Bernstein, Anne, 147–148
Berscheid, Ellen, 90, 91
Bible: homosexuality and, 58, 60; intersexuality and, 45
biology, homosexuality and, 66–75, 76–79
birth control: lack of use of, by "thirtynothings," 163. *See also* specific type
birth control pills, 104, 107, 108–109, 196
bisexuality, 198
blacks. *See* African Americans
bonobos, 77
Brahma, 7
brain, gender differences in, 213–219
breasts, 48–49

breech births, cesarean section and, 137, 138
Bright, Susie, 40
Buddhism, 6, 8, 9, 10
Byne, William, 70, 79

cancer, sex and, 181
Catholics: abortion and, 123–124; sexuality education and, 14, 15
cervical cap, 106, 108
cervical mucus method, 110
cesarean sections, unnecessary, 136–141
child abuse. *See* sexual abuse
China, love in, 89
chlamydia, 201
Clinton, Michael, 31, 33
Clomid, 128, 129
cloning, 131
coitus reservatus, 8
college students: AIDS and, 189–193; date rape and, 224–225; sexual experience in, and sexual arousal, 53–56
Commonwealth of Independent States. *See* Soviet Union
condoms, 104, 105, 106, 195–197; female, 104, 105, 106, 111–112
Confucius, 6, 8
congenital adrenal hyperplasia (CAH), 71–72, 223
congenital syphilis, 200
consumers, men as, 31–35
contraceptive sponge, 105, 106
contraceptives. *See* birth control; specific type
Contraction and Betrayal Stage, in relationships, 99–100
corpus callosum, 217
Czechoslovakia, sexuality education in former, 15, 18, 19

date rape, feminism and, 224–225
Davis, Keith, 86–87
de LaCoste, Marie-Christine, 69, 217, 218
demographic study, of men as consumers, 31–35
Denmark: sexuality education in, 15–16, 19. *See also* Scandinavia
Depo-Provera, 104, 107, 109, 196
Dewhurst, Christopher I., 46
dharma, 7
diabetes, sex and, 181
Diagnostic and Statistical Manual of Mental Disorders (DSM), on homosexuality, 68, 76
diaphragms, 105, 106, 108
doulas, 138
Durga, 7
dystocia, 137

Eals, Marion, 213
Eastern traditions, sexuality and spirituality and, 6–12
egg donors, 131
ejaculations, aging and, 176, 180
elderly, sexuality of, 176–178, 179–183
ensoulment, abortion and, 123
erections, aging and, 176–177, 180, 182
Eros lover, 93, 94
eroticism, gender differences in, 85
estrogen replacement therapy (ERT), 181
evolutionary psychology, mating strategies, 82–85
Expansion Stage, in relationships, 99, 100
external cephalic version, 138

family policy, comparison of, in United States, Scandinavia, and former Soviet Union, 21–23
fathers, single-parent families and, 159–160
Fausto-Sterling, Anne, 70
Fehr, Beverly, 92–93
female condom, 104, 105, 106, 111–112
female pseudohermaphrodites (ferms), 44–47
Femidom, 112. *See also* female condom
feminism: date rape and, 224–225; equality of women and, 226, 228
fetal distress, cesarean sections and, 137, 138
Finland: parental leave in, 21. *See also* Scandinavia
Fisher, Helen, 89
Fithian, Marilyn, 10
Freud, Sigmund, 76, 145, 220
friendships: between men and women, 227; love and, 92–93

Galaburda, Albert M., 217–218
gender differences: in brain, 213–219; in eroticism, 85; relationship of sexual arousal with sexual experience in college students and, 53, 56
gender equality, comparison of, in Scandinavia, United States, and former Soviet Union, 20–24
genital abnormalities, 44–47
Gerschwind, Norman, 217–218
global politics, of abortion, 123–127
gonorrhea, 198, 201
Gordon, Ronald R., 46
Gorski, Roger, 68–69, 70–71, 214, 217
Greece, homosexuality and, 58
Greenberg, David, 58–59

Haffner, Deborah, 144, 145, 147
Hall, Jeffrey, 73, 74–75
Hartman, William, 10
Hatfield, Elaine, 88–89, 91
Hazan, Cindy, 86, 87, 95–96
heart attacks, sex after, 180, 182
heart disease, sex and, 181
Hendrick, Clyde and Susan, 93–95
hermaphrodites, 44–47
high blood pressure, effect of medications for, on sex, 181
Hinduism, as Eastern sexual philosophy, 6, 7, 8, 10
Hines, Melissa, 216
hippocampus, 215
"homophobia," 64–65
homosexuality, 8, 50–51, 52, 198, 215; in ancient cultures, 58–59; biology and, 66–75, 76–79; children and, 147; effect of, on women, 60; Judaism and, 57–65; marriage and, 63–64; incidence of, 36, 37
Hooker, Evelyn, 67–68, 75
hormones, 213–219, 221
hysterectomies, sex after, 181
hypothalamus: differences in, of homosexuals, 50, 70, 78–79; gender differences in, 214, 215

ideological feminism, 226, 228
in vitro fertilization (IVF): ethical issues involved in, 128, 131; risks of, 128–130
incest, 9
India, love in, 89

infidelity, 37, 174–175; accidental, 169; emotionally retarded men and, 173–174; marital arrangements and, 171–172; mating strategies and, 84–85; myths regarding, 170–171; philandering and, 172–173; romantic, 169–171; spider women and, 173; among "thirtynothings," 164
intersexuality, 44–47
intimacy, attachment style and, 86–87
Isay, Richard, 67
IUD (intrauterine device), 104, 107, 109–110, 124, 196

Jankowiak, William, 89, 91
Jaspers, Karl, 6, 11
Judaism: homosexuality and, 57–65; intersexuality and, 45

Kali, 7, 9
Kama Sutra, 7
Kinsey, Alfred, 36, 37, 50, 67, 77–78
Krafft-Ebing, Richard von, 51
Kundalini, 9–10

labelling, prevention ability of contraceptives and, 196
Lao Tzu, 6, 7
latency period, 145–146
Laumann, Edward, 51–52
lesbians, 51, 72–73, 77. *See also* homosexuals
LeVay, Simon, 50, 69–70, 73, 75, 76, 78–79, 218
lingam, 7
lordosis, 217
love, study of, by science, 88–89, 90–96
Ludus lover, 93, 94
luteinizing-hormone (LH) feedback, 72

male pseudohermaphrodites (merms), 44–47
Mania lover, 93, 94
mantra, 9
marital arrangements, infidelity and, 171–172
marriage: for homosexuals, 63–64; love and, 89; sex among "thirtynothings" inside, 164; ways to make for sexier, 166–168
mating strategies, evolution and, 82–85
memories, recovery of, childhood sexual abuse, 203–206
men: abortion and, 126; changing role of, 31–35; conflict between women and, 220–223, 229–232; date rape and, 224–225; effect of sexual abuse on, 207–209; equality movement and, 226, 228; eroticism and, 85; friendship of, with women, 227; mating strategies of, 82, 83–85; physiological changes in, during aging process, 176–177, 180; survey of sexual practices of, 36–38
menopause, sexuality and, 179
men's movement, 230, 231
Mexico, love in, 89
Meyre-Bahlburg, Heino, 71, 72
minilaparotomy, 110
minipill, 108–109, 124. *See also* birth control pill
miscarriage, coping with, 133–135
mithuna, 7
moksha, 7
mothers, "attachment theory" of love and, 95–96

Müllerian regression factor, 72–73, 214

natural family planning, 107, 110
Near East, homosexuality in, 58
Netherlands, sexuality education in, 13–14, 18, 19
neurosyphilis, 200–201
Norplant, 107, 109, 113–115, 160–161; prevention of sexually transmitted diseases and, 104, 197
"not-thirtysomethings," sexual patterns of, 162–165

Oedipus conflict, homosexuality and, 76
Oneida community, 8
orgasms, physiological changes during aging process and, 180
original sin, 6, 7, 10
ovary transplants, 131
parental leave, in Scandinavia, 21

Parvati, 7
passionate love, 91
Passionate Love Scale (PLS), 91
Pattatucci, Angela, 74, 75
penile implants, 183
penile injection therapy, 182–183
Perganol, 128, 129
periodic abstinence, 107, 110
philandering, 172; among women, 172–173
physical attractiveness: effect of aging on sexuality and, 163–164, 177; male mating strategies and, 82, 83
physiology, changes in, during aging process, 176–177, 180, 182
Pilliard, Richard, 72, 73, 75, 78
Planned Parenthood, 124
politics, global, of abortion, 123–127
post traumatic stress disorder (PTSD), 204; sexual abuse of boys and, 208
Pragma lover, 93, 94
pregnancy: syphilis, 200, 201; teenage, 157–160
primary syphilis, 199
pro-choice movement: transforming, 25–30. *See also* abortion
prophetic feminism, 226
prostate surgery, sex after, 180–181
Proxmire, William, 90
puberty, 146–147
Putnam, Frank, 204–205
pygmy chimpanzees, 77

quickening, 123

rape, 120–121, 210–212, 224–225
Reality Female Condom, 104, 105, 106, 111–112
recovered-memory phenomenon, in childhood sexual abuse cases, 205–206
refractory period, aging and, 179, 182
Reinisch, June, 215–216
relationships, cycles of, 97–101
repression, recovered-memory phenomenon in childhood sexual abuse cases and, 203–206
reproductive technology: ethics of, 128, 131–132; risks of, 128–130
rhythm method, of birth control, 110
Rite of the Five Essences, 9
Roe v. Wade, 119–122
Roman Empire, homosexuality in, 59
romantic infidelity, 169–171

romantic love, 88, 89, 91–92
Roussel Uclaf, 116, 117, 118
RU 486, 116–118

Sanders, Stephanie A., 37
Scandinavia, gender equality in, vs. United States and former Soviet Union, 20–24
science, study of love by, 88–89
secondary syphilis, 199, 200
Seinfeld, Jerry, 41
self-esteem, breast size in women and, 48–49
sex hormones, 221; brain development and, 213–219
sex positive movement, 40
sex-organ defects, 44–47
sexual abuse: effect of, on boyhood victims in adulthood, 207–209; recovered-memory phenomenon in, 203–206
sexual arousal, relationship of, with sexual experiences of college students, 53–56
sexual healing, 40
sexual violence, 210–212. *See also* rape
sexuality, aging and, 176–178, 179–183
sexuality education, cross-cultural perspectives on, 13–19
sexually transmitted diseases, 51; preventing, 104, 195–197; teenagers and, 151–152, 160–161. *See also* specific diseases
Shakti, 7, 9
Shaver, Phillip R., 86, 87, 95–96
Shiva, 7, 9
Silverman, Irwin, 213
single parents, children and, 156–161
social attitudes: AIDS and, 188; sexuality in aging and, 177–178
Soviet Union: gender equality in former, vs. Scandinavia and United States, 20–24; sexuality education in former, 17–18, 19
sperm competition, 84
spermicides, 104, 106
spider women, infidelity and, 173
spirituality, sexuality in Eastern philosophies and, 6–12
splenium, 69, 217
sponge, contraceptive, 105, 106
sports, men and, 34
Sprecher, Susan, 91
Spur Posse, 40
state-dependent learning, repressed memories and, 204–205
sterilization, 110
Storge lover, 93, 94
Supreme Court, abortion and, 119–122
surgical sterilization, 110
Sweden: parental leave in, 21; sexuality education in, 16–17, 19. *See also* Scandinavia
syphilis, 198–202

Tantrism, as Eastern sexual philosophy, 6–12
Taoism, as Eastern sexual philosophy, 6–12
"teen conception index," 154
teenagers: abortion and, 119–122, 150–151, 158; causes of pregnancy in, 158; costs of pregnancy in, 150–152, 158–159; Norplant and, 160–161; pregnancy and unmarried, 151,

157–158; prevention of pregnancy in, 160; prevention of sexually transmitted diseases in, 151–152, 195–197; sexual activity in, 52, 149–150; sexuality education and, 13–19; welfare and, 151
Terr, Lenore, 204
tertiary syphilis, 199
"thirtynothings," sexual patterns of, 162–165
trial of labor, 138
tubal ligation, 110
Tucker, Raymond, 93

United States, gender equality in, vs. Scandinavia and former Soviet Union, 20–24
Upanishads, 9

vaccines, for AIDS, 187

vagina, changes in, during aging process, 179, 180, 182
vaginal birth after cesarean (VBAC), 137
vasectomy, 110
VBAC (vaginal birth after cesarean), 137
violence: sexual, 210–212; toward homosexuals, 64. *See also* rape
Vishnu, 7

Way of Knowledge, Hindu, 7
Webster v. Reproductive Health Services, 121
Where the Boys Are, 225
Wilson, Pamela, 144, 146, 147, 148
Williams, Robin, 154
women: breasts and, 48–49; conflict of, with men, 220–223, 229–232; date rape and, 224–225; effect of homosexuality on, 60–61; equality movement and, 226, 228; eroticism and, 85; friendship of, with men, 227; mating strategies of, 82–85; physical changes in, during aging process, 176, 177, 180; sexuality of in "thirtynothings," 163–165; unnecessary cesarean sections for, 136–141
work policies, comparisons of, in Scandinavia, United States, and former Soviet Union, 20–21

yabyum, 9
yantras, 7
Yin and Yang, 8, 11
yoga, as Eastern sexual philosophy, 9–10
yonti, 7
Young, Hugh H., 46

Zilbergeld, Bernie, 145

Credits/Acknowledgments

Cover design by Charles Vitelli

1. Sexuality and Society
Facing overview—United Nations photo by L. Barns.

2. Sexual Biology, Behavior, and Orientation
Facing overview—AP/Wide World photo.

3. Interpersonal Relationships
Facing overview—United Nations photo by John Isaac.

4. Reproduction
Facing overview—Sipa Press photo by Win McNamee. 111—Photo by Wisconsin Pharmacal Company.

5. Sexuality Through the Life Cycle
Facing overview—Dushking Publishing Group, Inc., photo by Marcuss Oslander.

6. Old/New Sexual Concerns
Facing overview—Picture Group photo by Rick Brownie. 195-196—Photos courtesy of FDA Consumer.

PHOTOCOPY THIS PAGE!!!*

ANNUAL EDITIONS ARTICLE REVIEW FORM

■ NAME: _____ DATE: _____

■ TITLE AND NUMBER OF ARTICLE: _____

■ BRIEFLY STATE THE MAIN IDEA OF THIS ARTICLE: _____

_____

_____

_____

_____

■ LIST THREE IMPORTANT FACTS THAT THE AUTHOR USES TO SUPPORT THE MAIN IDEA:

_____

_____

_____

_____

_____

_____

■ WHAT INFORMATION OR IDEAS DISCUSSED IN THIS ARTICLE ARE ALSO DISCUSSED IN YOUR TEXTBOOK OR OTHER READING YOU HAVE DONE? LIST THE TEXTBOOK CHAPTERS AND PAGE NUMBERS:

_____

_____

_____

_____

_____

_____

■ LIST ANY EXAMPLES OF BIAS OR FAULTY REASONING THAT YOU FOUND IN THE ARTICLE:

_____

_____

_____

_____

■ LIST ANY NEW TERMS/CONCEPTS THAT WERE DISCUSSED IN THE ARTICLE AND WRITE A SHORT DEFINITION:

_____

_____

_____

_____

*Your instructor may require you to use this Annual Editions Article Review Form in any number of ways: for articles that are assigned, for extra credit, as a tool to assist in developing assigned papers, or simply for your own reference. Even if it is not required, we encourage you to photocopy and use this page; you'll find that reflecting on the articles will greatly enhance the information from your text.

ANNUAL EDITIONS: HUMAN SEXUALITY 94/95
Article Rating Form

We Want Your Advice

Here is an opportunity for you to have direct input into the next revision of this volume. We would like you to rate each of the 49 articles listed below, using the following scale:

1. **Excellent: should definitely be retained**
2. **Above average: should probably be retained**
3. **Below average: should probably be deleted**
4. **Poor: should definitely be deleted**

Your ratings will play a vital part in the next revision. So please mail this prepaid form to us just as soon as you complete it.
Thanks for your help!

Annual Editions revisions depend on two major opinion sources: one is our Advisory Board, listed in the front of this volume, which works with us in scanning the thousands of articles published in the public press each year; the other is you—the person actually using the book. Please help us and the users of the next edition by completing the prepaid article rating form on this page and returning it to us. Thank you.

Rating	Article	Rating	Article
	1. Sexuality and Spirituality: The Relevance of Eastern Traditions		25. The Global Politics of Abortion
	2. Cross-Cultural Perspectives on Sexuality Education		26. Making Babies: Miracle or Marketing Hype? Risks, Caveats and Costs
	3. Family, Work, and Gender Equality: A Policy Comparison of Scandinavia, the United States, and the Former Soviet Union		27. Reproductive Revolution Is Jolting Old Views
			28. Coping with Miscarriage
	4. Beyond Abortion: Transforming the Pro-Choice Movement		29. Unnecessary Cesarean Sections: Halting a National Epidemic
	5. The Brave New World of Men		30. Raising Sexually Healthy Kids
	6. Sex in the Snoring '90s		31. Truth and Consequences: Teen Sex
	7. The New Sexual Revolution: Liberation at Last? Or the Same Old Mess?		32. Single Parents and Damaged Children: The Fruits of the Sexual Revolution
	8. The Five Sexes: Why Male and Female Are Not Enough		33. Let the Games Begin: Sex and the Not-Thirtysomethings
	9. What Is It with Women and Breasts?		34. 25 Ways to Make Your Marriage Sexier
	10. Why Do We Know So Little About Human Sex?		35. Beyond Betrayal: Life After Infidelity
			36. What Doctors and Others Need to Know: Six Facts on Human Sexuality and Aging
	11. Sexual Arousal of College Students in Relation to Sex Experiences		37. Sexuality and Aging: What It Means to Be Sixty or Seventy or Eighty in the '90s
	12. Homosexuality, the Bible, and Us—A Jewish Perspective		38. The Future of AIDS
	13. Homosexuality and Biology		39. Campuses Confront AIDS: Tapping the Vitality of Caring and Community
	14. The Gay Debate: Is Homosexuality a Matter of Choice or Chance?		40. Preventing STDs
	15. The Mating Game		41. Syphilis in the '90s
	16. Ways of Loving: A Matter of Style		42. A Question of Abuse
	17. What Is Love?		43. The Sexually Abused Boy: Problems in Manhood
	18. The Lessons of Love		44. Who's to Blame for Sexual Violence?
	19. Forecast for Couples		45. Sex Differences in the Brain
	20. Choosing a Contraceptive		46. Women and Men: Can We Get Along? Should We Even Try?
	21. The Female Condom		47. It's a Jungle Out There, So Get Used to It!
	22. The Norplant Debate		
	23. New, Improved and Ready for Battle		48. The Blame Game
	24. Desperation: Before *Roe v. Wade*, After *Roe* Is Reversed		49. Ending the Battle Between the Sexes

(Continued on next page)

ABOUT YOU

Name _____ Date _____

Are you a teacher? ☐ Or student? ☐
Your School Name _____
Department _____
Address _____
City _____ State _____ Zip _____
School Telephone # _____

YOUR COMMENTS ARE IMPORTANT TO US!

Please fill in the following information:

For which course did you use this book? _____
Did you use a text with this Annual Edition? ☐ yes ☐ no
The title of the text? _____
What are your general reactions to the Annual Editions concept?

Have you read any particular articles recently that you think should be included in the next edition?

Are there any articles you feel should be replaced in the next edition? Why?

Are there other areas that you feel would utilize an Annual Edition?

May we contact you for editorial input?

May we quote you from above?

ANNUAL EDITIONS: HUMAN SEXUALITY 94/95

BUSINESS REPLY MAIL
First Class   Permit No. 84   Guilford, CT

Postage will be paid by addressee

**The Dushkin Publishing Group, Inc.
Sluice Dock
DPG Guilford, Connecticut 06437**

No Postage
Necessary
if Mailed
in the
United States